深圳植物名录

深圳市城市管理局
深圳市园林科学研究所　编著
华南农业大学林学院

中国林业出版社

图书在版编目(CIP)数据

深圳植物名录/深圳市城市管理局等编著. —北京:中国林业出版社,2007.7
 ISBN 978-7-5038-4914-5

Ⅰ.深… Ⅱ.深… Ⅲ.植物—深圳—名录 Ⅳ.Q948.526.53

中国版本图书馆 CIP 数据核字(2007)第 109192 号

出 版	中国林业出版社(100009 北京市西城区德内大街刘海胡同7号)	
网 址	http://www.cfph.com.cn 电话:(010)66184477	
	E-mail:cfphz@public.bta.net.cn	
	www.zgly book.com	
发 行	新华书店北京发行所	
印 刷	北京地质印刷厂	
版 次	2007年11月第1版	
印 次	2007年11月第1次	
开 本	148mm×210mm 1/32	
印 张	11	
字 数	400千字	
定 价	36.00元	

深圳植物名录

编辑委员会

编辑单位

深圳市城市管理局
深圳市园林科学研究所
华南农业大学林学院

编辑委员会

主　任：叶果
副主任：周远松　谢良生
编　委：（按姓氏笔画为序）
　　　　王晓明　叶　果　刘德荣　庄雪影　朱伟华
　　　　邢福武　张业光　李秉滔　陈德华　周远松
　　　　郑明轩　唐跃琳　曹　华　谢良生　谢锐星
　　　　韩俊永　雷江丽　谭一凡

编著者：谢良生　雷江丽　庄雪影　谭一凡
　　　　曹　华　郑明轩

序

  植物是地球生态系统的第一生产者，也是构成地球生物圈的基石。在广阔的自然界里生活着各种各样的植物，是它们共同构成了地球上绚丽多彩、生机勃勃的植物世界。植物多样性给人类提供了各种粮食、能源、纤维、蔬果、工业原料及药物等资源；同时，植物在自然界中固定太阳能、净化环境、涵养水源、防治水土流失、调节气候以及园林美化、绿化上具有巨大的作用。因此，合理利用和保护植物和植物多样性研究是人类可持续发展的重大课题；植物多样性编目，出版植物名录，是生物多样性研究的基础项目。

  深圳地处热带、亚热带的过渡地带，是广东南部经济最发达的沿海城市，自然条件优越且特殊，植物种类丰富，植被类型比较复杂，保存有较古老和特有植物，包括亚热带季风常绿阔叶林，海边还有红树林、湿地植被等。深圳有近3000种的维管束植物，其中有许多植物不仅具有重要的科研价值，而且也有较高的经济价值。这些丰富的植物资源对于深圳及周边地区的生态平衡、植物资源的保护和可持续利用均有重要的科学意义和利用价值。

  作者在多年野外调查、标本采集和参考有关文献资料的基础上，编辑出版《深圳植物名录》。该书共收载了2979种维管束植物，隶属于257科1304属。该书的出版对深圳市及珠江三角洲地区的自然保护、生物多样性研究、园林植物资源的开发利用具有很大的实用价值，可供科研、教育、环境保护、医药、政府管理部门和旅游业等参考使用。

华南农业大学
李秉滔
2007年1月15日

Foreword

Plants are not only the primary producer of the earth's ecosystem, but also the basic elements of the biosphere. There are various plants on the earth. They build up the marvelous and vigorous plants kingdom. Plant-diversity can provide people with foods, energy, fiber, vegetables and fruits, industrial and medicinal materials. Meanwhile, plants can fix the solar energy, improve the environment, preserve water catchments, decrease soil erosion and water loss, beautify and enrich the urban landscape. Therefore, rational utilization and preservation of the plant resources and biodiversity study are the most important issues for the sustainable development of human beings. This check list is the basic reference for biodiversity study.

Shenzhen city lies between the tropical and sub-tropical areas. It is the best developed coastal city in Guangdong. Shenzhen has warm and humid climate, diverse flora and vegetation, including some ancient and endemic species. There are subtropical evergreen board-leaved forest, mangrove, and wetland vegetation. Local vascular plants are of high importance in science, but also in economy. They play an important role in maintaining the ecological balance and sustainable utilization of natural resources.

Based on field investigation, specimen collection, and literature review, it records a total of 2979 species and varieties of local vascular plants, belonging to 1304 genera and 257 families. Publication of the plant check list is a practical reference for natural conservation, biodiversity study and landscape construction. It is also useful information for scientific studies, education, environmental protection, medicine, governmental management, and tourism.

Li Bingtao
South China Agricultural University
January 15, 2007

前 言

《深圳植物名录》几点注释

此深圳植物名录共记载了本地维管植物 2979 种及变种，分别隶属 257 科 1304 属。

蕨类植物各科是依照秦仁昌（1978）系统排列，裸子植物各科按郑万钧（1990）系统排列，被子植物各科按 Hutchinson（1959）系统排列，属和种则按照拉丁学名的字母顺序排列。部分别名或异名分别列入括号。名录列出了每种植物的中文名、学名及英文名，外来植物前加"＊"号。

深圳的地理及气候

深圳市位于东经 113°6′~114°37′、北纬 22°27′~22°52′，全市总面积 1952.84 平方公里，其中特区内 395.81 平方公里。它东临大鹏湾，西连珠江口，南与香港新界接壤，北与东莞市、惠州市为邻。深圳市全境地势东南高，西北低，大部分为低丘陵地，间以台地和滨海平原。境内最高山峰为梧桐山，海拔 943.7 米。海拔在 700 米以上的山峰还有七娘山、大燕山和排牙山。深圳海岸线全长 230 公里，有 6 处港湾建有深水港。现有河流 160 余条，但集雨面积和流量不大。流域面积大于 100 平方公里的河流有深圳河、茅洲河、龙岗河、观澜河和坪山河，地表水资源相对短缺。

深圳属亚热带海洋性季风气候。冬季季风从每年 9 月开始由北方或东北方吹来，由 10 月延续至翌年 3 月中旬，冬季天气较凉和干燥。初春期间当风向由东北转至东南时，会间有浓雾和微雨。从 4 月中旬到 9 月，夏季季风由南方或西南方吹来，夏季多雨，天气炎热潮湿。每年 5~9 月是深圳台风季节，会有烈风和暴雨。

深圳年平均气温 22.4℃。最高月平均气温出现在 7 月，为

28.1℃；最低月平均气温在1月，为12.1℃。极端最高气温38.7℃，极端低气温0.2℃，全年无霜期为355天。深圳市雨量充沛，年平均降水量为1933.3毫米。降雨主要集中在每年的5～9月，而期间也是台风季节。年平均相对湿度为77%；年平均蒸发量高达1755.4毫米；年平均辐射热达5225兆焦耳/平方米；年平均日照时数为2120.5小时。

深圳的地质和土壤

在大地构造上，深圳地区属于震旦纪华南地台的一部分，曾发生几次海陆变迁。山地的母岩主要为花岗岩、砂页岩、凝灰岩以及部分变质岩和砂岩。土壤可分为山地土壤和冲积土两大类型。主要的山地土壤广泛分布于丘陵和山地，根据成土母质可为红壤、赤红壤和黄壤等三大类。其中赤红壤面积最大，占80%左右，主要分布于海拔300米以下的地方；红壤和黄壤约占20%，分别分布于海拔300～500米和海拔500米以上。冲积土分布在河流溪涧两岸的泛滥平原及沿海各滩地。

深圳的植被

深圳曾遍布森林，但近百年来不断受人为破坏，原生林早已毁坏殆尽，现存的植被都是近40年才恢复起来的次生林和人工林。

根据深圳地区自然植被的生境条件特点和群落的组成成分、外貌和结构特征，可划分为低地常绿季雨林、山地常绿阔叶林、沟谷雨林、红树林和人工林等5个类型。

由于长期人为活动的干扰，原生的常绿季雨林已遭全面砍伐，目前只在村落附近保留零星分布的次生林，由于村民一般有需要保护当地风水的信仰，这些次生林一般不会被樵伐，因而俗称"风水林"。"风水林"的建群种多为榕属（Ficus）、杜英属（Elaeocarpus）、五月茶属（Antidesma）、樟属（Cinnamomum）、润楠属（Machilus）、木姜子属（Litsea）、重阳木属（Bischofia）、鹅掌柴属（Schefflera）、桂木属（Artocarpus）、橄榄属（Canarium）等植物。林下灌木层种类较多，有九节属（*Psychotria*）、冬青属（*Ilex*）、紫金牛属（*Ardisia*）、卫矛属（*Euonymus*）、叶下珠属（*Phyllanthus*）、山竹子属（*Garcinia*）等植物。

山地常绿阔叶林是南亚热带丘陵山地的植被类型，分布于梧桐山、七娘山等地海拔 500～850 米的山地。以栲属（*Castanopsis*）、青冈属（*Cyclobalanopsis*）、润楠属的植物为主。林下灌木层常见的九节属、紫金牛属、蒲桃属（*Syzygium*）、木姜子属、野桐属（*Mallotus*）等种类。藤本植物常见的有买麻藤属（*Gnetum*）、崖豆藤属（*Millettia*）、锡叶藤属（*Tetracera*）、酸藤子属（*Emblia*）植物。

在梧桐山、七娘山、塘朗山的一些沟谷地段，零散分布有沟谷雨林乔木层，主要以榕属、杜英属、苹婆属（*Sterculia*）、野桐属为主。林下灌层以棕榈科的省藤属（*Calamus*）、黄藤属（*Daemonorops*）植物，及九节属、水团花属（*Adina*）、紫金牛属、山油柑属（*Acronychia*）、榕属、山竹子属、银柴属（*Aporosa*）等植物组成。

红树林多分布于沿海滩涂，但由于近年开发工业和住宅用地，红树林面积急剧减少。天然红树林通常由马鞭草科（Verbenaceae）、紫金牛科（Myrsinaceae）、大戟科（Euphorbiaceae）和红树科（Rhizophoraceae）少数植物组成。在红树林附近的陆生次生林主要植物有黄槿（*Hibiscus tiliaceus*）、血桐（*Macaranga tanarius*）、海杧果（*Cerbera manghas*）、露兜树（*Pandanus tectorius*）、草海桐（*Scaevola sericea*）等。海滩上常见的种类为海滩牵牛（*Ipomoea pes-caprae*）、单叶蔓荆（*Vitex trifolia* var. *simplicifolia*）、鬣刺（*Spinifex littoreus*）以及卤地菊（*Wedelia prostrata*）。

人工林多为防风固沙和防止水土流失的防护林，分布于海边和公路两旁。主要树种有木麻黄（*Casuarina equisetifolia*）、台湾相思（*Acacia confusa*）、马占相思（*Acacia mangium*）、尾叶桉（*Eucalyptus teretⅰcornis*）等。深圳还有很多果树，主要以荔枝（*Litchi chinensis*）、龙眼（*Dimocarpus longan*）和柑（*Citrus microcarpa*）、橙（*C. sinensis*）为主。

编著者
2007 年 1 月

Introduction

The *Check List of Shenzhen Plants* is edited based on field collection and reference reviews. A total of 2979 species and varieties of local vascular plants, belonging to 257 families and 1304 genera, are recorded.

The arrangement of the families of Pteridophytes follows the system proposed by Ching Ren-chang (Ching, 1978); that of the Gymnosperm follows that of Zheng Wanjun (Zheng et al., 1990); and that of the Angiosperm follows that of Hutchinson (Hutchinson, 1959). The arrangements of the genera and species within a family follow their scientific names alphabetically. Some common synonyms are listed in brackets following the scientific names. Each plant name includes scientific name, Chinese name, and English name. The exotic taxa are indicated with the asterisk (*). In this Check List, all the following plants are included as the local plants in the list:

(1) the plants naturally distribute in Shenzhen;

(2) the aged trees older than 100 years in Shenzhen;

(3) the plant species naturally distributed in the Pearl River Delta area, Hong Kong, and Macau;

(4) the traditional crops, fruits, and vegetables cultivated in Shenzhen and the Pearl River Delta area for more than 100 years.

Geography and Climate of Shenzhen

Shenzhen lies between latitudes 22°27'N and 22°52' N, and longitudes 113°6'E and 114°37'E to the east of the Pearl River (Zhujiang) estuary, with a total area of 1952.84 km^2. It is just south to the Tropic of Cancer. Its east is to the Mirs Bay (Dapeng Wan), west to the estuary of the Pearl River, south adjacent to the New Territory of Hong Kong, and neighbors Dongguan and Huizhou on the north. The topography is rugged in southeast and flat in northwest. Most areas are hilly areas with scattered tables and coastal plains. The highest point is Wutong Shan (943.7m). The peaks higher than 700 m altitude are Qiniang Shan, Dayan Shan, and Paiya Shan. There are a total of 230 km of the coastline, 6 deep-water ports. There are 160 rivulets, but the catchment area and discharge are not large. The catchment areas over 100 km^2 are Shenzhen River, Maozhou River, Longgang River, Guanlan River, Pingshan River. Their surface runoffs are low.

Shenzhen has monsoon climate. The winter monsoon, blowing from the north or northeast, dominated from September to mid-March, which brings cool and dry air to Shenzhen. The summer monsoon prevails from April to September, which results in hot and humid weather. Summer is also the main typhoon season, with heavy rain and thunderstorm.

The annual mean temperature is 22.4℃, ranging from 12.1℃ in January to 28.1℃ in July. The extremely high and low temperature are 38.7℃ and 0.2℃ respectively. The frost-free period is over 355 days. The mean annual rainfall is 1933.3 mm, most falling between May and September. Relative humidity is 77% and mean evaporation is 1755.4 mm; The mean radiant heat is 522.5 kJ/cm^2; The annual mean sunshine time is 2120.5 hours.

Geology and Soils

Shenzhen is a part of the South China sub-continent in the Sinian Period. Its sea levels have experienced several changes in the geological past. Rocks in the mountain areas are granite, arenaceous shale, tuff and some metamorphic rocks and sandstone rocks. There are two major types of soils in Shenzhen: hill soils and alluvial soils. Hill soils are widespread in upland and hilly areas, which can be subdivided into red earth, lateritic red earth, and yellow earth based on the parent material. Lateritic red earth is the commonest soil below 300m altitude, which occupies 80% of the area. Red earth and yellow earth make up the other 20%, which are mainly found in upland and hilly area between 300m altitude and 500m altitude respectively. The alluvium soils are only confined to the river banks, river plains, and seashore.

The Vegetation of Shenzhen

Shenzhen once covered by forest and the climatic vegetation is an evergreen or semi-deciduous forest. However, the vegetation has undergone great changes over the centuries under the influences of human beings' disturbances. Most of the existing forest is secondary forest developed in the past decades or man-made plantations. Five major forest types can be divided by floristic composition, community characteristics and structure: lowland evergreen monsoon forest, montane evergreen board-leaved forest, ravine rainforests, mangroves, and man-made plantations.

All the original lowland monsoon forest has been destroyed due to long term of human disturbances. Fung Shui Woods (Fung Shui means wind and water in Chinese) are the relics stands of the oldest lowland monsoon forest. They are near some aged villages and usually protected from logging for the ancient tradition by local residents. They are usually dominated by the trees of *Ficus*, *Elaeocarpus*, *Antidesma*, *Cinnamomum*, *Machilus*, *Litsea*, *Bischofia*, *Schefflera*, *Artocarpus*, and *Canarium*. The common understorey species in Fung Shui Woods belong to *Psychotria*, *Ilex*, *Ardisia*, *Euonymus*, *Garcinia* and *Phyllanthus*.

The montane evergreen board-leaved forests are confined to the upland areas above 500 m, such as Wutong Shan and Qiniang Shan. They are dominated by the species of *Castanopsis*, *Cyclobalanopsis*, and *Machilus*. The common understorey species in montane forest belong to *Psychotria*, *Ilex*, *Ardisia*, *Syzygium*, and *Litsea*.

The ravine rainforests are only scattered to a few ravines of Wutong Shan, Qiniang Shan, and Tanglang Shan. The dominant genera of the canopy trees are *Ficus*, *Elaeocarpus*, *Sterculia*, and *Mallotus*. The common understorey species belong to *Calamus*, *Daemonorops*, *Psychotria*, *Adina*, *Ardisia*, *Acronychia*, *Ficus*, *Garcinia*, and *Aporosa*.

The mangroves usually grow on muddy shores, but the areas rapidly declined due to extension of the industrial and residential development. The typical mangroves usually consist of a few species from Verbenaceae, Myrsinaceae, Euphorbiaceae and Rhizophoraceae. At the back belts of the mangroves, *Hibiscus tiliaceus*, *Macaranga tanarius*, *Cerbera manghas*, *Pandanus tectorius*, and *Scaevola sericea* are also very common. On the sand beaches, the common species are *Ipomoea pes-caprae*, *Vitex trifolia* var. *simplicifolia*, *Spinifex littoreus* and *Wedelia prostrata*.

The man-made plantations are planted for ecological purpose to prevent soil erosion and landslides. They are usually planted as roadside trees and along the river banks. The common species are *Casuarina equisetifolia*, *Acacia confusa*, *Acacia mangium*, and *Eucalyptus urophylla*. Most of the plantation species are exotic fast-growing trees. On the other hand, there are many orchards in Shenzhen. The common fruit species are *Litchi chinensis*, *Dimocarpus longan*, *Citrus microcarpa*, and *C. sinensis*.

The authors
2007 – 01

目 录

序
前言

蕨类植物 PTERIDOPHYTA ……………………………………… (1)
    松叶蕨科 Psilotaceae ……………………………………… (1)
    石杉科 Huperziaceae ……………………………………… (1)
    卷柏科 Selaginellaceae …………………………………… (1)
    木贼科 Equisetaceae ……………………………………… (2)
    瓶耳小草科 Ophioglossaceae …………………………… (2)
    观音座莲科 Angiopteridaceae …………………………… (2)
    紫萁科 Osmundaceae ……………………………………… (2)
    瘤足蕨科 Plagiogyriaceae ………………………………… (3)
    里白科 Gleicheniaceae …………………………………… (3)
    海金沙科 Lygodiaceae …………………………………… (3)
    膜蕨科 Hymenophyllaceae ……………………………… (4)
    蚌壳蕨科 Dicksoniaceae ………………………………… (4)
    桫椤科 Cyatheaceae ……………………………………… (5)
    碗蕨科 Dennstaedtiaceae ………………………………… (5)
    鳞始蕨科 Lindsaeaceae …………………………………… (5)
    姬蕨科 Hypolepidaceae …………………………………… (6)
    蕨科 Pteridiaceae ………………………………………… (6)
    凤尾蕨科 Pteridaceae …………………………………… (6)
    卤蕨科 Acrostichum ……………………………………… (7)
    中国蕨科 Sinopteridaceae ………………………………… (7)
    铁线蕨科 Adiantaceae …………………………………… (7)
    水蕨科 Parkeriaceae ……………………………………… (7)
    裸子蕨科 Hemionitidaceae ……………………………… (7)
    书带蕨科 Vittaria ………………………………………… (8)
    蹄盖蕨科 Athyriaceae …………………………………… (8)
    金星蕨科 Thelypteridacea ……………………………… (8)
    铁角蕨科 Aspleniaceae ………………………………… (10)
    睫毛蕨科 Pleurosoriopsidaceae ………………………… (10)

目 录

球子蕨科 Onocleaceae ……………………………………… (10)
乌毛蕨科 Blechnaceaeu ……………………………………… (10)
鳞毛蕨科 Dryopteridaceae …………………………………… (11)
三叉蕨科 Aspidiaceae ………………………………………… (12)
实蕨科 Bolbitidaceae ………………………………………… (13)
舌蕨科 Elaphoglossaceae ……………………………………… (13)
肾蕨科 Nephrolepidacea ……………………………………… (13)
条蕨科 Oleandraceae ………………………………………… (13)
骨碎补科 Davalliaceae ………………………………………… (13)
雨蕨科 Gymnogrammitidaceae ……………………………… (14)
双扇蕨科 Dipteridaceae ……………………………………… (14)
水龙骨科 Polypodiaceae ……………………………………… (14)
槲蕨科 Drynariaceae ………………………………………… (15)
鹿角蕨科 Platyceriaceae ……………………………………… (15)
禾叶蕨科 Grammitidaceae …………………………………… (16)
剑蕨科 Loxogrammaceae ……………………………………… (16)
苹科 Marsileaceae ……………………………………………… (16)
槐叶苹科 Salviniaceae ………………………………………… (16)
满江红科 Azollaceae …………………………………………… (16)

裸子植物 GYMNOSPERMAE …………………………… (17)

苏铁科 Cycadaceae …………………………………………… (17)
银杏科 Ginkgoaceae …………………………………………… (17)
松科 Pinaceae ………………………………………………… (17)
杉科 Taxodiaceae ……………………………………………… (18)
柏科 Cupressaceae …………………………………………… (18)
南洋杉科 Araucariaceae ……………………………………… (19)
罗汉松科 Podocarpaceae ……………………………………… (19)
三尖杉科 Cephalotaxaceae …………………………………… (19)
红豆杉科 Taxaceae …………………………………………… (20)
买麻藤科 Gnetaceae …………………………………………… (20)

被子植物 ANGIOSPERMAE ……………………………… (21)

双子叶植物 DICOTYLEDONEAE …………………… (21)

木兰科 Magnoliaceae ………………………………………… (21)
八角科 Illiciaceae ……………………………………………… (23)
五味子科 Schisandraceae ……………………………………… (23)

番荔枝科 Annonaceae	(23)
樟科 Lauraceae	(24)
青藤科 Illigeraceae	(27)
毛茛科 Ranunculaceae	(27)
金鱼藻科 Ceratophyllum	(28)
睡莲科 Nymphacaceae	(28)
小檗科 Berberidaceae	(29)
木通科 Lardizabalaceae	(29)
大血藤科 Sargentodoxaceae	(30)
防己科 Menispermaceae	(30)
马兜铃科 Aristolochiaceae	(31)
猪笼草科 Nepenthaceae	(31)
胡椒科 Piperaceae	(31)
三白草科 Saururaceae	(32)
金粟兰科 Chloranthaceae	(32)
罂粟科 Papaveraceae	(33)
白花菜科 Capparidaceae	(33)
辣木科 Moringaceae	(33)
十字花科 Cruciferae	(33)
堇菜科 Violaceae	(35)
远志科 Polygalaceae	(35)
景天科 Crassulaceae	(35)
虎耳草科 Saxifragaceae	(36)
茅膏菜科 Droseraceae	(37)
瓶子草科 Sarraceniaceae	(37)
沟繁缕科 Elatinaceae	(37)
石竹科 Caryophyllaceae	(37)
粟米草科 Molluginaceae	(38)
番杏科 Aizoaceae	(38)
马齿苋科 Portulacaceae	(38)
蓼科 Polygonaceae	(39)
商陆科 Phytolaccaceae	(40)
藜科 Chenopodiaceae	(40)
苋科 Amaranthaceae	(40)

落葵科 Basellaceae …… (41)
亚麻科 Linaceae …… (41)
牻牛儿苗科 Geraniaceae …… (41)
酢浆草科 Oxalidaceae …… (42)
旱金莲科 Tropaeolaceae …… (42)
凤仙花科 Balsaminaceae …… (42)
千屈菜科 Lythraceae …… (42)
安石榴科 Punicaceae …… (43)
柳叶菜科 Onagraceae …… (43)
菱科 Trapaceae …… (44)
小二仙草科 Haloragaceae …… (44)
水马齿科 Callitrichaceae …… (44)
瑞香科 Thymelaeaceae …… (44)
紫茉莉科 Nyctaginaceae …… (45)
山龙眼科 Proteaceae …… (45)
五桠果科 Dilleniaceae …… (45)
海桐花科 Pittosporaceae …… (45)
红木科 Bixaceae …… (46)
大风子科 Flacourtiaceae …… (46)
天料木科 Samydaceae …… (46)
西番莲科 Passifloraceae …… (46)
葫芦科 Cucurbitaceae …… (47)
秋海棠科 Begoniaceae …… (48)
番木瓜科 Caricaceae …… (49)
仙人掌科 Cactaceae …… (49)
茶科 Theaceae …… (50)
五列木科 Pentaphylacaceae …… (52)
猕猴桃科 Actinidiaceae …… (52)
水冬哥科 Saurauiaceae …… (52)
龙脑香科 Dipterocarpaceae …… (52)
桃金娘科 Myrtaceae …… (52)
野牡丹科 Melastomataceae …… (54)
使君子科 Combretaceae …… (55)
红树科 Rhizophoraceae …… (55)

金丝桃科 Hypericaceae	(56)
藤黄科 Guttiferae	(56)
椴树科 Tiliaceae	(56)
杜英科 Elaeocarpaceae	(57)
梧桐科 Sterculiaceae	(57)
木棉科 Bombacaceae	(58)
锦葵科 Malvaceae	(58)
金虎尾科 Malpighiaceae	(60)
粘木科 Ixonanthaceae	(60)
大戟科 Euphorbiaceae	(60)
交让木科 Daphniphyllaceae	(66)
小盘木科 Pandaceae	(66)
鼠刺科 Escalloniaceae	(66)
绣球花科 Hydrangeaceae	(66)
蔷薇科 Rosaceae	(67)
含羞草科 Mimosaceae	(69)
苏木科 Caesalpiniaceae	(71)
蝶形花科 Papilionaceae	(73)
旌节花科 Stachyuraceae	(80)
金缕梅科 Hamamelidaceae	(80)
黄杨科 Buxaceae	(81)
杨柳科 Salicaceae	(81)
杨梅科 Myricaceae	(81)
壳斗科 Fagaceae	(81)
木麻黄科 Casuarinaceae	(83)
榆科 Ulmaceae	(83)
桑科 Moraceae	(84)
荨麻科 Urticaceae	(87)
冬青科 Aquifoliaceae	(88)
卫矛科 Celastraceae	(89)
翅子藤科 Hippocarateaceae	(90)
茶茱萸科 Icacinaceae	(90)
铁青树科 Olacaceae	(90)
山柑子科 Opiliaceae	(90)

桑寄生科 Loranthaceae …… (90)
檀香科 Santalaceae …… (91)
蛇菰科 Balanophoraceae …… (91)
鼠李科 Rhamnaceae …… (91)
胡颓子科 Elaeagnaceae …… (92)
葡萄科 Vitaceae …… (92)
芸香科 Rutaceae …… (94)
苦木科 Simaroubaceae …… (95)
橄榄科 Burseraceae …… (95)
楝科 Meliaceae …… (96)
无患子科 Sapindaceae …… (96)
槭树科 Aceraceae …… (97)
清风藤科 Sabiaceae …… (98)
省沽油科 Staphyleaceae …… (98)
漆树科 Anacardiaceae …… (98)
牛栓藤科 Connaraceae …… (99)
胡桃科 Juglandaceae …… (99)
山茱萸科 Cornaceae …… (99)
八角枫科 Alangiaceae …… (100)
蓝果树科 Nyssaceae …… (100)
五加科 Araliaceae …… (100)
伞形花科 Umbelliferae …… (101)
杜鹃花科 Ericaceae …… (102)
越橘科 Vacciniaceae …… (103)
柿科 Ebenaceae …… (103)
山榄科 Sapotaceae …… (103)
肉实科 Sarcospermaceae …… (104)
紫金牛科 Myrsinaceae …… (104)
安息香科 Styracaceae …… (105)
山矾科 Symplocaceae …… (106)
马钱科 Loganiaceae …… (107)
木犀科 Oleaceae …… (107)
夹竹桃科 Apocynaceae …… (108)
萝藦科 Asclepiadaceae …… (110)

杠柳科 Periplocaceae ……（112）
茜草科 Rubiaceae ……（112）
忍冬科 Caprifoliaceae ……（117）
菊科 Compositae ……（118）
龙胆科 Gentianaceae ……（126）
睡菜科 Menyanthaceae ……（126）
报春花科 Primulaceae ……（126）
白花丹科 Plumbaginaceae ……（126）
车前草科 Plantaginaceae ……（127）
桔梗科 Campanulanceae ……（127）
半边莲科 Lobeliaceae ……（127）
草海桐科 Goodeniaceae ……（127）
花柱草科 Stylidiaceae ……（128）
田基麻科 Hydrophyllaceae ……（128）
紫草科 Boraginaceae ……（128）
茄科 Solanaceae ……（128）
旋花科 Convolvulaceae ……（130）
玄参科 Scrophulariaceae ……（132）
列当科 Orobanchaceae ……（134）
狸藻科 Lentibulariaceae ……（134）
苦苣苔科 Gesneriaceae ……（134）
紫葳科 Bignoniaceae ……（135）
爵床科 Acanthaceae ……（137）
马鞭草科 Verbenaceae ……（139）
唇形科 Labiatae ……（143）

单子叶植物 MONOCOTYLEDONEAE ……（146）

花蔺科 Butomaceae ……（146）
水鳖科 Hydrocharitaceae ……（146）
泽泻科 Alismataceae ……（147）
水蕹科 Aponogetonaceae ……（147）
眼子菜科 Potamogetonaceae ……（147）
鸭跖草科 Commelinaceae ……（147）
黄眼草科 Xyridaceae ……（149）
谷精草科 Eriocaulaceae ……（149）

凤梨科 Bromeliaceae …… (149)

芭蕉科 Musaceae …… (150)

旅人蕉科 Strelitziaceae …… (151)

姜科 Zingiberaceae …… (151)

美人蕉科 Cannaceae …… (153)

竹芋科 Marantaceae …… (153)

百合科 Liliaceae …… (153)

延龄草科 Trilliaceae …… (156)

雨久花科 Pontederiaceae …… (156)

菝葜科 Smilacaceae …… (156)

假叶树科 Ruscaceae …… (157)

天南星科 Araceae …… (157)

浮萍科 Lemnaceae …… (160)

香蒲科 Typhaceae …… (160)

石蒜科 Amaryllidaceae …… (160)

鸢尾科 Iridaceae …… (161)

百部科 Stemonaceae …… (162)

薯蓣科 Dioscoreaceae …… (162)

龙舌兰科 Agavaceae …… (162)

棕榈科 Arecaceae …… (164)

露兜树科 Pandanaceae …… (167)

仙茅科 Hypoxidaceae …… (167)

田葱科 Philydraceae …… (168)

水玉簪科 Burmanniaceae …… (168)

兰科 Orchidaceae …… (168)

灯心草科 Juncaceae …… (172)

莎草科 Cyperaceae …… (173)

禾本科 Gramineae …… (178)

索　引

中文名称 …… (191)

拉丁学名 …… (221)

英文名称 …… (281)

参考文献 …… (323)

蕨类植物 PTERIDOPHYTA

松叶蕨科 Psilotaceae

松叶蕨	*Psilotum nudum* (L.) Beauv.	Nude Fern

石杉科 Huperziaceae

蛇足石杉	*Huperzia serrata* (Thunb.) Trev.	Serrate Clubmoss
华南马尾杉	*Phlegmariurus fordii* (Baker) Ching	Ford's Phlegmariurus
石子藤(藤石松)	*Lycopodiastrum casuarinoides* (Spring) Holub	Climbed Clubmoss
石松	*Lycopodium japonicum* Thunb. ex Murray	Clubmoss
铺地蜈蚣(灯笼草)	*Palhinhaea cernua* (L.) A. Franco et Vasc.	Nodding Clubmoss

卷柏科 Selaginellaceae

二形卷柏	*Selaginella biformis* A. Br. ex Kuhn	Biform Spikemoss
缘毛卷柏	*Selaginella ciliaris* (Retz.) Spring	Ciliate Spikemoss
薄叶卷柏	*Selaginella delicatula* (Desv. ex Pior.) Alston	Delicate Spikemoss
深绿卷柏	*Selaginella doederleinii* Hieron	Doederlein's Spikemoss
异穗卷柏	*Selaginella heterostachys* Baker	Different Spikemoss
还魂草	*Selaginella involvens* (Sw.) Spring	Involute Spikemoss

小翠云	*Selaginella kraussiana* (Kunze) A. Br.	Mate Spikemoss
耳基卷柏	*Selaginella limbata* Alston	Limbated Spikemoss
江南卷柏	*Selaginella moellendorffii* Hieron	Moellendorff Spikemoss
卷柏	*Selaginella tamariscina* (Beauv.) Spring	Tamarisklike Spikemoss
翠云草	*Selaginella uncinata* (Desv.) Spring	Hooked Spikemoss
剑叶卷柏	*Selaginella xipholepis* Baker	Sword-leaved Spikemoss

木贼科 Equisetaceae

问荆	*Equisetum arvense* L.	Field Horsetail
纤弱木贼	*Equisetum debile* Roxb.	Frail Horsetail
节节草	*Equisetum ramosissimum* Desf. ex Vauch.	Ramose Scouring Rush

瓶耳小草科 Ophioglossaceae

尖头瓶尔小草	*Ophioglossum pedunculosum* Desv.	Pedunculated Adder's Tongue
带状瓶尔小草	*Ophioglossum pendula* Presl	Old World Adder's-tongue
瓶尔小草	*Ophioglossum vulgatum* L.	Common Adder's Tongue

观音座莲科 Angiopteridaceae

| 福建观音座莲 | *Angiopteris fokiensis* Hieron | Fokien Angiopteris |

紫萁科 Osmundaceae

狭叶紫萁	*Osmunda angustifolia* Ching	Narrow-leaf Osmanda
粗齿紫萁	*Osmunda banksiifolia* (Presl) Kuhn	Gross-dentate Osmanda
紫萁	*Osmunda japonica* Thunb.	Japanese Osmanda
粤紫萁	*Osmunda mildei* C. Chr.	Guangdong Osmanda

华南紫萁	*Osmunda vachellii* Hook.	Vachell's Interrupted Fern
	瘤足蕨科 Plagiogyriaceae	
瘤足蕨	*Plagiogyria adnata* (Bl.) Bedd.	Adnate Plagiogyria
华中瘤足蕨	*Plagiogyria euphlebia* Mett.	Fine-nerved Plagiogyria
耳形瘤足蕨(小牛肋巴)	*Plagiogyria stenoptera* (Hance) Diels	Auriform Plagiogyria
薄叶瘤足蕨	*Plagiogyria tenuifolia* Cop.	Tenuous-leaved Plagiogyria
	里白科 Gleicheniaceae	
里白	*Dicranopteris glaucum* (Thunb. ex Houtt.) Naikai	Glaucous Diplopterygium
铁芒萁(小里白)	*Dicranopteris linearis* (Burm. f.) Underw.	Linear Forked Fern
芒萁(铁狼萁)	*Dicranopteris pedata* (Houtt.) Nakai [*D. dichotoma* Bernh.]	Dichotomy Forked Fern
大芒萁(大羽芒萁)	*Dicranopteris splendida* (Hand.-Mazz.) Tagawa	Splendid Forked Fern
粤里白	*Diplopterygium cantonensis* (Ching) Nakai	Canton Diplopterygium
中华里白	*Diplopterygium chinense* (Ros.) De Vol	Chinese Hicriopteris
光里白	*Diplopterygium laevissimium* (Christ) Nakai	Flat Hicriopteris
假芒萁	*Sticherus laevigatus* Presl	Smooth Sticherum
	海金沙科 Lygodiaceae	
掌叶海金沙(海南海金沙)	*Lygodium conforme* C. Chr.	Conform Climbing Fern

曲轴海金沙(大叶海金沙)	*Lygodium flexuosum* (L.) Sw.	Flexuose Climbing Fern
海金沙	*Lygodium japonicum* (Thunb.) Sw.	Japanese Climbing Fern
狭叶海金沙	*Lygodium microstachyum* Desv.	Small-spiked Climbing Fern
小叶海金沙	*Lygodium scandens* (L.) Sw.	Small-leaf Climbing Fern

膜蕨科 Hymenophyllaceae

南洋假脉蕨	*Crepidomanes bipunctatum* (Poir.) Cop.	Two-dotted Crepidomanes
多脉假蕨(多脉假脉蕨)	*Crepidomanes insigne* (v. d. Bosch) Fu	Obvious Crepidomanes
华南假脉蕨(长柄假膜蕨)	*Crepidomanes racemulosum* (v. d. Bosch) Ching [*C. latealatum* auct. non (v. d. Bosch) Cop.]	Longstalk Crepidomanes
团扇蕨(圆扇蕨)	*Gonocormus minutus* (Bl.) v. d. Bosch	Minute Gonocormus
华东膜蕨(膜蕨)	*Hymenophyllum barbatum* (v. d. Bosch) Baker	Barbate Filmy Fern
小果蕗蕨(小果露蕨)	*Mecodium microsorum* (v. d. Bosch) Ching	Small-sorus Mecodium
华南长筒蕨(广西长筒蕨)	*Selenodesmium siamense* (Christ) Ching et C. H. Wang	Siamense Selenodesmium
管苞瓶蕨	*Trichomanes birmanicum* Bedd. [*Vandenboschia birmanica* Ching]	Tube-bract Bristle Fern

蚌壳蕨科 Dicksoniaceae

金毛狗(鲸口蕨)	*Cibotium barometz* (L.) J. Sm.	East Asian Tree Fern

蕨类植物 PTERIDOPHYTA

桫椤科 Cyatheaceae

桫椤（树蕨）	*Alsophila spinulosa* (Wall. ex Hook.) Tryon	Spiny Alsophila
大黑桫椤	*Gymnosphaera gigantea* (Wall. ex Hook.) J. Sm.	Large Gymnosphaera
细齿桫椤（韩氏桫椤）	*Gymnosphaera hancockii* (Cop.) Ching [*Alsophila denticulata* Baker]	Toothed Dentate Black Tree-fern
黑桫椤	*Gymnosphaera podophylla* (Hook.) Cop.	Black Tree-fern
*笔筒树	*Sphaeropteris lepifera* (J. Sm. ex Hook.) R. Tryon	Scaly Tree-fern

碗蕨科 Dennstaedtiaceae

华南鳞盖蕨	*Microlepia hancei* Prantl	Hance's Scaly-fern
虎克鳞盖蕨	*Microlepia hookeriana* (Wall. ex Hook.) Presl	Hooker's Scaly-fern
边缘鳞盖蕨（小叶山鸡尾巴草）	*Microlepia marginata* (Houtt.) C. Chr.	Marginate Scaly-fern

鳞始蕨科 Lindsaeaceae

剑叶鳞始蕨	*Lindsaea ensifolia* Sw.	Sword-leaved Lindsaea
异叶双唇蕨（异叶林蕨）	*Lindsaea heterophylla* Dry.	Different-leaved Lindsaea
鳞始蕨（陵齿蕨）	*Lindsaea odorata* Roxb. [*Lindsaea cultrata* (Willd.) Sw.]	Fragrant Lindsaea
团叶鳞始蕨（圆叶林蕨）	*Lindsaea orbiculata* (Lam.) Mett. ex Kuhn	Orbicular Lindsaea
阔叶乌蕨	*Sphenomeris biflora* (Kaulf.) Akas. [*Stenolomum biflorum* Ching]	Broad-pinna Wedgelet Fern

乌蕨(乌韭)	*Sphenomeris chinensis* (L.) Maxon [*Stenoloma chusanum* Ching]	Fairy Fern
	姬蕨科 Hypolepidaceae	
姬蕨(岩姬蕨)	*Hypolepis punctata* (Thunb.) Mett.	Downy Ground Fern
	蕨科 Pteridiaceae	
欧洲蕨	*Pteridium aquilinum* (L.) Kuhn var. *latiusculum* (Desv.) Underw. ex Heller	Bracken Fern
蕨(蕨菜)	*Pteridium esculentum* (Forst.) Cokayne	Esculent Bracken
密毛蕨(毛轴蕨)	*Pteridium revolutum* (Bl.) Nakai	Woolly-axle Bracken
	凤尾蕨科 Pteridaceae	
栗蕨(北投羊齿)	*Histiopteris incisa* (Thunb.) J. Sm.	Incised Histiopteris
条纹凤尾蕨(修纹凤尾蕨)	*Pteris cadieri* Christ	Carier Brake
刺齿凤尾蕨(天草凤尾蕨)	*Pteris dispar* Kunze	Disperate Brake
剑叶凤尾蕨	*Pteris ensiformis* Burm. f.	Sword Brake
溪边凤尾蕨(溪凤尾蕨)	*Pteris excelsa* Gaud.	Tall Brake
金钗凤尾蕨(傅氏凤尾蕨)	*Pteris fauriei* Hieron.	Faurie's Brake
*疏裂凤尾蕨	*Pteris finotii* Christ	Distant-cleft Brake
线羽凤尾蕨(三角脉凤尾蕨)	*Pteris linearis* Poir.	Lineate Brake
井栏边草(凤尾草)	*Pteris multifida* Poir.	Chinese Brake

凤尾蕨	*Pteris cretica* L. var. *nervosa* (Thunb.) Ching et S. H. Wu	Cretan Brake
半边旗	*Pteris semipinnata* L.	Semi-pinnated Brake
蜈蚣草(长叶甘草蕨)	*Pteris vittata* L.	Ladder Brake

卤蕨科 Acrostichum

卤蕨(金蕨)	*Acrostichum aureum* L.	Mangrove Brake

中国蕨科 Sinopteridaceae

毛轴碎米蕨	*Cheilosoria chusana* (Hook.) Ching et K. H. Shing	Chusan Lip-fern
薄叶碎米蕨	*Cheilosoria tenuifolia* (Burm. f.) Trev.	Narrow-leaved Lip-fern
隐囊蕨	*Notholaena hirsuta* (Poir.) Desv.	Hirsute Notholaena
野鸡尾(金粉蕨)	*Onychium japonicum* (Thunb.) Kunze	Japanese Clave Fern

铁线蕨科 Adiantaceae

铁线蕨	*Adiantum capillus-veneris* L.	Maidenhair Fern
鞭叶铁线蕨(有尾铁线蕨)	*Adiantum caudatus* L.	Walking Maidenhair Fern
扇叶铁线蕨	*Adiantum flabellulatum* L.	Fan-leaved Maidenhair Fern
半月铁线蕨(菲岛铁线蕨)	*Adiantum philippense* L. [*A. lunulatum* Burm. f.]	Philippine Maidenhair Fern

水蕨科 Parkeriaceae

水蕨	*Ceratopteris thalictroides* (L.) Brongn.	Water Fern

裸子蕨科 Hemionitidaceae

粉叶蕨(山苏花)	*Pityrogramma calomelanos* (L.) Link	Dixie Silver-back Fern

书带蕨科 Vittaria

广东书带蕨	*Vittaria chingii* B. S. Wang	Ching Grass Fern

蹄盖蕨科 Athyriaceae

毛柄短肠蕨（贯众）	*Allantodia dilatata* （Bl.） Ching	Widened Twin-sorus Fern
阔叶短肠蕨	*Allantodia matthewii* （Cop.） Ching	Matthew Twin-sorus Fern
淡绿短肠蕨	*Allantodia virescens* （Kunze） Ching	Grenish Twin-sorus Fern
假蹄盖蕨（东洋蹄盖蕨）	*Athyriopsis japonica* （Thunb.） Ching	Japanese Athyriopsis
毛轴假蹄盖蕨	*Athyriopsis petersenii* （Kunze） Ching	Petersen Athyriopsis
菜蕨（过沟菜蕨）	*Callipteris esculenta* （Retz.） J. Sm. ex Moore et Houlst.	Fleshy Lady-fern
双盖蕨（大羽双盖蕨）	*Diplazium donianum* （Mett.） Tard. -Blot	Twin-sorus Fern
内伶丁双盖蕨	*Diplazium neilingdingensis* Miau et W. B. Liao	Neilingding Twin-sorus Fern
单叶双盖蕨	*Diplazium subsinuatum* （Wall. ex Hook. et Grev.） Tagawa	Spear-Leaved Lady-fern
介蕨	*Dryoathyrium boryanum* （Willd.） Ching	Common Dryoathyrium

金星蕨科 Thelypteridacea

渐尖毛蕨（小叶凤凰尾巴草）	*Cyclosorus acuminatus* （Houtt.） Nakai	Acuminate Cyclosorus
齿牙毛蕨	*Cyclosorus dentatus* （Forsk.） Ching	Tapering Cyclosorus

异果毛蕨	*Cyclosorus heterocarpus* (Bl.) Ching	Heterocarpous Cyclosorus
毛蕨	*Cyclosorus interruptus* (Willd.) H. Ito	Interrupted Cyclosorus
宽羽毛蕨	*Cyclosorus latipinnus* (Benth.) Tard. -Blot	Broad-pinna Cyclosorus
华南毛蕨	*Cyclosorus parasiticus* (L.) Farw.	Parasitic Cyclosorus
截裂毛蕨	*Cyclosorus truncatus* (Poir.) Farw.	Truncate Cyclosorus
羽裂圣蕨	*Dictyocline wilfordii* (Hook.) J. Sm. [*D. griffithii* Moore]	Griffith's Dictyocline
普通针毛蕨	*Macrothelypteris torresiana* (Gaud.) Ching	Mariana Maiden Fern
金星蕨	*Parathelypteris glanduligera* (Kunze) Ching	Glandular Parathelypteris
毛脚金星蕨	*Parathelypteris angulariloba* (Ching) Ching	Hirsute-foot Parathelypteris
新月蕨	*Pronephrium aspera* (Presl) W. C. Shieh et J. L. Tsai	Rugged Pronephrium
红色新月蕨	*Pronephrium lakhimpurense* (Ros.) Holtt.	Red Pronephrium
单叶新月蕨	*Pronephrium simplex* (Hook.) Holtt.	Simple Pronephrium
二羽新月蕨	*Pronephrium triphyllum* (Sw.) Holtt.	Three-leaved Pronephrium
溪边假毛蕨	*Pseudocyclosorus ciliatus* (Wall. ex Benth.) Ching	Ciliate Pseudocyclosorus
镰形假毛蕨	*Pseudocyclosorus falcilobus* (Hook.) Ching	Falcate-lobed Pseudocyclosorus

铁角蕨科 Aspleniaceae

华南铁角蕨	*Asplenium austro-chinense* Ching	Southern China Spleenwort
毛轴铁角蕨	*Asplenium crinicaule* Hance	Hairy-stemmed Spleenwort
切边铁角蕨	*Asplenium excisum* Presl	Excised Spleenwort
镰叶铁角蕨(尖叶铁角蕨)	*Asplenium falcatum* Lam.	Falcate-leaved Spleenwort
大羽铁角蕨(野鸡尾)	*Asplenium neolaserpitii-folium* Tard.-Blot et Ching	Wedge-shaped Spleenwort
倒挂铁角蕨(倒挂草)	*Asplenium normale* Don	Normal Spleenwort
绿杆铁角蕨	*Asplenium obscurum* Bl.	Obscure Spleenwort
长叶铁角蕨	*Asplenium prolongatum* Hook.	Prolongated Spleenwort
假大羽铁角蕨	*Asplenium pseudolaserpitiifolium* Ching	Large-pinna Spleenwort
*狭基巢蕨(真武剑)	*Neottopteris antrophyoides* (Christ) Ching	Antrophyum-like Bird-nest Fern
巢蕨(鸟巢蕨)	*Neottopteris nidus* (L.) J. Smith	Bird-nest Fern
*长叶巢蕨	*Neottopteris phyllitidis* (Don) J. Smith	Long-leaved Bird-nest Fern

睫毛蕨科 Pleurosoriopsidaceae

| 睫毛蕨 | *Pleurosoriopsis makinoi* (Maxim.) Fomin | Makino Pleurosoriopsis |

球子蕨科 Onocleaceae

| 东方荚果蕨 | *Matteuccia orientalis* (Hook.) Trev. | Oriental Ostrich Fern |

乌毛蕨科 Blechnaceaeu

| 乌毛蕨(龙船蕨) | *Blechnum orientale* L. | Oriental Blechnum |

蕨类植物 PTERIDOPHYTA

苏铁蕨	*Brainea insignis* (Hook.) J. Sm.	Cycad-fern
崇澍蕨(哈氏狗脊)	*Chieniopteris harlandii* (Hook.) Ching	Harland's Chien Fern
狗脊蕨	*Woodwardia japonica* (L. f.) J. Sm.	Japanese Chain Fern
东方狗脊蕨	*Woodwardia orientalis* Sw.	Oriental Chain Fern
胎生狗脊(珠芽狗脊蕨)	*Woodwardia prolifera* Hook. et Arn.	Prolific Chain Fern
单牙狗脊蕨(顶芽狗脊)	*Woodwardia unigemmata* (Makino) Nakai	Unigemmate Chain Fern

鳞毛蕨科 Dryopteridaceae

中华复叶耳蕨(中华芒蕨)	*Arachniodes chinensis* (Ros.) Ching	Chinese Hollyfern
刺头复叶耳蕨(芒蕨)	*Arachniodes exilis* (Hance) Ching [*A. aristata* (G. Forst) Trindale]	East Indian Holly Fern
镰羽贯众(巴兰贯众)	*Cyrtomium balansae* (Christ) C. Chr.	Falcate Holly Fern
刺齿贯众(叶兰黄)	*Cyrtomium caryotideum* (Wall. ex Hook. et Grev.) Presl	Spiny-tooth Holly Fern
全缘贯众	*Cyrtomium falcatum* (L. f.) Presl [*Phanerophlebia falcata* (L. f.) Cop.]	Asian Holly Fern
*贯众(昏鸡头)	*Cyrtomium fortunei* J. Sm.	Fortune Holly Fern
阔鳞鳞毛蕨(多鳞毛蕨)	*Dryopteris championii* (Benth.) C. Chr. ex Ching	Champion Wood Fern
中华鳞毛蕨	*Dryopteris chinensis* (Baker) Koidz	Chinese Wood Fern
能高鳞毛蕨	*Dryopteris costalisora* Tagawa	Midribsorus Wood Fern

迷人鳞毛蕨(稀羽鳞毛蕨)	*Dryopteris decipiens* (Hook.) O. Kuntze	Deceive Wood Fern
黑足鳞毛蕨	*Dryopteris fuscipes* C. Chr.	Autumn Fern
柄叶鳞毛蕨	*Dryopteris podophylla* (Hook.) O. Kuntze	Stalk-leaf Wood Fern
无盖鳞毛蕨	*Dryopteris scottii* (Bedd.) Ching ex C. Chr.	Scott Wood Fern
华南鳞毛蕨	*Dryopteris tenuicula* Matthew et Christ.	South China Wood Fern
变异鳞毛蕨	*Dryopteris varia* (L.) O. Kuntze	Variant Wood Fern
灰缘耳蕨	*Polystichum eximium* (Mett.) C. Chr.	Eximious Shield Fern
*汝蕨	*Rumohra adiantiformis* (G. Forst.) Tindale	Adiantiform Rumohra

三叉蕨科 Aspidiaceae

三相蕨(厚叶轴脉蕨)	*Ataxipteris sinii* (Ching) Holttum	
靠脉肋毛蕨	*Ctenitis costulisora* Ching	Ribbed-sorus Ctenitis
虹鳞肋毛蕨(小金毛狗脊)	*Ctenitis rhodolepis* (Clarke) Ching	Red Scale Ctenitis
下延沙皮蕨(沙皮蕨)	*Hemigramma decurrens* (Hook.) Cop.	Decurrent Hemigramma
黄腺羽蕨	*Pleocnemia winitii* Holtt.	Winit Pleocnemia
毛轴芽蕨	*Pteridrys australis* Ching	Southern Pteridrys
下延三叉蕨	*Tectaria decurrens* (Presl) Cop.	Decurrent Halberd Fern
条裂三叉蕨	*Tectaria phaeocaulis* (Ros.) C. Chr. [*T. laciniata* Ching]	Brown-stalk Halberd Fern

蕨类植物 PTERIDOPHYTA

三叉蕨(叉蕨)	*Tectaria subtriphylla* (Hook. et Arn.) Cop.	Threeleaved Halberd Fern
	实蕨科 Bolbitidaceae	
*长叶实蕨	*Bolbitis heteroclita* (Presl) Ching	Longleaf Bolbitis
华南实蕨(海南实蕨)	*Bolbitis subcordata* (Cop.) Ching	Subcordate Bolbitis
刺蕨(恩蕨)	*Egenolfia appendiculata* (Willd.) J. Sm.	Appendiculed Egenolfia
中华刺蕨	*Egenolfia sinensis* (Baker) Maxon	Chinese Egenolfia
	舌蕨科 Elaphoglossaceae	
华南舌蕨(舌蕨)	*Elaphoglossum yoshinagae* (Yatabe) Makino	South China Elaphoglossum
	肾蕨科 Nephrolepidacea	
长叶肾蕨(尖羊齿)	*Nephrolepis biserrata* (Sw.) Schott	Broad Sword-fern
肾蕨(圆羊齿)	*Nephrolepis auriculata* (L.) Trimen	Tuberous Sword Fern
毛叶肾蕨(毛绒肾蕨)	*Nephrolepis hirsutula* (Forst.) Presl	Rough Sword Fern
	条蕨科 Oleandraceae	
华南条蕨	*Oleandra cumingii* J. Sm.	Cuming's Oleandra
	骨碎补科 Davalliaceae	
华南骨碎补	*Davallia austro-sinica* Ching	Southern China Davallia
大叶骨碎补(硬骨碎补)	*Davallia formosana* Hayata	South China Hare's-foot Fern
阴石蕨	*Humata repens* (L. f.) Diels	Large Hare's-foot Fern

| 圆盖阴石蕨(毛石蚕) | *Humata tyermannii* T. Moore | Repent Humate |

雨蕨科 Gymnogrammitidaceae

| 雨蕨 | *Gymnogrammitis dareiformis* (Hook.) Ching ex Tard.-Blot et C. Chr. | Common Gymnogrammitis |

双扇蕨科 Dipteridaceae

| *中华双扇蕨(八爪蕨) | *Dipteris chinensis* Christ | Chinese Dipteris |

水龙骨科 Polypodiaceae

掌叶线蕨(石壁莲)	*Colysis digitata* (Baker) Ching	Digitate Colysis
线蕨	*Colysis elliptica* (Thunb.) Ching	Snake's-eye Fern
断线蕨(石韦)	*Colysis hemionitidea* (Wall. ex Mett.) C. Presl	Interrupted Colysis
宽羽线蕨	*Colysis pothifolia* (D. Don) C. Presl	Broad Colysis
抱树莲(抱石莲)	*Drymoglossum piloselloides* (L.) C. Presl	Piloselike Drymoglossum
伏石蕨(瓜子草)	*Lemmaphyllum microphyllum* C. Presl	Little-leaf Lemmaphyllum
披针骨牌蕨	*Lepidogrammitis diversa* (Ros.) Ching	Diverse Lepidogrammitis
抱石莲(金龟藤)	*Lepidogrammitis drymoglossoides* (Baker) Ching	Drymoglossum-like Lepidogrammitis
骨牌蕨(骨牌草)	*Lepidogrammitis rostrata* (Bedd.) Ching	Beak-leaf Lepidogrammitis
瓦韦	*Lepisorus thunbergianus* (Kaulf.) Ching	Thunberg's Lepisorus

蕨类植物 PTERIDOPHYTA

攀援星蕨(东南星蕨)	*Microsorium buergerianum* (Miq.) Ching	Climbing Microsorium
江南星蕨(七星剑)	*Microsorium fortunei* (T. Moore) Ching	Fortune Microsorium
有翅星蕨	*Microsorium pteropus* (Bl.) Cop.	Java Fern
星蕨	*Microsorium punctatum* (L.) Cop.	Climbing Bird's Nest Fern
褐叶星蕨	*Microsorium superficiale* (Bl.) Ching	Brownleaf Microsorium
显脉星蕨	*Microsorium zippelii* (Bl.) Ching	Zippel Microsorium
多羽瘤蕨	*Phymatodes longissima* (Bl.) J. Smith	Many-pinnate Phymatodes
瘤蕨(茀蕨)	*Phymatodes scolopendria* (Burm. f.) Ching	Common Phymatodes
贴生石韦(上树咳)	*Pyrrosia adnascens* (Sw.) Ching	Tougue-fern
石韦(石兰)	*Pyrrosia lingua* (Thunb.) Farw.	Japanese Felt Fern

槲蕨科 Drynariaceae

槲蕨	*Drynaria fortunei* (Kunze) J. Sm. [*Drynaria roosii* Nakai]	Fortune's Drynaria
崖姜(穿石剑)	*Pseudodrynaria coronans* (Wall. ex Mett.) Ching	Rock-ginger Fern

鹿角蕨科 Platyceriaceae

*二歧鹿角蕨	*Platycerium bifurcatum* (Cav.) C. Chr.	Common Staghorn
鹿角蕨(蝙蝠蕨)	*Platycerium wallichii* Hook.	Indian Staghorn

*长叶鹿角蕨	*Platycerium willinckii* T. Moore [*P. bifurcatum* (Cav.) C. Chr. ssp. *willinckii* (T. Moore) Hennipman et M. C. Roos]	Java Staghorn
	禾叶蕨科 **Grammitidaceae**	
短柄禾叶蕨	*Grammitis dorsipila* (Christ) C. Chr. et Tard.	Short-stalk Grammitis
两广禾叶蕨	*Grammitis lasiosora* (Bl.) Ching	Hairysorus Grammitis
红毛禾叶蕨	*Grammitis hirtella* (Bl.) Tuyama	Red Hair Grammitis
	剑蕨科 **Loxogrammaceae**	
柳叶剑蕨(肺痨草)	*Loxogramme salicifolia* (Makino) Makino	Narrow-leaved Sword Fern
	苹科 **Marsileaceae**	
田字草(苹)	*Marsilea quadrifolia* L.	Water Shamrock
	槐叶苹科 **Salviniaceae**	
槐叶苹	*Salvinia natans* (L.) All.	Water Spangles
	满江红科 **Azollaceae**	
满江红(蒲桥)	*Azolla imbricata* (Roxb.) Nakai	Mosquito Fern

裸子植物 GYMNOSPERMAE

苏铁科 Cycadaceae

*越南篦齿苏铁	*Cycas elongata* (Leandri) D. Y. Yang et T. Chen	Elongated Cycad
仙湖苏铁	*Cycas fairylakea* D. Y. Wang	Fairylake Cycad
*软鳞苏铁	*Cycas furfuracea* W. V. Fitzg.	Scurfy Cycad
*海南苏铁	*Cycas hainanensis* C. J. Chen	Hainan Cycad
*元江苏铁	*Cycas parvulus* S. L. Yang	Yuanjiang Cycad
*蓖齿苏铁(凤尾蕉)	*Cycas pectinata* Griff.	Nepal Cycad
*苏铁	*Cycas revoluta* Thunb.	Sago-palm
*石山苏铁	*Cycas sexseminifera* F. N. Wei	Six-seeded Cycad
台湾苏铁(铁树)	*Cycas taiwaniana* Carruth.	Taiwan Cycad
*大型双子铁	*Dioon spinulosum* Dyer	Giant Dioon
*刺叶非洲铁	*Encephalartos ferox* Bertol. f.	Zululand Cycad
*鳞秕泽米苏铁(南美苏铁)	*Zamia furfuracea* L. f.	Cardboard Palm

银杏科 Ginkgoaceae

*银杏(白果树)	*Ginkgo biloba* L.	Ginkgo

松科 Pinaceae

*银杉(杉公子)	*Cathaya argyrophylla* Chun et Kuang	Cathay Silver Fir
油杉(海罗松)	*Keteleeria fortunei* (Murr.) Carr.	Fortune Keteleeria
*湿地松	*Pinus elliottii* Engelm.	Slash Pine

*华南五针松	*Pinus kwangtungensis* Chun ex Tsiang	Kwangtung Pine
南亚松(海南松)	*Pinus latteri* Mason	Latter Pine
马尾松(山松)	*Pinus massoniana* Lamb.	Chinese Red Pine
*油松(短叶马尾松)	*Pinus tabulaeformis* Carr.	Chinese Pine
	杉科 Taxodiaceae	
杉木(杉)	*Cunninghamia lanceolata* (Lamb.) Hook.	China Fir
水松	*Glyptostrobus pensilis* (Staunt.) K. Koch	Water Pine
*水杉	*Metasequoia glyptostroboides* Hu et Cheng	Dawn Redwood
*落羽杉(落羽松)	*Taxodium distichum* (L.) Rich.	Swamp Cypress
*池杉(池柏)	*Taxodium distichum* (L.) Rich. var. *imbricatum* (Bongn.) Parl. [*T. ascendens* Brongn.]	Pond Cypress
	柏科 Cupressaceae	
*翠柏	*Calocedrus macroleps* Kurz var. *macrolepis* Kurz	Chinese Incense Cedar
*日本扁柏(扁柏)	*Chamaecyparis obtusa* (Sieb. et Zucc.) Endl.	Japanese Cedar
*柏木(垂柏)	*Cupressus funebris* Endl.	Weeping Cypress
福建柏(建柏)	*Fokienia hodginsii* (Dunn) Henry et Thomas	Fokien Cypress
*侧柏(柏树)	*Platycladus orientalis* (L.) Franco [*Thuja orientalis* L.]	Chinese Arborvitae

*圆柏(桧柏)	*Sabina chinensis* (L.) Ait.	Chinese Juniper
*龙柏	*Sabina chinensis* (L.) Ait. 'Kaizuca'	Dragon Juniper
*铺地柏(爬地柏)	*Sabina procumbens* (Sieb. ex Endl.) Iwata et Kusaka	Creeping Juniper

南洋杉科 **Araucariaceae**

*贝壳杉(菲律宾贝壳杉)	*Agathis dammara* (Lamb.) Rich.	Queensland Kauri
*大叶南洋杉	*Araucaria bidwillii* Hook.	Bunya-bunya
*南洋杉(肯氏南洋杉)	*Araucaria cunninghamii* Sweet	Hoop Pine
*异叶南洋杉	*Araucaria heterophylla* (Salisb.) Franco	Norfolk Island Pine

罗汉松科 **Podocarpaceae**

长叶竹柏	*Nageia fleuryi* (Hickel) Laubenf.	Fleury Podocarpus
罗汉松	*Podocarpus macrophyllus* (Thunb.) D. Don	Buddhist Pine
*短叶罗汉松(小罗汉松)	*Podocarpus macrophyllus* (Thunb.) D. Don var. *maki* Endl.	Maki Podocarpus
竹柏	*Nageia nagi* (Thunb.) O. Kuntze [*Podocarpus nagi* (Thunb.) Zoll. et Mor. ex Zoll.]	Nagai Podocarpus
大叶罗汉松(百日青)	*Podocarpus neriifolius* D. Don	Thitmin

三尖杉科 **Cephalotaxaceae**

*三尖杉(粗榧)	*Cephalotaxus fortunei* Hook. f.	Fortune Plumyew

红豆杉科 Taxaceae

穗花杉	*Amentotaxus argotaenia* (Hance) Pilger	Common Amentotaxus

买麻藤科 Gnetaceae

罗浮买麻藤(买麻藤)	*Gnetum lofuense* C. Y. Cheng	Luofushan Joint-fir
小叶买麻藤(麻骨风)	*Gnetum parvifolium* (Warb.) C. Y. Cheng ex Chun	Small-leaved Joint-fir

被子植物 ANGIOSPERMAE

双子叶植物 DICOTYLEDONEAE

木兰科 Magnoliaceae

*鹅掌楸(马褂木)	*Liriodendron chinense* (Hemsl.) Sarg.	Chinese Tuliptree
*北美鹅掌楸(马褂木)	*Liriodendron tulipifera* L.	Tuliptree
香港木兰	*Magnolia championii* Benth.	Champion Magnolia
夜香木兰(夜合花)	*Magnolia coco* (Lour.) DC.	Coco Magnolia
*山玉兰	*Magnolia delavayi* Franch.	Delavay Magnolia
*玉兰(玉堂春)	*Magnolia denudata* Desr.	Yulan Magnolia
*荷花玉兰(洋玉兰)	*Magnolia grandiflora* L.	Laucel Magnolia
*紫玉兰(辛夷)	*Magnolia liliflora* Desr.	Lily Magnolia
*凹叶厚朴(温朴)	*Magnolia officinalis* Rehd. et Wils. ssp. *biloba* (Rehd. et Wils.) Cheng et Law	Twolobed Officinal Magnolia
长叶木兰	*Magnolia paenetalauma* Dandy	Long-leaf Magnolia
*二乔木兰	*Magnolia soulangeana* Soul.-Bod.	Saucer Magnolia
桂南木莲(仁昌木莲)	*Manglietia chingii* Dandy	Ching Manglietia
木莲(绿楠)	*Manglietia fordiana* Oliv.	Ford's Manglietia
*灰木莲(落叶木莲)	*Manglietia glauca* Bl.	Grey Manglietia

*大果木莲	*Manglietia grandis* Hu et Cheng	Large-fruit Manglietia
*海南木莲	*Manglietia hainanensis* Dandy	Hainan Manglietia
*红花木莲(木莲花)	*Manglietia insignis* (Wall.) Bl.	Red-flower Manglietia
*马关木莲	*Manglietia maguanica* Chang et B. L. Chen	Maguan Manglietia
*大叶木莲	*Manglietia megaphylla* Hu et Cheng	Large-leaf Manglietia
毛桃木莲	*Manglietia kwangtungensis* (Merr.) Dandy [*M. moto* Dandy]	Moto Manglietia
*乳源木莲	*Manglietia yuyuanensis* Y. W. Law	Yuyuan Manglietia
*白兰(缅桂)	*Michelia alba* DC.	White Champak
苦梓含笑(苦梓)	*Michelia balansae* (A. DC.) Dandy	Balanse Michelia
*黄兰(黄缅桂)	*Michelia champaca* L.	Champaca
乐昌含笑(景烈含笑)	*Michelia chapensis* Dandy	Tso Michelia
含笑(含笑花)	*Michelia figo* (Lour.) Spreng.	Banana Shrub
金叶含笑(金叶玉兰)	*Michelia foveolata* Merr. ex Dandy	Foveolate Michelia
*香籽含笑	*Michelia hedyosperma* Y. W. Law	Fragrant Michelia
醉香含笑(火力楠)	*Michelia macclurei* Dandy	McClure's Michelia
深山含笑(莫氏含笑)	*Michelia maudiae* Dunn	Maud's Michelia
*石碌含笑	*Michelia shiluensis* Chun et Y. F. Wu	Shilu Michelia

野含笑	*Michelia skinneriana* Dunn	Skinner Michelia
*云南含笑	*Michelia yunnanensis* Franch.	Yunnan Michelia
*乐东拟单性木兰（乐东木兰）	*Parakmeria lotungensis* (Chun et Tsoong) Y. W. Law	Lotung Parakmeria
*云南拟单性木兰	*Parakmeria yunnanensis* Hu	Yunnan Parakmeria
*合果木(山桂花)	*Paramichelia baillonii* (Pierre) Hu	Baillon Paramichelia
观光木(香花木)	*Tsoongiodendron odorum* Chun	Tsoong's Tree

八角科 Illiciaceae

厚皮香八角	*Illicium ternstroemioides* A. C. Smith	Ternstroemialike Anisetree

五味子科 Schisandraceae

黑老虎(冷饭团)	*Kadsura coccinea* (Lem.) A. C. Smith	Scarlet Kadusra
海风藤(异型南五味子)	*Kadsura heteroclita* (Roxb.) Craib	Curious Kadsura
南五味子(小号风沙藤)	*Kadsura longipedunculata* Finet et Gagnep.	Longpeduncule Kadsura

番荔枝科 Annonaceae

*刺果番荔枝	*Annona muricata* L.	Guananbana
*番荔枝	*Annona squamosa* L.	Sugar-apple
*鹰爪花	*Artabotrys hexapetalus* (L. f.) Bhandari	Eagle's Claw
香港鹰爪花	*Artabotrys hongkongensis* Hance	Hongkong Eagle's Claw
假鹰爪(酒饼叶)	*Desmos chinensis* Lour.	Chinese Desmos
白叶瓜馥木	*Fissistigma glaucescens* (Hance) Merr.	White-leaved Fissistigma

瓜馥木(钻山风)	*Fissistigma oldhamii* (Hemsl.) Merr.	Oldham's Fissistigma
多花瓜馥木(黑风藤)	*Fissistigma polyanthum* (Hook. f. et Thoms.) Merr.	Manyflower Fissistigma
香港瓜馥木(打鼓藤)	*Fissistigma uonicum* (Dunn.) Merr.	Hong Kong Fissistigma
*暗罗	*Polyalthia suberosa* (Roxb.) Thw.	Suberous Greenstar
嘉陵花	*Popowia pisocarpa* (Bl.) Endl.	Pea-like Fruit Popowia
光叶紫玉盘	*Uvaria boniana* Finet et Gagnep.	Glabrous-leaved Uvaria
山椒子(大花紫玉盘)	*Uvaria grandiflora* Roxb. ex Hornem.	Large-flowered Uvaria
紫玉盘(油饼木)	*Uvaria macrophylla* Roxb.	Largeleaf Uvaria

樟科 Lauraceae

美脉琼楠	*Beilschmiedia delicata* S. Lee et Y. T. Wei	Prettynerved Slugwood
网脉琼楠	*Beilschmiedia tsangii* Merr.	Tsang's Beilschmiedia
无根藤(罗纲藤)	*Cassytha filiformis* L.	Filiform Cassytha
毛桂(华南樟)	*Cinnamomum appelianum* Schewe	Haired-twig Cinnamon
肉桂(玉桂)	*Cinnamomum aromaticum* Nees [*C. cassia* Presl]	Cassia Bark Tree
阴香(香胶叶)	*Cinnamomum burmannii* (C. G. et Th. Nees) Bl.	Batavia Cinnamon
樟树(香樟)	*Cinnamomum camphora* (L.) Presl	Camphor Tree
*天竺桂(浙江樟)	*Cinnamomum japonicum* Sieb.	Japanese Cassia

野黄桂(桂皮树)	*Cinnamomum jensenianum* Hand.-Mazz.	Jensen's Cinnamon
*沉水樟	*Cinnamomum micranthum* (Hayata) Hayata	Small-flower Camphor Tree
黄樟(大叶樟)	*Cinnamomum porrectum* (Roxb.) Kosterm. [*C. parthenoxylon* (Jack) Meissn.]	Yellow Camphor Tree
粗脉樟	*Cinnamomum validinerve* Hance	Thick-nerved Cinnamon
*锡兰肉桂(锡兰檀)	*Cinnamomum zeylanicum* Nees	Cerlon Cinnamon
厚壳桂(白桂)	*Cryptocarya chinensis* (Hance) Hemsl.	Chinese Cryptocarya
黄果厚壳桂(生虫树)	*Cryptocarya concinna* Hance	Elegant Cryptocarya
乌药(香桂樟)	*Lindera aggregata* (Sims) Kosterm.	Tien-tai Spicebush
小叶乌药(小乌药)	*Lindera aggregata* (Sims) Kosterm. var. *playfairii* (Hemsl.) H. P. Tsui	Playfair's Spicebush
香叶树(香果树)	*Lindera communis* Hemsl.	Chinese Spicebush
山胡椒(油金楠)	*Lindera glauca* (Sieb. et Zucc.) Bl.	Greyblue Spicebush
华南山胡椒	*Lindera nacusua* (D. Don) Merr.	Hairy Spicebush
尖脉木姜子	*Litsea acutivena* Hayata	Sharp-veind Litsea
山苍子(山鸡椒)	*Litsea cubeba* (Lour.) Pers.	Fragrant Litse
潺槁木姜子(潺槁树)	*Litsea glutinosa* (Lour.) C. B. Rob.	Gluey Bark Litse

广西木姜子(广西新木姜子)	*Litsea kwangsiensis* H. T. Chang	Guangdong Litse
剑叶木姜子	*Litsea lancifolia* (Roxb. ex Nees) Benth. ex Hook. f.	Lanceleaf Litse
假柿木姜子(假柿树)	*Litsea monopetala* (Roxb.) Pers.	Rustyhairy Litse
圆叶豹皮樟	*Litsea rotundifolia* (Nees) Hemsl.	Roundleaf Litse
豹皮樟	*Litsea rotundifolia* (Nees) Hemsl. var. *oblongifolia* (Nees) Allen	Leopard Camphor
桂北木姜子	*Litsea subcoriacea* Yang et P. H. Huang	Leathery Leaf Litse
轮叶木姜子(槁树)	*Litsea verticillata* Hance	Whorl-leaf Litse
短序润楠(短花楠)	*Machilus breviflora* (Benth.) Hemsl.	Short-inflorescence Machilus
浙江润楠(长序润楠)	*Machilus chekiangensis* S. K. Lee	Chekiang Machilus
中华润楠(香港楠)	*Machilus chinensis* (Champ. ex Benth.) Hemsl.	Hong Kong Machilus
黄绒润楠(黄楠)	*Machilus grijsii* Hance	Yellow Machilus
薄叶润楠(华东楠)	*Machilus leptophylla* Hand.-Mazz.	Thin-leaved Machilus
多脉润楠(刨花润楠)	*Machilus pauhoi* Kanehira [*M. polyneura* H. T. Chang]	Pauho Machilus
柳叶润楠	*Machilus salicina* Hance	Willowleaf Machilus
红楠(红润楠)	*Machilus thunbergii* Sieb. et Zucc.	Red Machilus
绒毛润楠(绒楠)	*Machilus velutina* Champ. ex Benth.	Tomentose Machilus

香港新木姜	*Neolitsea cambodiana* Lec. var. *glabra* Allen	Glabrous Newlitse
鸭公树（大香籽）	*Neolitsea chuii* Merr.	Chu's Newlitse
广西新木姜	*Neolitsea kwangsiensis* Liou	Guangxi Newlitse
假肉桂（土肉桂）	*Neolitsea levinei* Merr.	Levine's Newlitse
显脉新木姜	*Neolitsea phanerophlebia* Merr.	Conspicuous-nerved Newlitse
南亚新木姜	*Neolitsea zeylanica* (Nees) Merr.	Ceylon Newlitse
*鳄梨（牛油果）	*Persea americana* Mill.	Avocado
*竹叶楠（小叶桢楠）	*Phoebe faberi* (Hemsl.) Chun	Faber's Phoebe
*檫木（檫树）	*Sassafras tzumu* (Hemsl.) Hemsl.	Chinese Sassafras

青藤科 Illigeraceae

青藤（宽叶青藤）	*Illigera celebica* Miq.	Illigera

毛茛科 Ranunculaceae

威灵仙（铁角威灵仙）	*Clematis chinensis* Osbeck	Chinese Clematis
厚叶铁线莲	*Clematis crassifolia* Benth.	Thick-leaved Clematis
甘木通（丝铁线莲）	*Clematis filamentosa* Dunn	Long-hairy Clematis
山木通	*Clematis finetiana* Levl. et Vant.	Finet Clematis
单叶铁线莲（老虎须）	*Clematis henryi* Oliv.	Henry Clematis
毛柱铁线莲（铜通美）	*Clematis meyeniana* Walp.	Meyen's Clematis
柱果铁线莲（黑木通）	*Clematis uncinata* Champ.	Hooked Clematis

*飞燕草(千鸟草)	*Consolida ajacis* (L.) Schur	Rocket Consolida
小回回蒜(自扣草)	*Ranunculus cantoniensis* DC.	Moslem Garlie
茴茴蒜(鸭脚板)	*Ranunculus chinensis* Bunge	Chinese Buttercup
毛茛	*Ranunculus japonicus* Thunb.	Japanese Buttercup
石龙芮(野堇菜)	*Ranunculus sceleratus* L.	Celery-leaved Crowfoot
尖叶唐松草(石笋还阳)	*Thalictrum acutifolium* (Hand.-Mazz.) Boivin	Meadow-rue
阴地唐松草	*Thalictrum umbricola* Ulbr.	Shade Meadow-rue

金鱼藻科 Ceratophyllum

金鱼藻(松藻)	*Ceratophyllum demersum* L.	Hornwort

睡莲科 Nymphacaceae

*莼菜(马蹄草)	*Brasenia schreberi* J. F. Gmel.	Common Watershield
*美国黄莲	*Nelumbo lutea* Pers.	American Lily
*荷花(莲花)	*Nelumbo nucifera* Gaertn.	Sacred Lily
*萍蓬草(萍蓬莲)	*Nuphar pumilum* (Timm.) DC.	Dwarf Cowlily
*白睡莲(睡莲)	*Nymphaea alba* L.	Pygmy Water-lily
*红睡莲	*Nymphaea alba* L. var. *rubra* Lonnr.	Red Water-lily
*齿叶睡莲	*Nymphaea lotus* L.	Egyptian Water Lily Lotus
*柔毛齿叶睡莲	*Nymphaea lotus* L. var. *pubescens* (Willd) Hook. f. et Thoms.	Pubescent Water-Lily
*黄睡莲	*Nymphaea mexicana* Zucc.	Yellow Water-Lily
*香睡莲	*Nymphaea odorata* Ait.	Fragrant Water-Lily
*印度红睡莲	*Nymphaea rubra* Roxb. ex Salisb.	Indian Red Water-Lily

*星花睡莲	*Nymphaea nouchali* Burm. f. [*N. stellata* Willd.]	Blue Indianlotus
*睡莲(子午莲)	*Nymphaea tetragona* Georgi	Pygmy Water Lily
*亚马孙王莲	*Victoria amazonica* Sowerby	Amazon Water Lily
*克鲁兹王莲	*Victoria cruziana* Orbign	Santa Cruz Water Lily

小檗科 Berberidaceae

*六角莲(独脚莲)	*Dysosma pleiantha* (Hance) Woodson	East-China Many-flowered May-apple
八角莲	*Dysosma versipellis* (Hance) M. Cheng ex Ying	Many-flowered May-apple
*阔叶十大功劳 (鸟不宿)	*Mahonia bealeibealii* (Fort.) Carr.	Beal's Mahonia
*小果十大功劳	*Mahonia bodinieri* Gagnep.	Bodinier Mahonia
*华南十大功劳	*Mahonia japonica* (Thunb.) DC.	Japanese Mahonia
海岛十大功劳	*Mahonia oiwakensis* Hayata	Island Mahonia
*南天竹	*Nandina domestica* Thunb.	Sacred Bamboo

木通科 Lardizabalaceae

白木通(三叶木通)	*Akebia trifoliata* (Thunb.) Koidz. ssp. *australis* (Diels) T. Shimizu	Austral Akebia
野木瓜(铁脚梨)	*Stauntonia chinensis* DC.	Chinese Stauntonia
牛藤果(那藤)	*Stauntonia brunoniana* (Wall. ex Hemsl.) Decne. ssp. *elliptica* (Hemsl.) H. N. Qin [*S. elliptica* Hemsl.]	Ovate-leaved Stauntonia
倒卵叶野木瓜	*Stauntonia obovata* Hemsl.	Ovate-leaved Stauntonia

大血藤科 Sargentodoxaceae

大血藤(血藤)	*Sargentodoxa cuneata* (Oliv.) Rehd. et Wils.	Sargent gloryvine

防己科 Menispermaceae

木防己	*Cocculus orbiculatus* (L.) DC.	Snail Seed
毛叶轮环藤	*Cyclea barbata* Miers	Barbate Cyclea
粉叶轮环藤	*Cyclea hypoglauca* (Schauer) Diels	Glaucousleaf Cyclea
轮环藤(百解藤)	*Cyclea racemosa* Oliv.	Racemose Cyclea
秤钩风	*Diploclisia affinis* (Oliv.) Diels	Similar Diploclisia
苍白秤钩风(穿墙风)	*Diploclisia glaucescens* (Bl.) Diels	Glaucescent Diploclisia
内伶仃秤钩风	*Diploclisia renincarpa* Miau et W. B. Liao	Neilingding Diploclisia
夜花藤	*Hypserpa nitida* Miers	Shining Hypserpa
细圆藤	*Pericampylus glaucus* (Lam.) Merr.	Gryblue Pericampylus
金线吊乌龟	*Stephania cepharantha* Hayata	Oriental Stephania
*千金藤	*Stephania japonica* (Thunb.) Miers	Japanese Stephania
粪箕笃	*Stephania longa* Lour.	Long Stephania
粉防己(汉防己)	*Stephania tetrandra* S. Moore	Fourstamen Stephania
青牛胆(山慈姑)	*Tinospora sagittata* (Oliv.) Gagnep.	Arrow-shaped Tinospora
华青牛胆(宽筋藤)	*Tinospora sinensis* (Lour.) Merr.	Chinese Tinospora

马兜铃科 Aristolochiaceae

广防己(滇防己)	*Aristolochia fangchi* Y. C. Wu ex L. D. Chow et S. M. Hwang	Fangchi
通城虎(五虎通城)	*Aristolochia fordiana* Hemsl.	Ford Dutchmanspipe
大叶马兜铃(圆叶马兜铃)	*Aristolochia kaempferi* Willd.	Curious Dutchmanspipe
香港马兜铃(威氏马兜铃)	*Aristolochia westlandii* Hemsl.	Westland's Birthwort
圆叶细辛(尾花细辛)	*Asarum caudigerum* Hance	Caudate Wildginger

猪笼草科 Nepenthaceae

猪笼草(猪笼入水)	*Nepenthes mirabilis* (Lour.) Druce	Pitcher Plant

胡椒科 Piperaceae

*西瓜皮椒草(豆瓣绿椒草)	*Peperomia argyreia* E. Morren	Watermelon Peperomia
石蝉草(火伤叶)	*Peperomia blanda* (Jacq.) Kunth [*P. dindygulensis* Miq.]	Dindygule Peperomia
*皱叶椒草(皱叶豆瓣绿)	*Peperomia caperata* Yunck.	Wrinkled-leaf Peperomia
硬毛草胡椒	*Peperomia cavaleriei* C. DC.	Cavaleri Peperomia
*灰绿椒草(银叶豆瓣绿)	*Peperomia incaua* A. Dietr.	Grey-green Peperomia
*钝叶椒草	*Peperomia obtusifolia* (L.) A. Dietr.	Obtuse-leaf Peperomia
草胡椒	*Peperomia pellucida* (L.) Kunth	Shiny Peperomia

*斑叶垂椒草(蔓性椒草)	*Peperomia serpens* C. DC. 'Variegata'	Spotted-leaf Creeping Peperomia
豆瓣绿(椒草)	*Peperomia tetraphylla*(G. Forst.) Hook. et Arn.	Four-leaf Peperomia
小叶爬崖香	*Piper arboricola* C. DC.	Small-leaf Pepper
华南胡椒	*Piper austrosinense* Y. C. Tseng	South China Pepper
蒌叶	*Piper betle* L.	Betel Pepper
山蒟	*Piper hancei* Maxim.	Hance's Pepper
香港蒟(毛蒟)	*Piper hongkongense* Hatusima [*P. puberulum* (Benth.) Maxim.]	Hong Kong Pepper
*荜拔	*Piper longum* L.	Long-leaf Pepper
*胡椒	*Piper nigrum* L.	Black Pepper
假蒟(哈蒟荜拔子)	*Piper sarmentosum* Roxb.	Short-spiked Pepper

三白草科 Saururaceae

鱼腥草(蕺菜)	*Houttuynia cordata* Thunb.	Heartleaf Houttuynia
三白草(水木通)	*Saururus chinensis*(Lour.) Baill.	Lizard's Tail

金粟兰科 Chloranthaceae

珠兰(珍珠兰)	*Chloranthus erectus*(Buch.-Ham.) Verdc.	Erect Chlorantus
四块瓦(四对叶)	*Chloranthus holostegius* Hand.-Mazz.	Integrifolious Chlorantus
单穗金粟兰	*Chloranthus monostachys* R. Br.	One-spike Chlorantus
多穗金粟兰	*Chloranthus multistachys* Pei	Manyspike Chlorantus
金粟兰	*Chloranthus spicatus*(Thunb.) Makino	Pearl-orchid

草珊瑚(鸡爪兰)	*Sarcandra glabra* (Thunb.) Nakai	Common Sarcandra
海南草珊瑚	*Sarcandra hainanensis* (Pei) Swamy et Bailey	Hainan Sarcandra

罂粟科 Papaveraceae

*博落回(号筒杆)	*Macleaya cordata* (Willd.) R. Br.	Pink Plumepoppy

白花菜科 Capparidaceae

广州槌果藤(广州山柑)	*Capparia cantoniensis* Lour.	Canton Caper
*黄醉蝶花	*Cleome lutea* Hook.	Yellow Spider-flower
*醉蝶花(凤蝶草)	*Cleome spinosa* Jacq.	Giant Spider-flower
臭矢菜(黄花草)	*Cleome viscosa* L.	Stink Grass
鱼木	*Crateva religiosa* G. Forst.	Spider Tree
单室鱼木(钝叶鱼木)	*Crateva trifoliata* (Roxb.) Sun	Obtuse-leaved Crateva

辣木科 Moringaceae

*象腿树(辣木)	*Moringa drouhardii* Jumelle	Bottle Tree

十字花科 Cruciferae

芥蓝(白花芥蓝)	*Brassica alboglabra* L. H. Bailey	Chinese Kale
擘蓝(芥兰头)	*Brassica caulorapa* DC. ex Laveille	Kohlrabi
青菜(白菜)	*Brassica chinensis* L.	Chinese White Cabbage
全叶芥(苦菜)	*Brassica integrifolia* (H. West) O. E. Schulz	Bitter Mustard
芥菜	*Brassica juncea* (L.) Czern. et Coss.	Leaf-mustard

根用芥菜(辣疙瘩)	*Brassica juncea* (L.) Coss. var. *megarrhiza* Tsen et Lee	Datoucai Leaf-mustard
* 大头菜(芜菁)	*Brassica napobrassica* Mill.	Rutabaga
* 羽衣甘蓝(叶牡丹)	*Brassica oleracea* L. var. *acephala* f. *tricolor* Hort.	Borecole
* 裂叶羽衣甘蓝	*Brassica oleracea* L. var. *acephala* f. *partita* Hort.	Lobed-leaf Borecole
甘蓝(椰菜)	*Brassica oleracea* L. var. *capitata* L.	Cabbage
菜心(菜薹)	*Brassica parachinensis* Bailey	Flowering Chinese Cabbage
白菜绍菜(黄牙白)	*Brassica pekinensis* (Lour.) Skeels	Peking Cabbage
荠菜	*Capsella bursa-pastoris* (L.) Medic.	Shepherd's Purse
弯曲碎米荠	*Cardamine flexuosa* With.	Bitter Cress
碎米荠(野荠菜)	*Cardamine hirsuta* L.	Pennsylvania Bittercress
圆齿碎米荠	*Cardamine scutata* Thunb.	Scutate-dentate Bittercress
* 香雪球(小白花)	*Lobularia maritima* (L.) Desv.	Spice snowball
* 紫罗兰	*Matthiola incana* (L.) R. Br.	Stock
西洋菜(豆瓣菜)	*Nasturtium officinale* R. Br.	Water Cress
萝卜(萝白)	*Raphanus sativus* L.	Garden Radish
长羽萝卜	*Raphanus sativus* L. var. *longipinnatus* Bailey	Chinese Radish
广东蔊菜(微子蔊菜)	*Rorippa cantoniensis* (Lour.) Ohwi	Guangzhou Yellowcress
无瓣蔊菜(塘葛菜)	*Rorippa dubia* (Pers.) Hara	Petalless Yellowcress
球果蔊菜(风花菜)	*Rorippa globosa* (Turcz.) Vassilcz.	Globate Yellowcress

葶菜(塘葛菜)	*Rorippa indica* (L.) Hiern	India Yellowcress

堇菜科 Violaceae

戟叶堇菜(紫花地丁)	*Viola betonicifolia* J. E. Smith	Wild Violet
蔓茎堇菜(七星莲)	*Viola diffusa* Ging	Spreading Violet
长萼堇菜(犁头草)	*Viola inconspicua* Bl.	Longsepal Violet
*香堇(香堇菜)	*Viola odorata* L.	Garden Violet
*紫花地丁(箭头草)	*Viola philippica* Cav. ssp. *munda* W. Beck	Purpleflower Violet
柔毛堇菜	*Viola principis* H. de Boiss.	Pubescent Violet
毛堇菜	*Viola thomsonii* Oudem.	Thomson Vionet
*大花三色堇(鬼脸花)	*Viola tricolor* L. var. *hortensis* DC.	Garden Pansy
堇菜(如意草)	*Viola verecunda* A. Gray	Common Violet

远志科 Polygalaceae

黄花远志(鸡根)	*Polygala arillata* Buch.-Ham. ex D. Don	Cionbag Milkwort
黄花倒水莲(屈头难)	*Polygala fallax* Hemsl.	Yellowflower Milkwort
金不换(臭苏)	*Polygala glomerata* Lour.	Southern China Milkwort
香港远志	*Polygala hongkongensis* Hemsl.	Hong Kong Milkwort
齿果草(莎萝莽)	*Salomonia cantoniensis* Lour.	Salomonia
蝉翼藤(蝉翼木)	*Securidacu inappendiculata* Hassk.	Securidaca

景天科 Crassulaceae

*明镜(盘叶莲花掌)	*Aeonium tabuliforme* Webb et Berth.	Tabletform Aeonium

*落地生根(灯笼花)	*Bryophyllum pinnatum* (L. f.) Oken	Air-plant
*玉树(景天树)	*Crassula arborescens* Willd.	Silver Jade Plant
*青锁龙(若绿)	*Crassula lycopodioides* Lam.	Moss Crassula
*燕子掌(肉质万年青)	*Crassula ovata* (Mill.) Druce	Jade Plant
*石莲花(八宝掌)	*Echeveria glauca* Baker	Glaucous Echeveria
*大叶落地生根(宽叶落地生根)	*Kalanchoe daigremontiana* Hamet et Perr.	Large-leaf Kalanchoe
*伽蓝菜	*Kalanchoe ceratophylla* Haw.	Kalanchoe
*条裂伽蓝菜(伽蓝花)	*Kalanchoe lanciniata* (L.) DC.	Laciniate Kalanchoe
*趣蝶莲(双飞蝴蝶)	*Kalanchoe synsepala* Baker	Cup Kalanchoe. Walking Kalanchoe
*唐印	*Kalanchoe thyrsiflora* Harv.	Paddle Plant
*棒花落地生根	*Kalanchoe tubiflora* (Harvey) Raym. -Hamet	Tubular Kalanchoe
*洋吊钟(玉吊钟)	*Kalanchoe verticillata* Elliot	Verticillate Kalanchoe
*万点星(雀䋟)	*Sedum acre* L.	Stonecrop
佛甲草(禾雀䋟)	*Sedum lineare* Thunb.	Buddhanail
*翡翠景天(串珠草)	*Sedum marganianum* Walth.	Malachite Stonecrop
*松叶景天	*Sedum mexicanum* Britt.	Mexican Stonecrop
*垂盆草(柔枝景天)	*Sedum sarmentosum* Bunge	Stringy Stonecrop

虎耳草科 Saxifragaceae

黄常山(鸡骨风)	*Dichroa febrifuga* Lour.	Antifebrile Dichroa

被子植物 ANGIOSPERMAE

鸡眼梅花草	*Parnassia wightiana* Wall. ex Wight et Arn.	Chicken eye Parnassia

茅膏菜科 Droseraceae

*捕蝇草(食虫草)	*Dionaea muscipula* J. Ellis	Venus Fly-trap
茅膏菜(锦地罗)	*Drosera burmannii* Vahl	Burmann Sundew
匙叶茅膏菜(小毛毡苔)	*Drosera spathulata* Labill.	Spathulate Sundew
宽苞茅膏菜	*Drosera spathulata* Labill. var. *loureirii* (Hook. et Arn.) Y. Z. Ruan	Spathulate Sundew

瓶子草科 Sarraceniaceae

*长叶瓶子草	*Sarracenia leucophylla* Raf.	Crimson Pitcherplant

沟繁缕科 Elatinaceae

田繁缕(密花草)	*Bergia ammanioides* Roxb. ex Roth	Common Bergia
大叶田繁缕	*Bergia capensis* L.	Bigleaf Bergia
三蕊沟繁缕	*Elatine triandra* Schkuhr	Threestamen Waterwort

石竹科 Caryophyllaceae

蚤缀(无心菜)	*Arenaria seropyllifolia* L.	Sandwort
*康乃馨(香石竹)	*Dianthus caryophyllus* L.	Carnation, Clove Pink
*石竹(洛阳花)	*Dianthus chinensis* L.	Chinese Pink
*日本石竹	*Dianthus japonicus* Thunb.	Japan Pink
*瞿麦	*Dianthus superbus* L.	Fringed Pink
荷莲豆	*Drymaria cordata* (L.) Willd.	Cordate Drymaria
*满天星(宿根霞草)	*Gypsophila paniculata* L.	Babys-breath
牛繁缕(鹅肠草)	*Malachium aquaticum* (L.) Fries	Aquatic Malachium

白鼓钉(声色草)	*Polycarpaea corymbosa* (L.) Lam.	Whitedrumnail
多荚草	*Polycarpon prostratum* (Forssk.) Aschers. et Schweinw.	Fruitful-grass
繁缕(鱼肠菜)	*Stellaria media* (L.) Cyr.	Common Chickweed
雀舌草(石灰草)	*Stellaria alsine* Grinum	Bog Starwort

粟米草科 Molluginaceae

| 簇花粟米草 | *Glinus oppositifolius* (L.) A. DC. [*Mollugo oppositifolia* L.] | Longstalk Glinus |
| 粟米草 | *Glinus stricta* L. [*Mollugo pentaphylla* L.] | Indian Chickweed |

番杏科 Aizoaceae

*宝绿(牛舌花)	*Glottiphyllum linguiforme* N. E. Br.	Tonguaeleaf Flower
海马齿(猪母菜)	*Sesuvium portulacastrum* L.	Seaside Purslane
番杏(新西兰菠菜)	*Tetragonia tetragonioides* (Pall.) O. Kuntze	New-Zealand Spinach

马齿苋科 Portulacaceae

*马齿苋树(绿玉树)	*Portulacaria afra* (L.) Jacq.	Elephant Bush
*半支莲(松叶牡丹)	*Portulaca grandiflora* Hook.	Rose-moss
马齿苋(瓜子菜)	*Portulaca oleracea* L.	Purslane
多毛马齿苋(毛马齿苋)	*Portulaca pilosa* L.	Pilous Purslane
土人参(假人参)	*Talinum paniculatum* (Jacq.) Gaertn.	Mock Ginseng

蓼科 Polygonaceae

金线草(朱砂七)	*Antenoron filiforme* (Thunb.) Rob. et Vant.	Goldthreadweed
*珊瑚藤(假菩提)	*Antigonon leptopus* Hook. et Arn.	Honolulu Vine
*竹节蓼(百足草)	*Homalocladium platycladum* (F. Muell.) Bailey	Centipede Plant
*绿帚(扫帚草)	*Kochia scoparia* Schrad. var. *sieversiana* (Pall.) Ulbr. f. *trichophylla* (Hort.) Schinz et Thell.	Broomsedge
扁蓄(粉节草)	*Polygonum aviculare* L.	Knotgrass
毛蓼(水辣蓼)	*Polygonum barbatum* L.	Hairy Polygonum
小毛蓼(小蓼子草)	*Polygonum barbatum* L. var. *gracile* (Danser) Steward	Slender Hairy Polygonum
红辣蓼(簇蓼)	*Polygonum caespitosum* Bl.	Clustered Knotweed
火炭母(五毒草)	*Polygonum chinense* L.	Smartweed
虎杖(苦杖)	*Polygonum cuspidatum* Sieb. et Zucc. [*Reynoutria japonica* Houtt.]	Tigerstick
光蓼(红辣蓼)	*Polygonum glabrum* Willd.	Smooth Knotweed
长剑叶蓼	*Polygonum hastato-sagittatum* Makino	Long Arrowleaf Knotweed
水蓼(辣蓼)	*Polygonum hydropiper* L.	Water Smartweed
蚕茧草(蓼子草)	*Polygonum japonicum* Meisn.	Japancsc Polygonum
山蓼(酸浆菜)	*Polygonum juncundum* Meisn.	Joyful Knotweed
大马蓼(白辣蓼)	*Polygonum lapathifolium* L.	White Smartweed
柳叶蓼(辣蓼草)	*Polygonum lapathifolium* L. var. *salicifolium* Sibth.	Dockleaved Knotweed

何首乌(首乌)	*Polygonum multiflorum* Thunb.	Many-flowered Polygonum
小蓼花	*Polygonum muricatum* Meissn.	Rough Knotweed
红蓼(东方蓼)	*Polygonum orientale* L.	Prince's Feather
杠板归(白草)	*Polygonum perfoliatum* L.	Perfoliate Knotweed
腋花蓼(珠仔草)	*Polygonum plebeium* R. Br.	Knotweed
香蓼(粘毛蓼)	*Polygonum viscosum* Ham. Buch. -Ham. ex D. Don	Fragrant Polygonum
假菠菜(假大黄)	*Rumex maritimus* L.	Golden Dock

商陆科 Phytolaccaceae

| 商陆(牛萝卜) | *Phytolacca acinosa* Roxb. | Indian Pokeberry |
| *美洲商陆(洋商陆) | *Phytolacca americana* L. | Common Pokeberry |

藜科 Chenopodiaceae

君达菜(牛皮菜)	*Beta vulgaris* L. var. *cicla* L.	Swiss Chard
藜(灰灰菜)	*Chenopodium acuminatum* Willd. ssp. *virgatum* (Thunb.) Kitam.	Narrowleaf Goosefoot
土荆芥(鹅脚草)	*Chenopodium ambrosioides* L.	Amercian Wormseed
小藜(灰苋头)	*Chenopodium ficifolium* Smith	Figleaf Goosefoot
*灰绿藜(小灰菜)	*Chenopodium glaucum* L.	Oakleaf Goosefoot
菠菜	*Spinacia oleracea* L.	Common Spinach
南方碱蓬	*Suaeda australis* (R. Br.) Moq.	South Sea-Blite

苋科 Amaranthaceae

土牛膝(倒扣草)	*Achyranthes aspera* L.	Common Achyranthes
牛膝(土牛膝)	*Achyranthes bidentata* Bl.	Twotooth Achyranthes
*红草(红苋草)	*Alternanthera bettzickiana* (Regel) Yoss	Garden Alternanthera

被子植物 ANGIOSPERMAE

空心莲子草(水花生)	*Alternanthera philoxeroides* (Mart.) Griseb.	Alligator Alternanthera
莲子草(虾钳菜)	*Alternanthera sessilis* (L.) DC.	Sessile Alternanthera
*红绿草(五色苋)	*Alternanthera tenella* Colla	Joyweed
三色苋(雁来红)	*Amaranthus tricolor* L.	Chinese Spinach
刺苋(簕苋菜)	*Amaranthus spinosus* L.	Spiny Amaranth
野苋(假苋菜)	*Amaranthus viridis* L.	Green Amaranth
青葙(野鸡冠)	*Celosia argentea* L.	Celosia
*圆绒鸡冠(鸡冠花)	*Celosia cristata* L.	Common Cockscomb
*凤尾鸡冠	*Celosia cristata* L. var. *plumosa* Hort.	Plumed Celosia
杯苋	*Cyathula prostrate* (L.) Bl.	Prostrate Cyathula
银花苋	*Gomphrena celosioides* Mart.	Silverflower Globeamaranth
千日红	*Gomphrena globosa* L.	Globeamaranth
*红花千日红	*Gomphrena globosa* L. var. *rubra* Hort.	Red Globeamaranth
*圆叶红苋(血苋)	*Iresine herbstill* Hook. f. ex Lindl.	Herbst Bloodleaf

落葵科 Basellaceae

落葵(潺菜)	*Basella alba* L.	Malabar-Nightshade

亚麻科 Linaceae

*石海椒	*Reinwardtia indica* Dumort	Yellow Flax

牻牛儿苗科 Geraniaceae

*香叶天竺葵	*Pelargonium graveolens* L'Her.	Fragrant-leaved Geranium
*天竺葵	*Pelargonium* × *hortorum* Bailey	Fish Geranium

酢浆草科 Oxalidaceae

*阳桃(杨桃)	*Averrhoa carambola* L.	Carambola
感应草	*Biophytum sensitivum* (L.) DC.	Reactiongrass
*大花酢浆草	*Oxalis bowiei* Lindl.	Largeflower Woodsorrel
酢浆草	*Oxalis corniculata* L.	Creeping Woodsorrel
*红花酢浆草(大酸味草)	*Oxalis corymbosa* DC.	Corymb Woodsorrel

旱金莲科 Tropaeolaceae

*旱金莲(金莲花)	*Tropaeolum majus* L.	Common Nasturtium

凤仙花科 Balsaminaceae

凤仙花(指甲花)	*Impatiens balsamina* L.	Garden Balsam
华凤仙	*Impatiens chinensis* L.	Chinese Snapweed
*新几内亚凤仙花(五彩凤仙花)	*Impatiens hawkeri* W. Bull	Hawker Snapweed
香港凤仙花	*Impatiens hongkongensis* Grey-Wilson	Hongkong Snapweed
*非洲凤仙花(矮凤仙)	*Impatiens walleriana* Hook. f.	Waller Snapweed

千屈菜科 Lythraceae

耳基水苋(耳叶水苋菜)	*Ammannia arenaria* H. B. K.	Sand Ammannia
泽水苋(水苋菜)	*Ammannia baccifera* L.	Common Water Ammannia
*紫雪茄花(雪茄花)	*Cuphea platycentra* Lemarie [*C. ignea* A. DC.]	Broadspur Cuphea
*香膏萼距花(香膏草)	*Cuphea balsamona* Cham. et Schlecht.	Balsam Cuphea

*萼距花(紫花满天星)	*Cuphea hookeriana* Walp.	Hooker Cuphea
*细叶萼距花(满天星)	*Cuphea hyssopifolia* H. B. K.	False Heather
*紫薇(痒痒树)	*Lagerstroemia indica* L.	Crape Myrtle
*大花紫薇(大叶紫薇)	*Lagerstroemia speciosa* (L.) Pers.	Queen Crape Myrtle
南紫薇	*Lagerstroemia subcostata* Koehne	South Crape Myrtle
*散沫花(指甲花)	*Lawsonia inermis* L.	Henna
*千屈菜(水柳)	*Lythrum salicaria* L.	Spiked Loosestrife
节节菜(水马齿苋)	*Rotala indica* (Willd.) Koehne	Indian Rotala
圆叶节节菜(水瓜子菜)	*Rotala rotundifolia* (Buch.-Ham. ex Roxb.) Koehne	Round-leaved Rotala

安石榴科 Punicaceae

*石榴(安石榴)	*Punica granatum* L.	Pomegranate
*重瓣安石榴	*Punica granatum* L. 'Multiplex'	Doublepetalous White Pomegranate

柳叶菜科 Onagraceae

*倒挂金钟(灯笼花)	*Fuchsia hybrida* Hort. ex Sieb. et Voss.	Common Fuchsia
假丁香蓼(过塘蛇)	*Ludwigia adscendens* (L.) Hara [*Jussiaea repens* L.]	Water-dragon
草龙	*Ludwigia hyssopifolia* (G. Don) Exell [*Jussiaea linifolia* Vahl]	Titimo
毛草龙	*Ludwigia octovalvis* (Jacq.) Raven	Primrose Willow

细花丁香蓼(水丁香)	*Ludwigia peploides* (Kunth) Raven [*L. caryophylla* (Lam.) Merr. et Metc.]	Pinkleaf Seedbox
细叶丁香蓼(小花水丁香)	*Ludwigia perennis* L.	Rote Stern Ludwigia
丁香蓼(小丁香)	*Ludwigia prostrata* Roxb.	Climbing Seedbox
滨海月见草(海边月见草)	*Oenothera drummondii* Hook.	Beach Primrose

菱科 Trapaceae

菱(菱角)	*Trapa bicornis* Osbeck	Horn Nut

小二仙草科 Haloragaceae

黄花小二仙草	*Haloragis chinensis* (Lour.) Merr.	China Seaberry
小二仙草	*Haloragis micrantha* (Thunb.) R. Br. ex Sieb. et Zucc.	Smallflower Seaberry

水马齿科 Callitrichaceae

广东水马齿	*Callitriche oryzetorum* Petr.	Guangdong Waterstarwort
水马齿	*Callitriche stagnalis* Scop.	Water Starwort

瑞香科 Thymelaeaceae

土沉香(牙香树)	*Aquilaria sinesis* (Lour.) Spreng.	China Eaglewood
白瑞香	*Daphne papyracea* Wall. ex Steud.	Papery Daphne
了哥王	*Wikstroemia indica* (L.) C. A. Mey.	Indian Wikstroemia
北江荛花	*Wikstroemia monnula* Hance	Lovely Stringbush
细轴荛花	*Wikstroemia nutans* Champ. ex Benth.	Nodding Wikstroemia

紫茉莉科 Nyctaginaceae

*光叶子花	*Bougainvillea glabra* Choisy	Paper Flower
*叶子花(簕杜鹃)	*Bougainvillea spectabilis* Willd.	Beautiful Bougainvillea
紫茉莉(胭脂花)	*Mirabilis jalapa* L.	Beauty-of-the-night

山龙眼科 Proteaceae

*银桦	*Grevillea robusta* Cunn. ex R. Br.	Silk Oak
小叶山龙眼(越南山龙眼)	*Helicia cochinchinensis* Lour.	Cochinchina Helicia
广东山龙眼	*Helicia kwangtungensis* W. T. Wang	Guangdong Helicia
网脉山龙眼	*Helicia reticulata* W. T. Wang	Reticulate Helicia
*全缘叶澳洲坚果	*Macadamia integrifolia* Maiden et Betche	Macadamianut
*澳洲坚果(夏威夷果)	*Macadamia terrifolia* F. Muell.	Terrateleaf Macadamia

五桠果科 Dilleniaceae

*五桠果(第伦桃)	*Dillenia indica* L.	Elephant Apple
大花五桠果(大花第伦桃)	*Dillenia turbinata* Finet et Gagnep.	Turbinate Dillenia
锡叶藤(雪藤)	*Tetracera asiatica* (Lour.) Hoogl.	Sandpaper Vine

海桐花科 Pittosporaceae

光叶海桐(崖子花)	*Pittosporum glabratum* Lindl.	Glabrous Pittosporum
狭叶海桐	*Pittosporum glabratum* var. *neriifolium* Rehd. et Wils.	Oleander-leaf Seatung
海桐	*Pittosporum tobira* (Thunb.) Ait.	Japanese Pittosporum

红木科 Bixaceae

*红木(胭脂树)	*Bixa orellana* L.	Arnotto Dye Plant

大风子科 Flacourtiaceae

*海南大风子	*Hydnocarpus hainanensis* (Merr.) Sleum.	Hainan Chaulmoogratree
箣柊	*Scolopia chinensis* (Lour.) Clos	Chinese Scolopia
广东箣柊	*Scolopia saeva* (Hance) Hance	Kwangtung Scolopia
长叶柞木	*Xylosma longifolium* Clos	Long-leaved Xylosma

天料木科 Samydaceae

嘉赐树	*Casearia glomerata* Roxb.	Clustered Casearia
毛叶嘉赐树	*Casearia villilimba* Merr.	Villous Casearia
华南天料木	*Homalium austro-chinense* G. S. Fan	Southern China Homalium
天料木	*Homalium cochinchinensis* (Lour.) Druce	Cochinchina Homalium
*红花天料木(母生)	*Homalium hainanense* Gagnep.	Hainan Homalium

西番莲科 Passifloraceae

*蓝翅西番莲	*Passiflora alatocaerulea* Lindl.	Skybluewing Passionflower
*蓝花西番莲	*Passiflora coerulea* L.	Bluecrown Passionflower
*蝙蝠西番莲	*Passiflora capsularis* L.	Capsule-fruited Passionflower
*蛇王藤	*Passiflora cochichinensis* Spreng.	Snakeking Vine
*鸡蛋果(紫果西番莲)	*Passiflora edulis* Sims	Purple Granadilla
*龙珠果(龙须果)	*Passiflora foetida* L.	Passion flower

葫芦科 Cucurbitaceae

中文名	学名	英文名
冬瓜	*Benincasa hispida* (Thunb.) Cogn.	Wax Gourd
节瓜(毛瓜)	*Benincasa hispida* (Thunb.) Cogn. var. *chieh-qua* How	Hairy Melon
西瓜	*Citrullus lanatus* (Thunb.) Mats. et Nakai	Water Melon
*木鳖	*Momordica cochinchinenses* (Lour.) Spreng.	Cochinchina Momordica
*甜瓜(香瓜)	*Cucumis melo* L.	Musk Melon
白瓜	*Cucumis melo* L. var. *conomon* (Thunb.) Makino	Oriental Pickling Melon
黄瓜(青瓜)	*Cucumis sativus* L.	Cucumber
南瓜(番瓜)	*Cucurbita moschata* (Duch. ex Lam.) Duch. ex Poir.	Crookneck Squash
*观赏瓜	*Cucurbita pepo* L. var. *ovifera* (L.) Harz	Ornamental Pumpkin
金瓜	*Gymnopetalum chinensis* (L.) Merr.	Goldenmelon
绞股蓝	*Gynostemma pentaphyllum* (Thunb.) Makino	Fiveleaf Gynostemma
*油渣果	*Hodgsonia macrocarpa* (Bl.) Cogn.	Large-fruit Hodgsonia
葫芦	*Lagenaria siceraria* (Molina) Standl.	Bottle Gourd
*蒲瓜	*Lagenaria siceraris* (Molina) Standl. var. *hispida* (Thunb.) Hara	Hispid Calabash
丝瓜	*Luffa acutangula* (L.) Roxb.	Angled Luffa
水瓜	*Luffa aegyptica* Mill.	Vegetable Sponge

*苦瓜(凉瓜)	*Momordica charantia* L.	Bitter Cucumber
木鳖子	*Momordica cochinchinensis* (Lour.) Spreng.	Wooden Tortoise
*佛手瓜	*Sechium edule* (Jacq.) Swartz	Chayote
茅瓜	*Solene amplexicaulis* (Lam.) Gandhi	Solena
大苞赤瓟(心叶赤瓟)	*Thladiantha cordifolia* (Bl.) Cogn.	Heartleaf Tubergourd
*蛇瓜(蛇丝瓜)	*Trichosanthes anguina* L.	Serpentgourd
王瓜	*Trichosanthes curcumeroides* (Ser.) Maxim.	Snakegourd
*栝楼	*Trichosanthes kirilowii* Maxim.	Mongolian Snakegourd
多型栝楼(全缘栝楼)	*Trichosanthes ovigera* Bl.	Entire Snakegourd
趾叶栝楼(瓜蒌)	*Trichosanthes pedata* Merr. et Chun	Pedateleaf Snakegourd
老鼠拉冬瓜	*Zehneria indica* (Lour.) Keraudren	Inida Zehneria
钮子瓜	*Zehneria maysorensis* (Wight et Arn.) Arn.	Button Melon

秋海棠科 Begoniaceae

*豹耳秋海棠	*Begonia bowerae* Ziesenh.	Eyelash Begonia
粗喙秋海棠	*Begonia crassirostris* Irmsch.	Thickrostrate Begonia
*丽格秋海棠(玫瑰海棠)	*Begonia elatior* Hort. ex Steud.	Elatior Begonia
紫背天葵	*Begonia fimbristipula* Hance	Mountain Begonia
*费氏秋海棠(绯氏秋海棠)	*Begonia fischeri* Schrank	Fischer Begonia

被子植物 ANGIOSPERMAE

*独活叶秋海棠（枫叶海棠）	*Begonia heracleifolia* Cham. et Schldl.	Star Begonia
*帝王秋海棠（地毡秋海棠）	*Begonia imperialis* Lem.	Green Imperial Begonia
*竹节秋海棠	*Begonia maculata* Raddi	Spotted Begonia
*铁十字秋海棠（马蹄秋海棠）	*Begonia masoniana* Iremsch.	Iron-cross Begonia
裂叶秋海棠	*Begonia palmata* D. Don	Palmate Begonia
蛤蟆叶秋海棠（蟆叶秋海棠）	*Begonia rex* Putz.	Assamking Begonia
*斑叶秋海棠	*Begonia rex-cultorum* Bailey	Spotted-leaf Begonia
*四季秋海棠（蚬肉秋海棠）	*Begonia semperflorens* Link et Otto	Perpetual Begonia
*球根秋海棠（茶花海棠）	*Begonia tuberhybrida* Voss.	Tuberous Begonia

番木瓜科 Caricaceae

*木瓜（番木瓜）	*Carica papaya* L.	Papaya

仙人掌科 Cactaceae

*天轮柱	*Cereus fernambucensis* Lem.	Madacaru
*仙人球	*Echinopsis multiplex* Pfeiff. et Otto	Thin Hedgehogcactus
*令箭荷花（红孔雀）	*Nopalxochia ackermennii* (Haw.) F. M. Knuth	Red Orchid Cactus
*昙花	*Epiphyllum oxypetalum* (DC.) Haw.	Orchid Cactus
*量天尺（三棱柱）	*Hylocereus undatus* (Haw.) Britt. et Rose	Pitaya
*仙人掌	*Opuntia stricta* (Haw.) Haw. var. *dillenii* (Ker-Gawl.) L. D. Benson	Prickly-pear

木麒麟	*Pereskia aculeata* Mill.	Leafy Cactus
*巴西蟹爪	*Schlumbergera bridgesii* (Lem.) Loefgr.	Christmas Cactus
*蟹爪兰	*Schlumbergera truncata* (Haw.) Moran. [*Zygocactus truncatus* (Haw.) K. Schum.]	Crab Cactus
*狮子头	*Thelocactus lophothele* Britt. et Rose	Lionhead Thelocactus

茶科 Theaceae

*毛杨桐	*Adinandra glischroloma* Hand.-Mazz.	Liangguang Adinandra
海南杨桐	*Adinandra hainanensis* Hayata	Hainan Adinandra
黄瑞木(杨桐)	*Adinandra millettii* (Hook. et Arn.) Benth. et Hook. f. ex Hance	Adinandra
普洱茶	*Camellia assamica* (Mast.) Chang	Puer Tea
香港毛蕊茶	*Camellia assimilis* Champ. et Benth.	Pointed-leaf Camellia
短柱茶	*Camellia brevistyla* (Hayata) Coh. St.	Shrotstyle Camellia
长尾毛蕊茶(尾叶山茶)	*Camellia caudata* Wall.	Tail-leaved Camellia
*红花油茶	*Camellia chekiang-oleosa* Hu	Zhejiang Camellia
红皮糙果茶(克氏茶)	*Camellia crapnelliana* Tutch.	Crapnell's Camellia
柃叶茶(绿青果)	*Camellia euryoides* Lindl.	Euryaleaf Tea
大苞白山茶(葛亮洪茶)	*Camellia granthamiana* Sealy	Grantham's Camellia

*山茶花(红山茶)	*Camellia japonica* L.	Japan Camellia
落瓣油茶	*Camellia kissi* Wall.	Kiss Camellia
金花茶(亮叶离蕊茶)	*Camellia nitidissima* Chi	Goldenflower Tea
油茶(茶子树)	*Camellia oleifera* Abel	Oil Tea
柳叶毛蕊茶(柳叶茶)	*Camellia salicifolia* Champ. ex Benth.	Willow-leaved Camellia
*茶梅	*Camellia sasanqua* Thunb.	Sassanqua
南山茶(红花油茶)	*Camellia semiserrata* Chi	Halftooth Camellia
茶(茶树)	*Camellia sinensis* (L.) O. Kuntze	Tea
红淡比(森氏杨桐)	*Cleyera japonica* Thunb.	Japan Cleyera
耳叶柃	*Eurya auriformis* Chang	Eared Eurya
米碎花(岗茶)	*Eurya chinensis* R. Br.	False Tea
华南毛柃	*Eurya ciliata* Merr.	Ciliata Eurya
二列叶柃	*Eurya distichophylla* Hemsl.	Two-ranked-leaf Eurya
岗柃(米碎木)	*Eurya groffii* Merr.	Groff's Eurya
细枝柃(亮叶柃)	*Eurya loquaiana* Dunn	Slenderbranch Eurya
黑柃	*Eurya macartneyi* Champ.	Macartney's Eurya
格药柃(隔药柃)	*Eurya muricata* Dunn	Muricate Eurya
细齿叶柃(柃木)	*Eurya nitida* Korth.	Shining Eurya
窄叶柃	*Eurya stenophylla* Mcrr.	Narrowleaf Eurya
毛果柃	*Eurya trichocarpa* Korth.	Hairyfruit Eurya
大头茶	*Gordonia axillaris* (Roxb.) Dietr.	Hongkong Gordonia
木荷(荷树)	*Schima superba* Gardn. et Champ.	Gugertree

西南木荷(红荷木)	*Schima wallichii* Choisy	Wallich Gugertree
厚皮香	*Ternstroemia gymnanthera* (Wight et Arn.) Bedd.	Nakedanther Ternstroemia
石笔木	*Tutcheria championii* Nakai	Champion Slatepenciltree
小果石笔木	*Tutcheria microcarpa* Dunn	Smallfruit Slatepentree

五列木科 Pentaphylacaceae

五列木	*Pentaphylax euryoides* Gardn. et Champ.	Common Pentaphylax

猕猴桃科 Actinidiaceae

灰毛猕猴桃	*Actinidia cinerascens* C. F. Liang	Greyhair Kiwifruit
猕猴桃	*Actinidia deliciosa* C. F. Liang [*A. chinensis* Planch. var. *hispida* C. F. Liang]	Hispid Actinidia
毛花猕猴桃	*Actinidia eriantha* Benth.	Hairyflower Kiwifruit
阔叶猕猴桃	*Actinidia latifolia* (Gardn. et Champ.) Merr.	Broad-leaved Actinidia

水冬哥科 Saurauiaceae

水东哥(水枇杷)	*Saurauia tristyla* DC.	Common Saurauia

龙脑香科 Dipterocarpaceae

*青梅(青皮)	*Vatica mangachapoi* Blanco	Stellatehair Vatica

桃金娘科 Myrtaceae

肖蒲桃(赛赤楠)	*Acmena acuminatissima* (Bl.) Merr. et Perry	Sharpleaf Acmena
岗松	*Baeckea frutescens* L.	Dwarf Mountain Pine
*松叶红千层	*Callistemon pinifolius* Sweet	Pine-leaf Bottle-brush
*红千层(红瓶刷)	*Callistemon rigidus* R. Br.	Stiff Bottle-brush

*串钱柳(垂枝红千层)	*Callistemon viminalis* G. Don ex Loud.	Tall Bottle-brush
水翁(水榕)	*Cleistocalyx operculatus* (Roxb.) Merr. et Perry	Operculate Waterfig
*赤桉	*Eucalyptus camaldulensis* Dehnh.	Longbeak Eucalyptus
*柠檬桉	*Eucalyptus citriodora* Hook. f.	Lemon-scented Gum
*蓝桉	*Eucalyptus globulus* Labill.	Blue Gum
*大叶桉(尤加利)	*Eucalyptus robusta* Smith	Swamp Mahogany
*细叶桉	*Eucalyptus tereticornis* J. E. Smith	Forest Gray Gum
*棱果蒲桃(红果仔)	*Eugenia uniflora* L.	Surinam Cherry
*细叶白千层	*Melaleuca parviflora* Lindl.	Littleflower Melaleuca
*白千层	*Melaleuca quinquenervia* (Cav.) S. T. Blake [*M. leucadendron* auct. non L.]	Paper-bark Tree
*番石榴(鸡矢果)	*Psidium guajava* L.	Guava
桃金娘(山稔)	*Rhodomyrtus tomentosa* (Ait.) Hassk.	Rose Myrtle
赤楠	*Syzygium buxifolium* Hook. ex Arn.	Boxleaf Syzygium
灶地乌骨木(子凌蒲桃)	*Syzygium championii* (Benth.) Merr. et Perry	Champion Syzygium
*乌墨(海南蒲桃)	*Syzygium cumini* (L.) Skeels	Duhat
卫矛蒲桃	*Syzygium euonymifolium* (Metc) Merr. et Perry	Euonymusleaf Syzygium
*水蒲桃(水石榴)	*Syzygium fluviatile* (Hemsl.) Merr. et Perry	Waterbamboo Syzygium

轮叶蒲桃(三叶赤楠)	*Syzygium grijsii* (Hance) Merr. et Perry	Whorlleaf Syzygium
红车(红鳞蒲桃)	*Syzygium hancei* Merr. et Perry	Hance Syzygium
蒲桃	*Syzygium jambos* (L.) Alston	Rose-apple
山蒲桃(白车)	*Syzygium levinei* (Merr.) Merr. et Perry	Wild Syzygium
香蒲桃	*Syzygium odoratum* (Lour.) DC.	Fragrant Syzygium
红枝蒲桃	*Syzygium rehderianum* Merr. et Perry	Rehder Syzygium
*洋蒲桃(莲雾)	*Syzygium samarangense* (Bl.) Merr. et Perry	Java Apple
*红胶木	*Lophostemon confertus* (R. Br.) P. G. Wilson et J. T. Waterhouse [*Tristania conferta* R. Br.]	Brisbane Box

野牡丹科 Melastomataceae

毛药花(棱果花)	*Barthea barthei* (Hance ex Benth.) Krass.	Southern China Barthea
柏拉木	*Blastus cochinchinensis* Lour.	Cochinchina Blastus
*小花柏拉木	*Blastus congniauxii* Stapf	Cogniaux Blastus
*宝莲花(宝莲灯)	*Medinilla magnifica* Lindl.	Magnificent Medinilla
多花野牡丹	*Melastoma affine* D. Don	Manyflower Melastoma
野牡丹	*Melastoma candidum* D. Don	Common Melastoma
地稔(铺地锦)	*Melastoma dodecandrum* Lour.	Twelvestamen Melastoma
细叶野牡丹	*Melastoma intermedium* Dunn	Intermediate Melastoma
展毛野牡丹	*Melastoma normale* D. Don	Patenthairy Melastoma
毛稔	*Melastoma sanguineum* Sims	Bloodred Melastoma

谷木	*Memecylon ligustrifolium* Champ. ex Benth.	Privetleaf Memecylon
黑叶谷木	*Memecylon nigrescens* Hook. et Arn.	Blackleaf Memecylon
金锦香	*Osbeckia chinensis* L.	Chinese Osbeckia
*银毛野牡丹	*Tibouchina aspera* Aubl. var. *asperrima* Cogn.	Rough Tibouchina
虎颜花	*Tigridiopalma magnifica* C. Chen	Magnific Tigridiopalma

使君子科 Combretaceae

风车子	*Combretum alfredii* Hance	Alfred Windmill
*使君子	*Quisqualis indica* L.	Rangoon Creeper
*阿江榄仁	*Terminalia arjuna* Wight et Arn.	Arjuna Myrobalan
*马尼拉榄仁(菲律宾榄仁)	*Terminalia calamansanai* (Blanco) Rolfe	Philippine Almond
榄仁树(法国枇杷)	*Terminalia catappa* L.	Indian Almond
*小叶榄仁(非洲榄仁树)	*Terminalia mantalyi* H. Perrier	Madagascar Terminalia
*美洲榄仁	*Terminalia muelleri* Benth.	Australian Almond
千果榄仁	*Terminalia myriocarpa* Van Huerck et Muell. Arg.	Bayberry Waxmyrtlefruit Terminalia

红树科 Rhizophoraceae

木榄	*Bruguiera gymnorrhiza* (L.) Lam.	Many-petaled Mangrove
竹节树	*Carallia brachiata* (Lour.) Merr.	India Carallia
秋茄	*Kandelia candel* (L.) Druce	Kandelia

金丝桃科 Hypericaceae

黄牛木	*Cratoxylum cochinchinense* (Lour.) Bl.	Common Oxwood
田基黄(地耳草)	*Hypericum japonicum* Thunb. ex Murray	Japanese St. Johnswort
金丝桃	*Hypericum monogynum* L.	Monostyle St. Johnswort

藤黄科 Guttiferae

横经席(薄叶胡桐)	*Calophyllum membranaceum* Gardn. et Champ.	Membranaceus Beautyleaf
多花山竹子	*Garcinia multiflora* Champ. ex Benth.	Many-flowered Garcinia
岭南山竹子	*Garcinia oblongifolia* Champ. et Benth.	Oblongleaf Garcinia
*菲岛福木(福树)	*Garcinia subelliptica* Merr.	Philippine Garcinia
*铁力木(铁梨木)	*Mesua ferrea* L.	Common Mesua

椴树科 Tiliaceae

甜麻	*Corchorus aestuans* L.	Sweet Jute
*黄麻	*Corchorus capularis* L.	Roundpod Jute
*小花扁担杆	*Grewia biloba* G. Don var. *parviflora* (Bunge) Hand.-Mazz.	Small-flower Grewia
扁担杆	*Grewia biloba* D. Don	Bilobed Grewia
破布叶(布渣叶)	*Microcos paniculata* L.	Paniculate Microcos
*文定果	*Muntingia colabura* L.	Common Muntingia
长钩刺蒴麻(毛刺蒴麻)	*Triumfetta cana* Bl.	Hair Triumfetta
刺蒴麻	*Triumfetta rhomboidea* Jacq.	Common Triumfetta

杜英科 Elaeocarpaceae

*尖叶杜英(长芒杜英)	*Elaeocarpus apiculatus* Mast.	Apiculate Elaeocarpus
中华杜英	*Elaeocarpus chinensis* (Gardn. et Champ.) Hook. f. ex Benth.	Chinese Elaeocarpus
杜英(胆八树)	*Elaeocarpus decipiens* Hemsl.	Common Elaeocarpus
拟杜英(显脉杜英)	*Elaeocarpus dubius* A. DC.	Mock Elaeocarpus
*水石榕(海南杜英)	*Elaeocarpus hainanensis* Oliv.	Hainan Elaeocarpus
日本杜英(薯豆)	*Elaeocarpus japonicus* Sieb. et Chun	Japanese Elaeocarpus
绢毛杜英	*Elaeocarpus nitentifolius* Merr. et Chun	Shiningleaf Elaeocarpus
长柄杜英	*Elaeocarpus petiolatus*(Jack) Wall. ex Kurz	Longpetiole Elaeocarpus
*锡兰橄榄(锯叶杜英)	*Elaeocarpus serratus* L.	Ceylon Elaeocarpus
山杜英	*Elaeocarpus sylvestris* (Lour.) Poir.	Woodland Elaeocarpus
薄果猴欢喜	*Sloanea leptocarpa* Diels	Thinfruit Monkeyjoy
猴欢喜	*Sloanea sinensis* (Hance) Hemsl.	China Monkeyjoy

梧桐科 Sterculiaceae

*槭叶瓶干树(槭叶桐)	*Brachychiton acerifolius* (Cunn.) F. Muell.	Flame Bottletree
*昆士兰瓶干树(酒瓶树)	*Brachychiton rupestris* (Lindl.) Schum.	Queensland Bottletree
刺果藤	*Byttneria aspera* Col.	Burvine

*海南梧桐	*Firmiana hainanensis* Kosterm.	Hainan Phoenix Tree
*梧桐	*Firmiana simplex* (L.) W. F. Wight	Chinese Parasol-tree
山芝麻	*Helicteres angustifolia* L.	Helicteres
*长柄银叶树	*Heritiera angustata* Pierre	Longstalk Silvertree
银叶树(大白叶仔)	*Heritiera littoralis* Dryand.	Looking-glass Tree
蝴蝶树(加卜)	*Heritiera parvifolia* Merr.	Smallleaf Silvertree
马松子	*Melochia corchorifolia* L.	Juteleaf Melochia
翻白叶树(半枫荷)	*Pterospermum heterophyllum* Hance	Heterophyllous Wingseedtree
窄叶翅子树	*Pterospermum lanceaefolium* Roxb.	Lanceleaf Wingseedtree
两广梭罗(梭罗树)	*Reevesia thyrsoideas* Lindl.	Bunch-like Reevesia
假苹婆	*Sterculia lanceolata* Cav.	Scarlet Sterculia
苹婆(凤眼果)	*Sterculia nobilis* Smith	Noble Bottle-tree
蛇婆子	*Waltheria americana* L.	Florida Waltheria
木棉科 Bombacaceae		
木棉(英雄花)	*Bombax ceiba* L. [*B. malabaricum* DC.]	Tree Cotton
*白花异木棉(美丽异木棉)	*Ceiba insignis* (Kunth) Gibbs et Semir [*Chorisia insignis* H. B. K.]	Pink Silk Cotton Tree
*吉贝(爪哇木棉)	*Ceiba pentandra* (L.) Gaertn.	Kapok Ceiba
*马拉巴栗(发财树)	*Pachira macrocarpa* (Cham. et Schlcht.) Walp.	Largefruit Pachira
锦葵科 Malvaceae		
黄葵(黄蜀葵)	*Abelmoschus moschatus* (L.) Medic.	Muskmallow

磨盘草	*Abutilon indicum* (L.) Sweet	India Abutilon
*金玲花	*Abutilon striatum* Dickson	Striped Abutilon
苘麻	*Abutilon theophrastii* Medic.	Chingma Abutilon
*蜀葵	*Althaea rosea* (L.) Cavan.	Hollyhock
*紫叶槿(红叶槿)	*Hibiscus acetosella* Welw. ex Hiern	Redleaf Hibiscus
木麻槿(红麻)	*Hibiscus cannabinus* L.	Kenaf Hibiscus
*红秋葵(槭叶秋葵)	*Hibiscus coccineus*(Medic.) Walt.	Scarlet Rosemallow
*海滨木槿(海槿)	*Hibiscus hamabo* Sieb. et Zucc.	Yellow Hibiscus
全叶美丽芙蓉	*Hibiscus indicus* (Burm. f.) Hochr.	Indian Hibiscus
木芙蓉	*Hibiscus mutabilis* L.	Changeable Rose-mallow
*吊灯扶桑(吊灯花)	*Hibiscus schizopetalus* (Mast.) Hook. f.	Coral Hibiscus
木槿	*Hibiscus syriacus* L.	Rose of Sharon
黄槿	*Hibiscus tiliaceus* L.	Cuban Bast
*锦葵	*Malva sinensis* Cav.	China Mallow
野葵(冬葵)	*Malva verticillata* L.	Musk Mallow
赛葵	*Malvastrum coromandelianum* (L.) Garcke	False Mallow
*悬铃花	*Malvaviscus arboreus* Cav. var. *penduliflorus* (DC.) Schery	Turk's Cap
黄花稔	*Sida acuta* Burm. f.	Acute Sida
桤叶黄花稔	*Sida alnifolia* L.	Alderleaf Sida
小叶黄花稔	*Sida alnifolia* L. var. *microphylla* (Cavan.) S. Y. Hu	Littleleaf Sida

中华黄花稔	*Sida chinensis* Retz.	China Sida
长梗黄花稔	*Sida cordata*（Burm. f.）Boiss.	Longstalk Sida
心叶黄花稔	*Sida cordifolia* L.	Heart-leaved Sida
白背黄花稔	*Sida rhombifolia* L.	Sida Hemp
榛叶黄花稔	*Sida subcordata* Span.	Subcordate Sida
杨叶肖槿(恒春黄槿)	*Thespesia populnea*（L.）Soland. ex Corr.	Portia Tree
地桃花(肖梵天花)	*Urena lobata* L.	Rose Mallow
中华地桃花	*Urena lobata* L. var. *chinensis*（Osbeck）S. Y. Hu	China Rose Mallow
梵天花(狗脚迹)	*Urena procumbens* L.	Procumbent Indian Mallow

金虎尾科 Malpighiaceae

风车藤(风筝果)	*Hiptage benghalensis*（L.）Kurz	Bengal Hiptage
*金英	*Thryallis gracilis* O. Kuntze	Slender Thryallis
*星果藤(三星果)	*Tristellateia australasiae* A. Rich.	Maiden's Jealousy

粘木科 Ixonanthaceae

| 粘木 | *Ixonanthes chinensis* Champ. | Chinese Ixonanthes |

大戟科 Euphorbiaceae

铁苋菜	*Acalypha australis* L.	Australian Acalypha
*狗尾红(红穗铁苋)	*Acalypha hispida* Burm. f.	Redhot Cat-tail
*红桑(红叶桑)	*Acalypha wilkesiana* Muell. -Arg.	Copper Leaf
山麻杆	*Alchornea rugosa*（Lour.）Muell. -Arg.	Pinnatevein Christmas Bush

被子植物 ANGIOSPERMAE

红背山麻杆	*Alchornea trewioides* (Benth.) Muell. -Arg.	Christmas Bush
*石栗	*Aleurites moluccana* (L.) Willd.	Candlenut Tree
五月茶	*Antidesma bunius* (L.) Spreng.	Salamander-tree
黄毛五月茶	*Antidesma fordii* Hemsl.	Salamander Tree
方叶五月茶	*Antidesma ghaesembilla* Gaertn.	Black Currant Tree
日本五月茶	*Antidesma japonicum* Sieb. et Zucc.	Dense-flowered China Laurel
柳叶五月茶	*Antidesma pseudomicrophyllum* Croiz.	Narrowleaf Laurel
小叶五月茶	*Antidesma venosum* E. Mey. ex Tul.	Small-leaved China Laurel
银柴(大沙叶)	*Aporosa dioica* (Roxb.) Muell. -Arg.	Chinese Aporusa
秋枫(常绿重阳木)	*Bischofia javanica* Bl.	Java Bishopwood
重阳木	*Bischofia polycarpa* (Levl.) Airy-Shaw	Squarelea China Laurel
黑面神	*Breynia fruticosa* (L.) Hook. f.	Waxy Leaf
*雪花木(白雪树)	*Breynia nivosa* (W. C. Sm.) Small	Snow Bush
小叶黑面神(小黑面叶)	*Breynia vitis-idaea* (Burm. f.) C. E. C. Fisch. [*B. officinalis* Hemsl.]	Officinal Breynia
禾串树	*Bridelia insulana* Hance	Insular Bridelia
土密藤	*Bridelia stipularis* (L.) Bl.	Stipulate Breynia

土密树（逼迫子）	*Bridelia tomentosa* Bl. [*B. monoica* Merr.]	Pop-gum Seed
白桐树（刁了棒）	*Claoxylon indicum* (Reinw. ex Bl.) Hassk.	Common Claoxylon
蝴蝶果	Cleidiocarpon cavaleriei (Levl.) Airy Shaw	Butterflyfruit
棒柄花	*Cleidion brevipetiolatum* Pax et Hoffm.	Shortpetiole Cleidion
*洒金榕（变叶木）	*Codiaeum variegatum* (L.) A. Juss. var. *pictum* (Lodd.) Muell.-Arg.	Croton
*长叶变叶木（变色叶）	*Codiaeum variegatum* (L.) A. Juss. var. *pictum* (Lodd.) Muell.-Arg. f. *ambigum* Pax	Longleaf Croton
*蜂腰洒金榕	*Codiaeum variegatum* (L.) A. Juss. var. *pictum* (Lodd.) Muell.-Arg. f. *appendiculatum* Pax	Appendiculate Croton
*宽叶变叶木	*Codiaeum variegatum* (L.) A. Juss. var. *pictum* (Lodd.) Muell. Arg. f. *platylla* Pax	Wideleaf Croton
鸡骨香	*Croton crassifolius* Geisel.	Thick-leaved Croton
毛果巴豆（桃叶双眼龙）	*Croton lachnocarpus* Benth.	Hairy-fruited Croton
巴豆（双眼龙）	*Croton tiglium* L.	Croton-oil Plant
钝叶核果木	*Drypetes obtusa* Merr. et Chun	Blunt-fruit Drypetes
黄桐	*Endospermum chinense* Benth.	Endospermum
*红麒麟	*Euphorbia aggregata* A. Berger	Aggregated Euphorbia
*肖黄栌（紫锦木）	*Euphorbia cotinifolia* L.	Caribbean Copper Plant

*猩猩草(草本一品红)	*Euphorbia cyathophora* Murray	Mexican Fire Plant
*桃叶猩猩草	*Euphorbia heterophylla* L.	Fireplant
飞扬草	*Euphorbia hirta* L.	Spurge
*地锦	*Euphorbia humifusa* Willd.	Humifuse Spurge
通奶草(光叶小飞杨)	*Euphorbia hypericifolia* L.	Garden Spurge
*银边翠	*Euphorbia marginata* Pursh	Silvemargin Spurge
*虎刺梅(铁海棠)	*Euphorbia milli* Ch. des Moul.	Crown of Thorns
*麒麟角(玉麒麟)	*Euphorbia neriifolia* L. 'Cristata'	Crested Neriumleaf Euphorbia
铺地草(伏生大戟)	*Euphorbia prostrata* Ait.	Prostrate Spurge
*一品红(圣诞花)	*Euphorbia pulcherrima* Willd. ex Klotzsch	Poinsettia
千根草(小飞扬草)	*Euphorbia thymifolia* L.	Thyme-leaved Spurge
*光棍树(绿玉树)	*Euphorbia tirucalli* L.	Milk-bush
*彩云阁(三角麒麟)	*Euphorbia trigona* Haw.	Triangular Euphorbia
海漆树	*Excoecaria agallocha* L.	Milky Mangrove
红背桂	*Excoecaria cochinchinensis* Lour.	Redback Osmanthus
毛果算盘子	*Glochidion eriocarpum* Champ. ex Benth.	Hairy-fruited Abacus Plant
厚叶算盘子	*Glochidion hirsutum* (Roxb.) Voigt [*G. dasyphyllum* K. Koch]	Thick-leaved Abacus Plant
大叶算盘子(艾胶算盘子)	*Glochidion lanceolarium* (Roxb.) Voigt [*G. macrophyllum* Benth.]	Large-leaved Abacus Plant

甜叶算盘子(菲岛算盘子)	*Glochidion philippicum* (Cav.) C. B. Rob.	Philippines Abacus Plant
算盘子(算珠树)	*Glochidion puberum* (L.) Hutch.	Abacus Plant
白背算盘子	*Glochidion wrightii* Benth.	Wright's Abacus Plant
香港算盘子(美短盘)	*Glochidion zeylanicum* (Gaertn.) A. Juss. [*G. hongkongense* Muell.-Arg.	Hong Kong Abacus Plant
*巴西橡胶树(橡胶树)	*Hevea brasiliensis* (Willd. ex A. Juss.) Muell.-Arg.	Brazil Rubbertree
麻疯树(黄肿树)	*Jatropha curcas* L.	Physic Nut
棉叶麻疯树	*Jatropha gossypifolia* L.	Belly-ache Bush
*琴叶珊瑚(琴叶樱)	*Jatropha integerrima* Jacq.	Spicy Jatropha
*佛肚花(玉树珊瑚)	*Jatropha podagrica* Hook.	Gout Stalk
刺果血桐	*Macaranga auriculata* (Merr.) Airy-Shaw	Earfruit Macaranga
鼎湖血桐	*Macaranga sampsonnii* Hance	Sampson Macaranga
血桐	*Macaranga tanarius* (L.) Muell.-Arg.	Elephant's Ear
白背叶	*Mallotus apelta* (Lour.) Muell.-Arg.	White-back Mallotus
粗毛野桐	*Mallotus hookerianus* (Seem.) Muell.-Arg.	Hooker Mallotus
圆叶野桐(山苦茶)	*Mallotus peltatus* (Geisel.) Muell.-Arg. [*M. oblongifolia* (Miq.) Muell.-Arg.]	Wild Bittertea
白楸	*Mallotus paniculatus* (Lam.) Muell.-Arg.	Turn-in-the-wind

粗糠柴	*Mallotus philippinensis* (Lam.) Muell.-Arg.	Kamala Tree
石岩枫	*Mallotus repandus* (Willd.) Muell.-Arg.	Creepy Mallotus
木薯	*Manihot esculentus* Crantz	Cassava
*红雀珊瑚(大银龙)	*Pedilanthus tithymaloides* (L.) Poit.	Redbird Cactus
越南叶下珠(乌蝇翼)	*Phyllanthus cochinchinensis* (Lour.) Spreng.	Vietnam Leaf-flower
余甘子(油甘树)	*Phyllanthus emblica* L.	Myrobalan
珠子草	*Phyllanthus niruri* L.	Necklace Plant
小果叶下珠(烂头钵)	*Phyllanthus reticulatus* Poir.	Reticulated Leaf-flower
无毛小果叶下珠	*Phyllanthus reticulatus* Poir. var. *glaber* Muell.-Arg.	Glabrous Reticulated Leaf-flower
叶下珠	*Phyllanthus urinaria* L.	Night-closing Leaf
蜜甘草	*Phyllanthus ussuriensis* Rupr. ex Maxim.	Longstalked Phyllanthus
黄珠子草	*Phyllanthus virgatus* Forst. f.	Twiggy Phyllanthus
*蓖麻	*Ricinus communis* L.	Castorbean
山乌桕	*Sapium discolor* (Champ. ex Benth.) Muell.-Arg.	Moutain Tallow
乌桕(蜡子树)	*Sapium sebiferum* (L.) Roxb.	Chinese Tallow-tree
艾堇(海滨守宫木)	*Sauropus bacciformis* (L.) Airy-Shaw	Berry-shaped Sauropus
*龙脷叶	*Sauropus spatulifolius* Beille	Spatulate leaf Sauropus
地阳桃	*Sebastiania chamaelea* (L.) Muell.-Arg.	Creeping Sebastiania

白饭树	*Securinega virosa* (Roxb. ex Willd.) Baill. [*Fluggea virosa* (Roxb. ex Willd.) Voigt]	Common Bushweed
中华三宝木	*Trigonostemon chinensis* Merr.	Chinese Trigonostemon
油桐(三年桐)	*Vernicia fordii* (Hemsl.) Airy-Shaw	Oiltung
木油桐(千年桐)	*Vernicia montana* Lour.	Muyou Oiltung
	交让木科 Daphniphyllaceae	
牛耳枫	*Daphniphyllum calycinum* Benth.	Calys-shaped Tigernanmu
交让木	*Daphniphyllum macropodum* Miq.	Macropodous Tigernanmu
虎皮楠	*Daphniphyllum oldhamii* (Hemsl.) Rosenth.	Oldham Tigernanmu
	小盘木科 Pandaceae	
小盘木	*Microdesmis caseraiifolia* Planch. ex Hook.	Saucerwood
	鼠刺科 Escalloniaceae	
鼠刺	*Itea chinensis* Hook. et Arn.	Chinese Sweetspire
矩形叶鼠刺	*Itea oblonga* Hand.-Mazz.	Oblong-leaf Sweetspire
	绣球花科 Hydrangeaceae	
常山(黄常山)	*Dichroa febrifuga* Lour.	Antifebrile Dichroa
*八仙花(绣球花)	*Hydrangea macrophylla* (Thunb.) Ser.	Largeleaf Hydrangea
星毛冠盖藤	*Pileostegia tomentella* Hand.-Mazz.	Tomentose Pileostegia
冠盖藤	*Pileostegia viburnoides* Hook. f. et Thoms.	Common Pileostegia

蔷薇科 Rosaceae

蛇莓	*Duchesnea indica* (Andr.) Focke	Snake Strawberry
山枇杷(大花枇杷)	*Eriobotrya cavaleriei* (Levl.) Rehd.	Bigflower Loquat
香花枇杷	*Eriobotrya fragrans* Champ. ex Benth.	Wild Loquat
枇杷	*Eriobotrya japonica* (Thunb.) Lindl.	Loquat
*草莓	*Fragaria ananassa* Duch.	Garden Strawberry
大叶桂樱	*Laurocerasus zippeliana* (Miq.) T. T. Yu et L. T. Lu	Bigleaf Cherrylaurel
闽粤石楠(边沁石斑木)	*Photinia benthamiana* Hance	Photinia
饶平石楠(细叶石斑木)	*Photinia raupingensis* Kuan	Yaoping Photinia
*紫叶李(红叶李)	*Prunus cerasifera* Ehrh. var. *atropurea* Jacq.	Myrobalan Plum
*冬樱花	*Cerasus cerasoides* (D. Don) S. Ya. Sokolov [*Prunus majestica* Koehne]	Wild Himalayan Cherry
全缘樱桃(镶边樱)	*Laurocerasus marginata* (Dunn) Yu et Lu [*Prunus marginata* Dunn]	Entire Cherry-laurel
梅	*Armeniaca mume* Sieb. [*Prunus mume* Sieb. et Zucc.]	Mume Plant
桃	*Amygdalus persica* L. [*Prunus persica* L.]	Peach
腺叶桂樱(腺叶野樱)	*Laurocerasus phaeosticta* (Hance) S. K. Schneid. [*Prunus phaeosticta* (Hance) Maxim.]	Brown Punctate Cherry

李	*Prunus salicina* Lindl.	Plum
臀果木(臀形果)	*Pygeum topengii* Merr.	Topeng Pygeum
*火棘	*Pyracantha fortuneana* (Maxim.) Li	Fortune Firethorn
豆梨(麻子梨)	*Pyrus calleryana* Decne.	Callery Pear
锲叶豆梨(棠梨)	*Pyrus calleryana* Decne. var. *koehnei* (Schneid.) Yu	Koehne Callery Pear
沙梨	*Pyrus pyrifolia* (Burm. f.) Nakai	Sand Pear
车轮梅(春花)	*Raphiolepis indica* (L.) Lindl. [*R. rugosa* Nakai]	Hongkong Hawthorn
细叶石斑木	*Raphiolepis lanceolata* Hu	Lanceolate Raphiolepis
柳叶石斑木	*Raphiolepis salicifolia* Lindl.	Willowleaf Raphiolepis
月季	*Rosa chinensis* Jacq.	China or Bengal Rose
*小型月季	*Rosa chinensis* Jacq. var. *mimima* Voss.	Roulett Rose
广东蔷薇	*Rosa kwangtungensis* Yu et Tsai	Guangdong Rose
金樱子	*Rosa laevigata* Michx.	Cherokee Rose
多花蔷薇(七姐妹)	*Rosa multiflora* Thunb.	Seven Sisters Rose
*香水月季	*Rosa odorata* Sweet	Tea Rose
*玫瑰	*Rosa rugosa* Thunb.	Rose
光叶蔷薇(魏氏蔷薇)	*Rosa wichuraiana* Crep.	Memorial Rose
粗叶悬钩子	*Rubus alceaefolius* Poir.	Roughleaf Raspberry
江西悬钩子	*Rubus gressitii* Metc.	Jiangxi Raspberry
白花悬钩子(白花箣藤)	*Rubus leucanthus* Hance	Whiteflower Raspberry

茅莓(蛇泡簕)	*Rubus parvifolius* L.	Herbaceous Myrica
梨叶悬钩子	*Rubus pirifolius* Smith	Pearleaf Raspberry
锈毛木莓	*Rubus reflexus* Ker-Gawl.	Raspberry
浅裂绣毛莓	*Rubus reflexus* Ker-Gawl. var. *hui* (Diels apud Hu) Metc.	Hu Rustyhair Raspberry
红泡刺(深裂叶锈毛莓)	*Rubus reflexus* Ker-Gawl. var. *lancelobus* Metc.	Lanceolobed Rustyhair Raspberry
空心泡(蔷薇梅)	*Rubus rosaefolius* Smith	Rose-leaved Strawberry
红脉悬钩子	*Rubus sumatranus* Miq.	Redglandular Raspberry
木莓	*Rubus swinhoei* Hance	Swinhoe Raspberry
光滑悬钩子	*Rubus tsangii* Merr.	Tsang Raspberry
*广东绣线菊	*Spiraea cantoniensis* Lour.	Bridal Wreath
中华绣线菊	*Spiraea chinensis* Maxim.	Chinese Spiraea

含羞草科 Mimosaceae

*大叶相思(耳叶相思)	*Acacia auriculiformis* A. Cunn. ex Benth.	Auriculate Acacia
藤金合欢	*Acacia concinna* (Willd.) DC.	Elegant Acacia
台湾相思(相思树)	*Acacia confusa* Merr.	Taiwan Acacia
*金合欢	*Acacia farnesiana* (L.) Willd.	Sweet Acacia
*马占相思	*Acacia mangium* Willd.	Mangium Acacia
蛇藤(羽叶金合欢)	*Acacia pennata* (L.) Willd.	Snake Acacia
海红豆(孔雀豆)	*Adenanthera pavonina* L. var. *microsperma* (Teijsm. et Binnend.) Nielsen	Red Sandalwood
楹树	*Albizia chinensis* (Osbeck) Merr.	Chinese Albizia

天香藤	*Albizia corniculata* (Lour.) Druce	Corniculate Albizia
*南洋楹	*Albizia falcataria* (L.) Fosberg	Molucca Albizia
合欢	*Albizia julibrissin* Durazz.	Silk-tree
山合欢	*Albizia kalkora* (Roxb.) Prain	Wild Siris
*阔荚合欢(大叶合欢)	*Albizia lebbeck* (L.) Benth.	Lebbek Tree
猴耳环(围诞树)	*Archidendron clypearia* (Jack.) Nielsen [*Pithecellobium clypearia* (Jack.) Benth.]	Common Apes Ear-ring
亮叶猴耳环(亮叶围诞树)	*Archidendron lucidum* (Benth.) Nielsen [*Pithecellobium lucidum* Benth.]	China Apes Ear-ring
薄叶猴耳环	*Archidendron utile* (Chun et How) Nielsen [*Pithecellobium utile* Chun et How]	Common Apes Ear-ring
*红绒球(美蕊花)	*Calliandra haematocephala* Hassk.	Red Powderpuff
*粉扑花(粉红合欢)	*Calliandra riparia* Pittier	Surinam Calliandra
榼藤子(眼镜豆)	*Entada phaceoloides* (L.) Merr.	Giant Bean
*银合欢(白合欢)	*Leucaena leucocephala* (Lam.) de Wit	White Popinac
*簕仔树(光荚含羞草)	*Mimosa bimucronata* (DC.) O. Kuntze [*M. sepiaria* Benth.]	Glabrous-pod Mimosa

*巴西含羞草(美洲含羞草)	*Mimosa diplotricha* C. Wrigth ex Sauvalle [*M. invisa* Mart. ex Colla]	Shame Plant
*含羞草(怕丑草)	*Mimosa pudica* L.	Sensitive Plant
*牛蹄豆(金龟树)	*Pithecellobium dulce* (Roxb.) Benth.	Manila Tamarind
*雨树	*Samanea saman* (Jacq.) Merr.	Raintree

苏木科 Caesalpiniaceae

阔裂叶羊蹄甲	*Bauhinia apertilobata* Merr. et Metc.	Broadlobed Bauhinia
*洋紫荆(红花羊蹄甲)	*Bauhinia blakeana* Dunn	Hongkong Orchid Tree
龙须藤	*Bauhinia championii* (Benth.) Benth.	Champion's Bauhinia
首冠藤(深裂叶羊蹄甲)	*Bauhinia corymbosa* Roxb.	Camel's Foot
光叶羊蹄甲(粉叶羊蹄甲)	*Bauhinia glauca* (Wall. ex Benth.) Benth.	Climbing Bauhinia
广东羊蹄甲	*Bauhinia kwangtungensis* Merr.	Guangdong Bauhinia
*羊蹄甲	*Bauhinia purpurea* L.	Purple Bauhinia
*宫粉羊蹄甲(洋紫荆)	*Bauhinia variegata* L.	Buddhist Bauhinia
*白花羊蹄甲(白花洋紫荆)	*Bauhinia variegata* L. var. *candida* (Roxb.) Voigt	White Bauhinia
刺果苏木(大托叶云实)	*Caesalpinia bonduc* (L.) Roxb.	Gray Nickers
华南云实(假老虎簕)	*Caesalpinia crista* L.	Wood Gossip Caesalpinia

云实	*Caesalpinia decapetala* (Roth.) Alston	Mysore Thorn
大叶华南云实(大叶云实)	*Caesalpinia magnifoliolata* Metc.	Largeleaf Caesalpinia
小叶云实	*Caesalpinia millettii* Hook. et Arn.	Smallleaf Caesalpinia
南蛇簕(喙荚云实)	*Caesalpinia minax* Hance	Snake Caesalpinia
*金凤花(洋金凤)	*Caesalpinia pulcherrima* (L.) Sw.	Parado Spride
苏木	*Caesalpinia sappan* L.	Sappan Caesalpinia
鸡嘴簕	*Caesalpinia sinensis* (Hemsl.) J. E. Vidal	Cockbillthorn
春云实(鸟爪簕藤)	*Caesalpinia vernalis* (L.) Champ. ex Benth.	Spring Caesalpinia
*翅荚决明	*Cassia alata* L.	Winded Cassia
*双荚决明	*Cassia bicapsularis* L.	Double-fruited Cassia
*腊肠树(阿勃勒)	*Cassia fistula* L.	Golden-shower
短叶决明(大叶山扁豆)	*Cassia leschenaultiana* DC.	Wild Sensitive-plant
含羞草决明(山扁豆)	*Cassia mimosoides* L.	Mimosa-leaved Cassia
望江南(野扁豆)	*Cassia occidentalis* L.	Coffee Senna
*铁刀木(暹罗决明)	*Cassia siamea* Lam.	Kassod Tree
*黄槐决明(黄花决明)	*Cassia surattensis* Burm. f.	Sunshine Tree
决明	*Cassia tora* L.	Sickle Senna
*紫荆	*Cercis chinensis* Bunge	China Redbud
*凤凰木	*Delonix regia* (Hook.) Raf.	Flame Tree

格木	*Erythrophleum fordii* Oliv.	Ford Checkwood
小果皂荚(南方皂荚)	*Gleditsia australis* Hemsl.	Small-fruited Honeylocust
华南皂荚	*Gleditsia fera* (Lour.) Merr.	South China Honeylocust
皂荚	*Gleditsia sinensis* Lam.	China Honeylocust
*李叶豆(李叶苏木)	*Hymenaea courbaril* L.	India Locust
*短萼仪花	*Lysidice brevicalyx* Wei	Shortcalyx Lysidice
*仪花(红花树)	*Lysidice rhodostegia* Hance	Redbracted Lysidice
*盾柱木(双翼豆)	*Peltophorum pterocarpum* (DC.) Baker ex K. Heyn.	Yellow Poinciana
*中国无忧花(火焰花)	*Saraca dives* Pierre	China Saraca
*油楠	*Sindora glabra* Merr.	Oil Sindora
*东京油楠	*Sindora tonkinensis* A. Cheval. ex K. et S. S. Larsen	Tonkin Sindora
*酸豆(罗望子)	*Tamarindus indica* L.	Tamarind

蝶形花科 Papilionaceae

广东相思子(鸡骨草)	*Abrus cantoniensis* Hance	Prayer-beads
毛相思子(毛鸡骨草)	*Abrus mollis* Hance	Hairy Rosary Pea
相思子(相思豆)	*Abrus precatorius* L.	Rosary Pea
链荚豆	*Alysicarpus vaginalis* (L.) DC.	Alyce Clover
*蔓花生(铺地黄金)	*Arachis duranensis* Krapov. et W. C. Gregory	Creeping Peanut
*落花生(花生)	*Arachis hypogaea* L.	Groundnut

藤槐	*Bowringia callicarpa* Champ. ex Benth.	Common Bowringia
木豆(三叶豆)	*Cajanus cajan*（L.）Millsp.	Pigeon Pea
蔓草虫豆	*Cajanus scarabaeoides*（L.）Thou.	Scarab-like Cajanus
小刀豆(白扁豆)	*Canavalia cathartica* Thou.	Small Knife Bean
海刀豆	*Canavalia maritima*（Aubl.）Thou.	Sea Sword Bean
＊栗豆树(绿元宝)	*Castanospermum australe* A. Cunn. ex Mudie	Black Bean
铺地蝙蝠草	*Christia obcordata*（Poir.）Bakh. f.	Obcordata Christia
蝙蝠草(蝴蝶草)	*Christia vespertilionis*（L. f.）Bahn. f.	Common Christia
蝴蝶花豆(蝶豆)	*Clitoria ternatea* L.	Butterfly Pea
舞草	*Codariocalyx motorius*（Houtt.）Ohashi	Common Danceweed
响铃豆	*Crotalaria albida* Heyne ex Roth	Whitish Rattlebox
凸尖野百合(大猪屎豆)	*Crotalaria assamica* Benth.	Assam Rattlebox
长萼猪屎豆(长萼野百合)	*Crotalaria calycina* Schrank	Long-calyx Rattlebox
猪屎豆	*Crotalaria pallida* Ait.	Pallid Rattlebox
吊裙草(凹叶野百合)	*Crotalaria retusa* L.	Yellow-flower Pea
野百合	*Crotalaria sessiliflora* L.	Purple-flowered Crotalaria
球果猪屎豆(椭圆叶野百合)	*Crotalaria uncinella* Lam.	Hooked Rattlebox

被子植物 ANGIOSPERMAE

光萼猪屎豆(光叶野百合)	*Crotalaria zanzibarica* Benth.	Glabrous Sepel Crotalaria
南岭黄檀(南岭檀)	*Dalbergia balansae* Prain	South China Rosewood
两广黄檀(两粤黄檀)	*Dalbergia benthamii* Prain	Behtham's Rosewood
弯枝黄檀(扭黄檀)	*Dalbergia candenatensis* (Dennst.) Prain	Twisted Rosewood
藤黄檀(藤檀)	*Dalbergia hancei* Benth.	Scandent Rosewood
香港黄檀	*Dalbergia millettii* Benth.	Hongkong Rosewood
含羞草黄檀(象鼻藤)	*Dalbergia mimosoides* Franch.	Trunk Rosewood
*降香黄檀(花梨母)	*Dalbergia odorifera* T. Chen	Fragrant Rosewood
白花鱼藤	*Derris alborubra* Hemsl.	White-flowered Derris
毛鱼藤(毒鱼藤)	*Derris elliptica* (Roxb.) Benth.	Elliptic Fishvine
中南鱼藤(霍氏鱼藤)	*Derris fordii* Oliv.	Ford Fishvine
亮叶霍氏鱼藤(亮叶中南鱼藤)	*Derris fordii* Oliv. var. *lucida* How	Ford's Derris
鱼藤	*Derris trifoliata* Lour.	Derris
小槐花山蚂蟥(小槐花)	*Desmodium caudatum* (Thunb.) DC.	Desmodium
大叶山蚂蝗	*Desmodium gangeticum* (L.) DC.	Big-leaved Desmodium
圆叶野百合(圆叶舞草)	*Codariocalyx gyroides* (Roxb. ex Link.) Hassk. [*Desmodium gyroides* (Roxb.) DC.]	Round Leaf Codariocalyx
假地豆	*Desmodium heterocarpon* (L.) DC.	False Groundnut

异叶山蚂蝗(异叶山绿豆)	*Desmodium heterophyllum* (Willd.) DC.	Heterophyllous Tick Clover
大叶拿身草(疏花山蚂蟥)	*Desmodium laxiflorum* DC.	Bigleaf Ticktrefoil
小叶三点金(小叶山绿豆)	*Desmodium microphyllum* (Thunb.) DC.	Small-leaved Desmodium
多花山蚂蝗	*Desmodium multiflorum* DC. [*D. floribundum* (D. Don) Sweet]	Many-flower Tickclover
显脉山蚂蝗(显脉山绿豆)	*Desmodium reticulatum* Champ. et Benth.	Distinct-nerved Tickclover
广东金钱草	*Desmodium styracifolium* (Osb.) Merr.	Snowbell-leaved Tick Clover
三点金	*Desmodium triflorum* (L.) DC.	Three-flowered Desmodium
绒毛山蚂蝗(绒毛山绿豆)	*Desmodium velutinum* (Willd.) DC.	Velvet Leaved Desmodium
黄毛野扁豆	*Dunbaria fusca* (Wall.) Kurz.	Yellow Hairy Dunbaria
长柄野扁豆	*Dunbaria podocarpa* Kurz.	Long Stipe Dunbaria
圆叶野扁豆	*Dunbaria punctata* (Wight et Arn.) Benth. [*D. rotundifoloa* (Lour.) Merr.]	Round-leaved Dunbaria
猪仔笠	*Eriosema chinensis* Vog.	Chinese Eriosema
*龙牙花(象牙红)	*Erythrina corallodendron* L.	Coralbean Tree
*美丽刺桐(鸡冠刺桐)	*Erythrina crista-galli* L.	Cockspur Coralbean
*刺桐	*Erythrina variegata* L.	India Coral Tree
*黄脉刺桐(斑叶刺桐)	*Erythrina variegata* L. var. *picta* Graf.	Yellow-vein Coralbean Tree

大叶千斤拔	*Flemingia macrophylla* (Willd.) Prain	Large-leaved Flemingia
蔓千斤拔(千斤拔)	*Flemingia prostrata* Roxb. f. ex Roxb.	Philippine Flemingia
*大豆(黄豆)	*Glycine max* (L.) Merr.	Soy Bean
庭藤	*Indigofera decora* Lindl.	Indigo Vine
刚毛木蓝(刚毛木蓝)	*Indigofera hirsuta* L.	Hirsute Indigo
穗序木蓝	*Indigofera spicata* Forsk.	Spike Indigo
野青树(假蓝靛)	*Indigofera suffruticosa* Mill.	Subshrub Indigo
三叶木蓝(地蓝根)	*Indigofera trifoliata* L.	Trifoliolate Indigo
鸡眼草(人字草)	*Kummerowia striata* (Thunb.) Schindl.	Striate Kummerowia
*扁豆(逸生)	*Lablab purpureus* (L.) Sweet	Hyacinth-bean
中华胡枝子(台湾胡枝子)	*Lespedeza chinensis* G. Don	Chinese Lespedeza
截叶胡枝子(铁扫帚)	*Lespedeza cuneata* (Dum.-Cours.) G. Don	Cuneate Lespedeza
美丽胡枝子(马扫帚)	*Lespedeza formosa* (Vog.) Koehne	Beautiful Lespedeza
绿花崖豆藤(硬骨藤)	*Millettia championi* Benth.	Champion's Millettia
香花崖豆藤(山鸡血藤)	*Millettia dielsiana* Harms ex Diels	Diel's Millettia
亮叶崖豆藤(亮叶鸡血藤)	*Millettia nitida* Benth.	Glittering-leaved Millettia
丰城崖豆藤(丰城鸡血藤)	*Millettia nitida* Benth. var. *hirsutissima* Z. Wei	Shiningleaf Millettia

香港崖豆藤(香港崖豆)	*Millettia oraria* Dunn	Hongkong Millettia
厚果崖豆藤(厚果鸡血藤)	*Millettia pachycarpa* Benth.	Thick-pericarped Millettia
美花崖豆藤(印度鸡血藤)	*Millettia pulchra* (Benth.) Kurz	Indian Millettia
鸡血藤	*Millettia reticulata* Benth.	Leatherleaf Millettia
牛大力藤(美丽崖豆藤)	*Millettia speciosa* Champ. ex Benth. (M. speciosa Champ.)	Showy Millettia
白花油麻藤(勃氏黧豆)	*Mucuna birdwoodiana* Tutch.	Birdwood' Mucuna
香港油麻藤(香港黧豆)	*Mucuna championi* Benth.	Hongkong Mucuna
海南黧豆(琼油麻藤)	*Mucuna hainanensis* Hayata	Hainan Mucuna
大果油麻藤(褐毛黧豆)	*Mucuna macrocarpa* Wall.	Bigfruit Mucuna
刺毛黧豆(刺蒺黧豆)	*Mucuna pruriens* (L.) DC.	Shag Mucuna
常春油麻藤(常绿黧豆)	*Mucuna sempervirens* Hemsl. ex Forb. et Hemsl.	Evergreen Mucuna
凹叶红豆(恒春红豆)	*Ormosia emarginata* (Hook. et Arn.) Benth.	Emarginate-leaved Ormosia
韧荚红豆	*Ormosia indurata* L. Chen	Hard-fruited Ormosia
海南红豆	*Ormosia pinnata* (Lour.) Merr.	Hainan Ormosia
软荚红豆(荔叶红豆)	*Ormosia semicastrata* Hance	Soft-fruited Ormosia
豆薯沙葛(凉瓜)	*Pachyrhizus erosus* (L.) Urb.	Yam Bean

毛排钱草	*Phyllodium elegans* (Lour.) Desv.	Elegant Phyllodium
排钱草	*Phyllodium pulchellum* (L.) Desv.	Beautiful Phyllodium
水黄皮	*Pongamia pinnata* (L.) Merr.	Pongam Tree
*四棱豆(翼豆)	*Psophocarpus tetragonolobus* (L.) DC.	Winged-bean
紫檀(印度紫檀)	*Pterocarpus indicus* Willd.	Burmese Rosewood
葛	*Pueraria lobata* (Willd.) Ohwi	Kudzu Vine
山葛藤	*Pueraria lobata* (Willd.) Ohwi var. *montana* (Lour.) Van der Maesen	Montane Kudzu
甘葛藤(粉葛)	*Pueraria lobata* (Willd.) Ohwi. var. *thomsonii* (Benth.) Van der Maesen	Thomson's Kudzu
三裂叶野葛	*Pueraria phaseoloides* (Roxb.) Benth.	Wild Kudzu Vine
密子豆	*Pycnospora lutescens* (Poir.) Schindl.	Leutescent Pycnospora
鹿藿(老鼠眼)	*Rhynchosia volubilis* Lour.	Rat's Eye Bean
刺田菁(多刺田菁)	*Sesbania bispinosa* (Jacq.) W. F. Wight	Spiny Sesbania
*田菁(碱青)	*Sesbania cannabina* (Retz.) Poir.	Common Sesbania
印度田菁(山菁)	*Sesbania sesban* (L.) Merr.	India Sesbania
密节坡油甘	*Smithia conferta* Smith	Dense-flowered Smithia
*龙爪槐	*Sophora japonica* L. var. *pendula* Loud.	Pendent Japanese Pagodatree
密花豆	*Spatholobus suberectus* Dunn	Suberet Spatholobus

葫芦茶	*Tadehagi triquetrum* (L.) Ohashi	Triquetrous Tadehagi
灰毛豆(红花灰叶)	*Tephrosia purpurea* (L.) Pers.	Purple Tephrosia
猫尾豆(猫尾草)	*Uraria crinita* (L.) Desv. ex DC.	Cat's Tail Bean
狸尾草(狗尾豆)	*Uraria lagopodioides* (L.) Desv. ex DC.	Dog's Tail Bean
滨豇豆(海豇豆)	*Vigna marina* (Burm.) Merr.	Marine Cowpea
山绿豆	*Vigna minima* (Roxb.) Ohwei et Ohashi	Small Cowpea
*绿豆	*Vigna radiata* (L.) Wilozek	Mung Bean
豆角(长豇豆)	*Vigna unguiculata* (L.) Walp. [*V. sinensis* (L.) Savi ex Hassk.]	Common Cowpea, Coppea
*紫藤	*Wisteria sinensis* (Sims) Sweet	Chinese Wisteria
丁癸草(人字草)	*Zornia gibbosa* Span.	Twinleaf Zornia

旌节花科 Stachyuraceae

| *中华旌节花(旌节花) | *Stachyurus chinensis* Franch. | China Stachyurus |

金缕梅科 Hamamelidaceae

蕈树(阿丁枫)	*Altingia chinensis* (Champ.) Oliv. ex Hance	Mountain Lichi
蚊母树(蚊子树)	*Distylium racemosum* Sieb. et Zucc.	Mother-of-mosquitoes Tree
秀柱花	*Eustigma oblongifolium* Gardn. et Champ.	Oblong-leaved Eustigma
缺萼枫香	*Liquidambar acalycina* H. T. Chang	Calyxless Sweetgum
枫香树(枫树)	*Liquidambar formosana* Hance	Beautiful Sweetgum

檵木	*Loropetalum chinense* (R. Br.) Oliv.	Chinese Loropetalum
*红花檵木(红檵木)	*Loropetalum chinense* (R. Br.) Oliv. f. *rubrum* H. T. Chang	Red-flower Loropetalum
*壳菜果(米老排)	*Mytilaria laosensis* Lec.	Lao Mytilaria
红花荷(红苞木)	*Rhodoleia championi* Hook. f.	Champion Rhodoleia
半荷枫	*Semiliquidambar cathayensis* H. T. Chang	Semiliquidambar
东方水丝梨(尖叶水丝梨)	*Sycopsis dunnii* Hemsl.	Sharpleaf Fighazel

黄杨科 Buxaceae

匙叶黄杨(雀舌黄杨)	*Buxus harlandii* Hance	Harland's Box
*桃叶黄杨	*Buxus henryi* Mayr.	Bigflower Box
*锦熟黄杨(尖叶黄杨)	*Buxus sempervirens* L.	Common Box
黄杨(豆瓣黄杨)	*Buxus sinica* (Rehd. et Wils.) Cheng	Chinese Box
*细叶黄杨(狭叶黄杨)	*Buxus stenophylla* Hance	Narrowleaf Box

杨柳科 Salicaceae

*垂柳(柳树、柳)	*Salix babylonica* L.	Weeping Willow

杨梅科 Myricaceae

杨梅	*Myrica rubra* (Lour.) Sieb. et Zucc.	Strawberry Tree

壳斗科 Fagaceae

板栗(栗子)	*Castanea mollissima* Bl.	Chinese Chestnut

小红栲(米锥、小叶栲)	*Castanopsis carlesii* (Hemsl.) Hayata	Carles's Chinkapin
中华锥(锥栗)	*Castanopsis chinensis* Hance	Chinese Oatchestnut
甜槠栲(甜槠树)	*Castanopsis eyrei* (Champ.) Tutch.	Eyre's Chinkapin
罗浮栲(白椽)	*Castanopsis fabri* Hance	Faber's Chestnut
川鄂栲(栲树)	*Castanopsis fargesii* Franch.	Farges's Chinkapin
黧蒴栲(裂斗锥栗)	*Castanopsis fissa* (Champ. ex Benth.) Rehd. et Wils.	Breakingfruit Chinkapin
红锥(赤稀黧)	*Castanopsis hystrix* A. DC.	Red Oatchestnut
吊皮锥(格氏栲)	*Castanopsis kawakamii* Hayata	Kawakami Chestnut
鹿角锥(鹿角栲)	*Castanopsis lamontii* Hance	Lamont Castanopsis
大叶锥栗(钩锥)	*Castanopsis tibetana* Hance	Tibet Oatchestnut
黄青冈栎(岭南青冈)	*Cyclobalanopsis championii* (Benth.) Oerst.	Champion's Oak
饭甑青冈	*Cyclobalanopsis fleuryi* (Hick. et A. Camus) Chun	Fleury Oak
小叶青冈(辽东栎)	*Cyclobalanopsis myrsinifolia* (Bl.) Oerst.	Myrsinaleaf Oak
竹叶青冈(竹叶青冈栎)	*Cyclobalanopsis neglecta* Schott.	Bamboo-leaved Oak
毛果青冈(赤栎)	*Cyclobalanopsis pachyloma* (Seem.) Schott.	Thickleaf Oak
烟斗石栎(杯果石栎)	*Lithocarpus corneus* (Lour.) Rehd.	Horny Tanoak
厚斗柯(斗柯)	*Lithocarpus elizabethae* (Tutch.) Rehd.	Thickcupsule Tanoak
柯(石栎)	*Lithocarpus glaber* (Thunb.) Nakai	Glabrous Tanoak

硬斗石栎(硬壳稠)	*Lithocarpus hancei* (Benth.) Rehd.	Hance's Tanoak
木姜叶柯(多穗柯)	*Lithocarpus litseifolius* (Hance) Chun	Litseleaf Tanoak
圆锥柯(圆锥石栎)	*Lithocarpus paniculatus* Hand.-Mazz.	Panicle Tanoak
栎叶柯	*Lithocarpus quercifolius* Huang et Y. T. Chang	Oakleaf Tanoak
紫玉盘柯(紫玉盘石栎)	*Lithocarpus uvariifolius* (Hance) Rehd.	Uvariformleaf Tanoak
麻栎(橡树)	*Quercus acutissima* Carruth.	Sawtooth Oak

木麻黄科 Casuarinaceae

*细枝木麻黄(银线木麻黄)	*Casuarina cunninghamiana* Miq.	River Oak
*木麻黄(短枝木麻黄)	*Casuarina equisetifolia* Forst.	Horsetail Tree

榆科 Ulmaceae

糙叶树	*Aphananthe aspera* (Thunb.) Planch.	Scabrous Aphanathe
光叶白颜树(滇糙叶树)	*Aphananthe cuspidata* (Bl.) Planch. [*Gironniera cuspidata* (Bl.) Kurz]	Yunnan Roughleaftree
*华南朴	*Celtis austro-sinensis* Chun	South China Hackberry
紫弹树	*Celtis biondii* Pamp.	Biond's Hackberry
朴树(相思树)	*Celtis sinensis* Pers.	Chinese Hackberry
樟叶朴(假玉桂)	*Celtis timorensis* Span [*C. cinnamonea* Lindl. ex Planch.]	Philippine Hackberry
白颜树	*Gironniera subaequalis* Planch.	White Gironniera

*青檀(翼朴)	*Pteroceltis tatarinowii* Maxim.	Wingceltis
狭叶山黄麻	*Trema angustifolia*(Planch.) Bl.	Narrowleaf Wildjute
光叶山黄麻	*Trema canabina* Lour.	Nakedleaf Wildjute
山油麻(假黄麻)	*Trema canabina* Lour. var. *dielsiana* (Hand.-Mazz.) C. J. Chen	Diels Trema
山黄麻(山油麻)	*Trema orientalis* (L.) Bl.	Mountain Hemp
榔榆(榆树)	*Ulmus parvifolia* Jacq.	Langyu Elm
*榉树(光叶榉)	*Zelkova serrata* (Thunb.) Makino	Japanese Zelkora

桑科 Moraceae

*面包树(罗蜜树)	*Artocarpus altilis* (Park.) Fosberg	Breadfruit
木波罗(波罗蜜)	*Artocarpus heterophyllus* Lam.	Jackfruit
白桂木(将军树)	*Artocarpus hypargyreus* Hance	Silver-back Artocarpus
桂木(红桂木)	*Artocarpus nitidus* ssp. *lingnanensis* (Merr.) Jarr.	Lingnan Artocarpus
二色波罗蜜(沙蕾木)	*Artocarpus styracifolius* Pierre	Styraxleaf Artocarpus
胭脂树	*Artocarpus tonkinensis* A. Chev. ex Gagnep.	Tonkin Artocarpus
葡蟠(藤构)	*Broussonetia kazinoki* Sieb. et Zucc.	Kazinoki Papermulberry
构树(楮树)	*Broussonetia papyrifera* (L.) L'Herit. ex Vent.	Paper Mulberry
*深裂号角树	*Cecropia adenopus* Mast. ex Miq.	Ambay Pumpwood
穿破石(畏芝)	*Cudrania cochinchinensis* (Lour.) Kudo et Masam.	False Custard

柘树(柘桑)	*Cudrania tricuspidata* (Carr.) Bureau ex Lavallee [*Maclura tricuspidata* Carr.]	Tricuspid Cudrania
水蛇麻(桑草)	*Fatoua villosa* (Thunb.) Nakai	Villous Fatoua
高山榕	*Ficus altissima* Bl.	Mountain Fig
大果榕	*Ficus auriculata* Lour.	Eared Strangler Fig
垂叶榕(柳叶榕)	*Ficus benjamina* L.	Weeping Fig
花叶垂榕	*Ficus benjamina* L. 'Golden Princess'	Variegated Benjamin Fig
*无花果	*Ficus carica* L.	Edible Fig
雅榕(万年青树)	*Ficus concinna* (Miq.) Miq.	Elegant Fig
*橡胶榕	*Ficus elasticar* Roxb. ex Hornem.	India-rubber Tree
天仙果	*Ficus erecta* Thunb. var. *beecheyana* (Hook. et Arn.) King	Fairy Fig
黄毛榕	*Ficus esquiroliana* Levl.	Fulvous Fig
水同木	*Ficus fistulosa* Reinw ex Bl.	Common Yellow Stem-fig
台湾榕	*Ficus formosana* Maxim.	Taiwan Fig
窄叶台湾榕	*Ficus formosana* Maxim. var. *shimadai* (Hayata) W. C. Chen	Narrow-leaved Taiwan Fig
藤榕(狭叶薜荔)	*Ficus hederacea* Roxb.	Climbing Fig
异叶榕	*Ficus heteromorpha* Hemsl.	Diverseleaf Fig
粗叶榕(五指毛桃)	*Ficus hirta* Vahl	Hairy Fig
对叶榕	*Ficus hispida* L. f.	Oppositeleaf Fig

*大琴叶榕(大琴榕)	*Ficus lyrata* Warb.	Big Fiddle-leaf Fig
小叶榕(榕树)	*Ficus microcarpa* L. f.	Smallfruit Banyan
黄金榕	*Ficus microcarpa* L. f. 'Golden Leaves'	Golden-leaved Banyan
九丁树(凸叶榕)	*Ficus nervosa* Heyne ex Roth.	Veined Fig
*琴叶榕(倒吊葫芦)	*Ficus pandurata* Hance	Fiddleleaf Fig
全缘榕	*Ficus pandulata* Hance var. *holophylla* Migo	Hololeaf Fig
薜荔(凉粉果)	*Ficus pumila* L.	Climbing Fig
舶梨榕(梨果榕)	*Ficus pyriformis* Hook. et Arn.	Ribbed Bush Fig
*菩提榕(菩提树)	*Ficus religiosa* L.	Peepul Tree
羊乳榕	*Ficus sagittata* Vahl	Arrowleaf Fig
爬藤榕	*Ficus samentosa* Buch. -Ham. ex J. E. Sm.	Sarmentose Fig
竹叶榕	*Ficus stenophylla* Hemsl.	Bambooleaf Fig
锥叶榕(假斜叶榕)	*Ficus subulata* Bl.	Sea Fig
笔管榕(雀榕)	*Ficus superba* Miq. var. *japonica* Miq.	Superb Fig
*地果(小叶铺地榕)	*Ficus tikoua* Bur.	Digua Fig
斜叶榕(水榕)	*Ficus tinctoria* ssp. *gibbosa* (Bl.) Corner [*F. gibbosa* Bl.]	Gibbous Fig
青果榕(牛奶树)	*Ficus variegata* Bl. var. *chlorocarpa* (Benth.) King	Greenfruit Fig
变叶榕	*Ficus variolosa* Lindl. ex Benth.	Variedleaf Fig

白肉黄果榕(突脉榕)	*Ficus vasculosa* Wall. ex Miq.	White Fig Tree
大叶榕(黄葛榕)	*Ficus virens* Ait. var. *sublanceolata*(Miq.) Corner	Big-leaved Fig
牛筋藤	*Malaisia scandens*(Lour.) Planch.	Strength-vine
桑(家桑)	*Morus alba* L.	White Mulberry

荨麻科 Urticaceae

舌柱麻(两广紫麻)	*Archiboehmeria atrata* (Gagnep.) C. J. Chen	Linguanramie
水苎麻(掌叶麻)	*Boehmeria macrophylla* Hornem.	Big-leaved False Nettle
苎麻(苧麻)	*Boehmeria nivea*(L.) Gaud.	Ramie
青叶苎麻(绿叶苎麻)	*Boehmeria nivea*(L.) Gaud. var. *tenacissima*(Gaudich) Miq.	Virid-leaved Boehmeria
线条楼梯草(狭叶楼梯草)	*Elatostema lineolatum* Wight var. *majus* Wedd	Narrowleaf Stairweed
多序楼梯草(石生楼梯草)	*Elatostema macintyrei* Dunn	Manyhead Stairweed
糯米团(糯米条)	*Gonostegia hirta*(Bl.) Miq.	Hairy Gonostegia
紫麻(野麻)	*Oreocnidea frutescens* (Thunb.) Miq.	Shrubby Wood Nettle
华南赤车	*Pellionia grijsii* Hance	South China Redcarweed
赤车	*Pellionia radicans*(Sieb. et Zucc.) Wedd.	Rooted Pellionia
*吐烟花(吐烟草)	*Pellionia repens*(Lour.) Merr.	Creeping Pellionia
蔓赤车(毛赤车)	*Pellionia scabra* Benth.	Rough Pellionia

冷水花(花叶荨麻)	*Pilea angulata* (Bl.) Bl.	Roundpetal Coldwaterflower
*花叶冷水花	*Pilea cadierei* Gagnep. et Guill.	Alluminum Plant
小叶冷水花(透明草)	*Pilea microphylla* (L.) Liebm.	Artillery Clearweed
*蛤蟆草(皱皮草)	*Pilea mollis* Hemsl.	Moon Valley
盾叶冷水花	*Pilea peltata* Hance	Peltate Coldwaterflower
*镜面草(翠屏草)	*Pilea peperomioides* Diels	Lens Coldwaterflower
*泡叶冷水花(毛虾蟆草)	*Pilea repens* Liebm.	Black-leaf Panamiga
红雾水葛	*Pouzolzia sanguinea* (Bl.) Merr.	Red Pouzolzia
雾水葛	*Pouzolzia zeylanica* (L.) Benn.	Ceylon Pouzolzia
藤麻	*Procris wightiana* Wall. ex Wedd.	Crenate Procris

冬青科 Aquifoliaceae

秤星树(梅叶冬青)	*Ilex asprella* (Hook. et Arn.) Champ. ex Benth.	Rough-leaved Holly
*枸骨	*Ilex cornuta* Lindl. et Paxt.	Horny Holly
榕叶冬青(台湾糊樗)	*Ilex ficoidea* Hemsl.	Fig-leaved Holly
细花冬青(纤花冬青)	*Ilex graciliflora* Champ. ex Benth.	Small-flowered Holly
谷木叶冬青(隐脉冬青)	*Ilex memecylifolia* Champ. ex Benth.	Dwarf Holly
毛冬青(茶叶冬青)	*Ilex pubescens* Hook. et Arn.	Downy Holly
铁冬青(救必应)	*Ilex rotunda* Thunb.	Chinese Holly

小果铁冬青(微果冬青)	*Ilex rotunda* Thunb. var. *microcarpa* (Lindl. et Paxt.) S. Y. Hu	Smallfruit Chinese Holly
三花冬青(小冬青)	*Ilex triflora* Bl.	Teafruit Holly
亮叶冬青(绿冬青)	*Ilex viridis* Champ. ex Benth.	Small-leaved Holly

卫矛科 Celastraceae

过山枫	*Celastrus aculeatus* Merr.	Aculeate Bittersweet
青江藤(南华南蛇藤)	*Celastrus hindsii* Benth.	Chinese Bitter-sweet
圆叶南蛇藤	*Celastrus kusanoi* Hayata	Roundleaf Bittersweet
独子藤(单子南蛇藤)	*Celastrus monospermus* Roxb.	Bentham's Bitter-sweet
紫刺卫矛(棱枝卫矛)	*Euonymus angustatus* Sprague	Narrow-leaved Euonymus
中华卫矛(华卫矛)	*Euonymus nitidus* Benth. [*E. chinensis* Lindl.]	Chinese Euonymus
常春卫矛(长春卫矛)	*Euonymus hederaceus* Champ. ex Benth.	Ivy-like Euonymus
疏花卫矛	*Euonymus laxiflorus* L.	Loose-flowered Euonymus
长圆叶卫矛	*Euonymus oblongifolius* Loes. et Rehd.	Oblongleaf Euonymus
双花假卫矛	*Microtropis biflora* Merr. et Freem.	Biflor Microtropis
福建假卫矛(福建赛卫矛)	*Microtropis fokienensis* Dunn	Fokien Microtropis
广州假卫矛	*Microtropis obscurinervia* Merr. et Freem.	Broad-leaved Microtropis
网脉假卫矛(细脉假棕矛)	*Microtropis reticulata* Dunn	Thin-nerved Microtropis

翅子藤科 Hippocarateaceae

短柄翅子藤(雅致翅子藤)	*Loeseneriella concina* A. C Smith	Fairy Webseedvine

茶茱萸科 Icacinaceae

定心藤(甜果藤)	*Mappianthus iodoides* Hand.-Mazz.	Common Mappianthus

铁青树科 Olacaceae

华南青皮木(碎骨仔树)	*Schoepfia chinensis* Gardn. et Champ.	Chinese Schoepfia

山柑子科 Opiliaceae

山柑藤(山柑)	*Cansjera rheedii* J. F. Gmelin	Rheed Cansjera

桑寄生科 Loranthaceae

五蕊寄生(乌榄寄生)	*Dendrophthoe pentandra* (L.) Miq.	Fivestamen Dendrophthoe
离瓣寄生(五瓣寄生)	*Helixanthera parasitica* Lour.	Fivepetal Helixanthera
油茶离瓣寄生(油茶寄生)	*Helixanthera sampsoni* (Hance) Danser	Sampson Coilthrum
栗寄生	*Korthalsella japonica* (Thunb.) Engl.	Japan Korthalsella
鞘花寄生(鞘花)	*Macrosolen cochinchinensis* (Lour.) Van Tregh	Sheathflower
红花寄生	*Scurrula parasitica* L.	Witches's Broom
广寄生(梧州寄生茶)	*Taxillus chinensis* (DC.) Danser	China Taxillus
桑寄生(桑上寄生)	*Taxillus sutchuenensis* (Lec.) Danser	Szechwan Taxillus
扁枝槲寄生(槲寄生)	*Viscum articulatum* Burm. f.	Mistletoe

被子植物 ANGIOSPERMAE

棱枝槲寄生(樟木寄生)	*Viscum diospyrosicolum* Hayata	Ebonyshoot Mistletoe
枫香槲寄生(枫树寄生)	*Viscum liquidambaricolum* Hayata	Sweetgumshoot Mistletoe
柄果槲寄生(刀叶槲寄生)	*Viscum multinerve* (Hayata) Hayata	Stipefruit Mistletoe
瘤果槲寄生(柚树寄生)	*Viscum ovalifolium* DC.	Oriental Mistletoe

檀香科 Santalaceae

| 寄生藤 | *Dendrotrophe varians* Miq. [*D. frutescens* (Champ. ex Benth.) Danser] | Shrubby Parasiticvine |

蛇菰科 Balanophoraceae

| 广东蛇菰(红冬蛇菰) | *Balanophora harlandii* Hook. f. | Harland's Balanophora |
| 香港蛇菰 | *Balanophora hongkongensis* K. M. Lau | Hong Kong Balanophora |

鼠李科 Rhamnaceae

多花勾儿茶(亮叶老鼠耳)	*Berchemia floribunda* (Wall.) Brongn. [*B. racemosa* Sieb. et Zucc.]	Japanese Supple-jack
铁包金(老鼠耳)	*Berchemia lineata* (L.) DC.	Lineate Supple-juck
光枝勾儿茶(铁包金)	*Berchemia polyphylla* Wall. ex Lawson var. *leioclada* Hand.-Mazz.	Smooth-branched Hooktea
枳椇(拐枣)	*Hovenia dulcis* Thunb.	Japan Turnjujube
马甲子	*Paliurus ramosissimus* (Lour.) Poir.	Thorny Wingnut
山绿柴(圆叶鼠李)	*Rhamnus brachypoda* C. Y. Wu ex Y. L. Chen	Twiggy Buckthorn

黄药(长叶冻绿)	*Rhamnus crenata* Sieb. et Zucc.	Oriental Buckthorn
尼泊尔鼠李	*Rhamnus napalensis* (Wall.) Lawson	Nepal Buckthron
亮叶雀梅藤(钩状雀梅藤)	*Sageretia lucida* Merr.	Lucidleaf Sageretia
雀梅藤(雀梅)	*Sageretia thea* (Osbeck) Johnst	Hedge Sageretia
翼核果	*Ventilago leiocarpa* Benth.	Smooth-fruited Ventilago
*枣树(枣)	*Ziziphus jujuba* Mill.	Common Jujube
*台湾青枣(滇刺枣)	*Ziziphus mauritiana* Lam.	Indian Jujube

胡颓子科 Elaeagnaceae

密花胡颓子(羊奶果)	*Elaeagnus conferta* Roxb.	Denseflower Elaeagnus
蔓胡颓子(藤胡颓子)	*Elaeagnus glabra* Thunb.	Glabrous Elaeagnus
角花胡颓子	*Elaeagnus gonyanthes* Benth.	Angularflower Elaeagnus
鸡柏胡颓子	*Elaeagnus loureiri* Champ.	Loureiro Elaeagnus
胡颓子	*Elaeagnus pungens* Thunb.	Thorny Elaeagnus
香港胡颓子	*Elaeagnus tucherii* Dunn	Hongkong Elaeagnus

葡萄科 Vitaceae

广东蛇葡萄(粤蛇葡萄)	*Ampelopsis cantoniensis* (Hook. et Arn.) Planch.	Canton Ampelopsis
大叶蛇葡萄(羽叶蛇葡萄)	*Ampelopsis chaffanjoni* (Levl.) Rehd.	Chaffanjon Snakegrape
蛇葡萄(山葡萄)	*Ampelopsis sinica* (Miq.) W. T. Wang	China Snakegrape

被子植物 ANGIOSPERMAE

角花乌蔹梅	*Cayratia corniculata* (Benth.) Gagnep.	Corniculata Cayratia
乌蔹梅	*Cayratia japonica* (Thunb.) Gagnep.	Japan Cayratia
翅茎白粉藤	*Cissus hexangularis* Thorel ex Planch.	Wingedstem Treebine
白粉藤(菱叶粉藤)	*Cissus repens* Lam.	Creeping Treebine
*光叶火筒树(光叶炎筒树)	*Leea glabra* C. L. Li	Glabrous Leea
*台湾火筒树(火筒树)	*Leea guineensis* G. Don	Taiwan Leea
异叶爬墙虎(爬墙虎)	*Parthenocissus dalzielii* Gagnep. [*P. heterophylla* (Bl.) Merr.]	Diverse-leaved Creeper
三叶爬墙虎(中国地锦)	*Parthenocissus semicordatus* (Wall.) Planch. [*P. himalayana* (Royle) Planch.]	Himalayas Creeper
地锦(爬山虎)	*Parthenocissus tricuspidata* (Sieb. et Zucc.) Planch.	Boston Ivy
三叶崖爬藤	*Tetrastigma hemsleyanum* Diels et Gilg	Hemsley's Rockvine
崖爬藤	*Tetrastigma obtectum* (Wall.) Planch. ex Franch.	Common Rockvine
扁担藤(过江扁龙)	*Tetrastigma plunicaule* (Hook.) Gagnep.	Norrow-leaved Grape
小果葡萄(野葡萄)	*Vitis balanseana* Planch.	Little-fruited Grape
蘡薁(华北葡萄)	*Vitis bryoniifolia* Bunge	North China Grape
绵毛葡萄	*Vitis retordii* Rom. du Caill. ex Planch.	Tomentose Grape

*葡萄	*Vitis vinifera* L.	Wine Grape
	芸香科 Rutaceae	
山油柑(降真香)	*Acronychia pedunculata* (L.) Miq.	Peduncle Acronychia
酒饼簕(狗橘)	*Atalantia buxifolia* (Poir.) Oliv. ex Benth.	Boxleaf Atalantia
酸橙(代代)	*Citrus aurantium* L.	Seville Orange
柚(文旦)	*Citrus grandis* (L.) Osbeck	Pummelo
*柠檬(洋柠檬)	*Citrus limon* (L.) Burm. f.	Lemon
*黎檬(广东柠檬)	*Citrus limonia* Osbeck	Chinese Lemon
佛手(佛手香橼)	*Citrus medica* L. var. *sarcodactylis* (Noot.) Swingle	Finger Citron
*香橼(枸橼)	*Citrus medica* L.	Medicinal Citron
橘(四季橘)	*Fortunella japonica* (Thunb.) Swingle [*Citrus madurensis* Lour.]	Round Kumquat
柑橘(广橘)	*Citrus reticulata* Blanco	Satsuma Orange
甜橙(橙)	*Citrus sinensis* (L.) Osbeck	Sweet Orange
黄皮	*Clausena lansium* (Lour.) Skeels	Wampi
三叉苦(三丫苦)	*Evodia lepta* (Spreng.) Merr.	Thin Evodia
楝叶吴茱萸	*Evodia glabrifolia* (Champ. ex Benth.) Huang [*E. meliaefolia* (Hance) Benth.]	Melia-leaved Evodia
山橘(猴子柑)	*Fortunella hindsii* (Champ. ex Benth.) Swingle	Mountain Kumquat
金橘(金弹)	*Fortunella margarita* (Lour.) Swingle	Oval Kumquat

山小橘(山柑橘)	*Glycosmis parviflora* (Sims) Kurz	Glycosmis
九里香(千里香)	*Murraya paniculata* (L.) Jack. [*M. exotica* L.]	Orange-jessamine
芸香(臭草)	*Ruta graveolens* L.	Common Rue
飞龙掌血	*Toddalia asiatica* (L.) Lam.	Lopez Root
椿叶花椒(樗叶花椒)	*Zanthoxylum ailanthoides* Sieb. et Zucc.	Ailanthus-like Prickly Ash
簕党花椒(簕党)	*Zanthoxylum avicennae* (Lam.) DC.	Prickly Ash
两面针(光叶花椒)	*Zanthoxylum nitidum* (Roxb.) DC.	Shiny-leaved Prickly Ash
*胡椒木(山椒)	*Zanthoxylum piperitum* DC.	
大叶臭花椒	*Zanthoxylum myriacanthum* Wall. ex Hook. f. [*Z. rhetsoides* Drake]	Big-leaved Prickly Ash
花椒簕(花椒)	*Zanthoxylum scandens* Bl.	Climbing Prickly Ash

苦木科 Simaroubaceae

常绿臭椿(福氏臭椿)	*Ailanthus fordii* Noot.	Green Ailanthus
岭南臭椿	*Ailanthus triphysa* (Dennst.) Alston	Hairyleaf South Ailanthus
鸦胆子(老鸦胆)	*Brucea javanica* (L.) Merr.	False Sumac
苦树	*Picrasma quassioides* (D. Don) Benn.	Indian Quassiawood

橄榄科 Burseraceae

橄榄(白榄)	*Canarium album* (Lour.) Raeusch.	Chinese White Olive

| 乌榄(黑榄) | *Canarium tramdenum* Dai et Yakovl. [*C. pimela* Leenh.] | Black Olive |

楝科 Meliaceae

*四季米仔兰(四季米兰)	*Aglaia duperreana* Pierre	
*米仔兰(米兰)	*Aglaia odorata* Lour.	Mock Lime
大叶山楝	*Aphanamixis grandifolia* Bl.	Largeleaf Wildmedia
山楝	*Aphanamixis polystachya* (Wall.) R. N. Parker	Common Wildmedia
麻楝	*Chukrasia tabularis* A. Juss.	Chittagong Chickrassy
*毛麻楝	*Chukrasia tabularis* A. juss. var. *velutina* (Wall.) King	Hairy Chickrassy
香港坚木	*Dysoxylum hongkongense* (Tutch.) Merr.	Hongkong Pencilwood
*非洲楝(塞楝)	*Khaya senegalensis* (Desr.) A. Juss.	Senegal Khaya
苦楝(楝树)	*Melia azedarach* L.	China-berry
*大叶桃花心木(洪都拉斯桃花心木)	*Swietenia macrophylla* King	Central America Mahogany
*桃花心木(西印度群岛桃花心木)	*Swietenia mahagoni* (L.) Jacq.	West Indies Mahogany
小果香椿(紫椿)	*Toona microcarpa* (C. DC.) Harms	Purple Toona
香椿(椿树)	*Toona sinensis* (A. Juss.) Roem.	Chinese Mahogany

无患子科 Sapindaceae

| 倒地铃 | *Cardiospermum halicacabum* L. | Balloon Vine |

被子植物 ANGIOSPERMAE

龙眼(桂圆)	*Dimocarpus longan* Lour.	Longan
车桑子坡柳(车桑子)	*Dodonaea viscosa* (L.) Jacq.	Clammy Hop Seed
*伞花木	*Eurycorymbus cavaleriei* (Levl.) Rehd. et Hand.-Mazz.	Cavaler Eurycorymbus
*复羽叶栾树(国庆花)	*Koelreuteria bipinnata* Franch.	Golden Rain Tree
*台湾栾树	*Koelreuteria elegans* (Seem.) A. C. Smith ssp. formosana (Hayata) Meyer	Taiwan Goldraintree
荔枝	*Litchi chinensis* Sonn.	Lychee
褐叶柄果木(柄果木)	*Mischocarpus pentapetalus* (Roxb.) Radlk.	Brownleaf Mischocarp
红毛丹(毛荔枝)	*Nephelium lappaceum* L.	Rambutan
无患子(肥皂树)	*Sapindus saponaria* L. [*S. mukorossi* Gaertn.]	Soap Berry

槭树科 Aceraceae

三角槭(三角枫)	*Acer buergerianum* Miq.	Buerger Maple
十蕊槭(阔翅槭)	*Acer decandrum* Merr.	Tenstamen Maple
罗浮槭(红翅槭)	*Acer fabri* Hance	Luofu Maple
亮叶槭	*Acer lucidum* Metc.	Lucidleaf Maple
马峦槭	*Acer maluanshanensis* X. M. Wang, R. H. Miau et W. B. Liao	Maluanshan Maple
*鸡爪槭(鸡爪枫)	*Acer palmatum* Thunb.	Japanese Maple
滨海槭(华南飞蛾树)	*Acer sino-oblongum* Metc.	South China Maple
岭南槭	*Acer tutcheri* Duthie	Tutcher's Maple

清风藤科 Sabiaceae

香皮树	*Meliosma fordii* Hemsl. ex Forb. et Hemsl.	Ford's Meliosma
笔罗子	*Meliosma rigida* Sieb. et Zucc.	Stiff-leaved Meliosma
樟叶泡花树(绿樟)	*Meliosma squamulata* Hance	Chinese Meliosma
山羡叶泡花树(花木香)	*Meliosma thorellii* Lec.	Buchananialeaf Meliosma
簇花清风藤	*Sabia fasciculata* Lec. ex L. Chen	Fascicled-flower Sabia
清风藤	*Sabia japonica* Maxim.	Japan Sabia
毛萼清风藤(黑风藤)	*Sabia limoniacea* Wall. ex Hook. f. var. *ardisoides* (Hook. et Arn.) L.	Hairy-sepal Lemon Sabia
尖叶清风藤(台湾清风藤)	*Sabia swinhoei* Hemsl.	Swinhoe's Sabia

省沽油科 Staphyleaceae

野鸦椿(鸡肾果)	*Euscaphis japonica* (Thunb.) Kanitz	Common Euscaphis
锐尖山香圆	*Turpinia arguta* (Lindl.) Seem.	Acute Fieldcitron
山香圆(高地山香圆)	*Turpinia montana* (Bl.) Kurz	Montane Turpinia
光山香圆	*Turpinia montana* (Bl.) Kurz var. *glaberrima* (Merr.) T. Z. Hsu	Glabrous Turpinia

漆树科 Anacardiaceae

*腰果(鸡腰果)	*Anacardium occidentale* L.	Cashew
南酸枣(五眼果)	*Choerospondias axillaris* (Roxb.) Burtt. et Hill	Hog Plum

*人面子(人面树)	*Dracontomelon duperreanum* Pierre	Yanmin
杧果	*Mangifera indica* L.	Mango
*扁桃	*Mangifera persiciformis* C. Y. Wu et T. L. Ming	Peachform Mango
黄连木	*Pistacia chinensis* Bunge	China Pistachio
盐肤木(五倍子树)	*Rhus chinensis* Mill.	Chinese Sumac
滨盐肤木(盐霜柏)	*Rhus chinensis* Mill. var. *roxburghii* (DC.) Rehd.	Seashore Sumac
白背盐肤木(白背漆)	*Rhus hypoleuca* Champ. ex Benth.	Whiteback Sumac
*大叶肉托果	*Semecarpus gigantifolia* Vidal	Largeleaf Markingnut
岭南酸枣(五眼果)	*Spondias lakonensis* Pierre [*S. lakonensis* (Pierre) Stapf]	Canton Mombin
野漆树	*Toxicodendron succedaneum* (L.) O. Kuntze	Field Lacquertree

牛栓藤科 Connaraceae

红叶藤	*Rourea microphylla* (Hook. et Arn.) Planch.	Littleleaf Rourea
牛栓藤(大叶红叶藤)	*Rourea minor* (Gaertn.) Alston	Redleaf Rourea

胡桃科 Juglandaceae

广东黄杞(少叶黄杞)	*Engelhardtia fenzelii* Merr.	Fenzel's Engelhardia
黄杞	*Engelhardtia roxburghiana* Lindl.	Yellow Basket-willow

山茱萸科 Cornaceae

桃叶珊瑚	*Aucuba chinensis* Benth.	Chinese Aucuba

| 狭叶桃叶珊瑚 | *Aucuba chinensis* Benth. var. *angusta* Wang | Narrowed-leaf Chinese Aucuba |
| 香港四照花(野荔枝) | *Dendrobenthamia hongkongensis* (Hemsl.) Hutch. | Hongkong Dogwood |

八角枫科 Alangiaceae

| 八角枫 | *Alangium chinense* (Lour.) Harms | Chinese Alangium |
| 毛八角枫 | *Alangium kurzii* Craib | Kurz Alangium |

蓝果树科 Nyssaceae

| *喜树(旱莲木) | *Camptotheca acuminata* Decne. | Common Camptotheca |

五加科 Araliaceae

五加(五花)	*Acanthopanax gracilistylus* W. W. Smith	Slenderstyle Acanthopanax
白簕花(白刺根)	*Acanthopanax trifoliatus* (L.) Merr.	Three-leaved Acanthopanax
虎刺楤木(小郎伞)	*Aralia armata* (Wall.) Seem.	Spine Aralia
楤木(刺老包)	*Aralia chinensis* L.	Chinese Aralia
黄毛楤木	*Aralia decaisneana* Hance	Yellow Hair Aralia
树参(半枫荷)	*Dendropanax dentigerus* (Harms) Merr.	Tree Ginseng
变叶树参(白半枫荷)	*Dendropanax proteus* (Champ. ex Benth.) Benth.	Biformed Tree-gingseng
马蹄参(大果五加)	*Diplopanax stachyanthus* Hand.-Mazz.	China Hoofrenshen
*八角金盘	*Fatsia japonica* (Thunb.) Decne. et Planch.	Japanese Fatsia
*洋常春藤(西洋常春藤)	*Hedera helix* L.	English Ivy

被子植物 ANGIOSPERMAE

常春藤(中华常春藤)	Hedera nepalensis K. Koch var. sinensis (Tobl.) Rehd.	China Ivy
幌伞枫	Heteropanax fragrans (Roxb.) Seem.	Fragrant Heteropanax
十蕊大参(波缘大参)	Macropanax decandrus Hoo.	Tenstamen Biggingseng
*圆叶福禄桐(圆叶南洋森)	Polyscias balfouriana (Hort. ex Sander) Bailey	Balfour Polyscias
*南洋参(福禄桐)	Polyscias fruticosa (L.) Harms	Indian Polyscias
福禄桐(银边南洋参)	Polyscias guilfoylei (Cogn. et March.) Bailey	Guilfoylei Polyscias
*辐叶鹅掌柴(澳洲鸭脚木)	Schefflera actinophylla (Endl.) Harms	Actinoleaf Schefflera
鹅掌藤	Schefflera arboricola Hayata	Miniature Umbrella Plant
*穗序鹅掌柴	Schefflera delavayi (Franch.) Harms ex Diels	Delavay Schefflera
*孔雀木(手树)	Schefflera elegantissima (Veitch. ex Mast.) Lowry et Frodin [Dizygotheca elegantissima (Veitch. ex Mast.) R. Veitch et Guillaumin]	Spider Aralia
鹅掌柴(鸭脚木)	Schefflera heptaphylla (L.) Frodin [S. octophylla (Lour.) Harms]	Ivy Tree
通脱木	Tetrapanax papyriferus (Hook.) K. Koch	Ricepaperplant

伞形花科 Umbelliferae

紫花前胡(土当归)	Angelica decursiva (Miq.) Franch.	Purpleflower Angelica

芹菜(旱芹)	*Apium graveolens* L.	Celery
积雪草(崩大腕)	*Centella asiatica* (L.) Urban	Moneywort
芫荽(香菜)	*Coriandrum sativum* L.	Coriander
胡萝卜(南鹤虱)	*Daucus carota* L. var. *sativa* Hoffm.	Carrot
刺芫荽(假芫茜)	*Eryngium foetidum* L.	Foetid Eryngo
*小茴香	*Foeniculum vulgare* Mill.	Common Fennel
珊瑚菜(海沙参)	*Glehnia littoralis* F. Schmidt ex Miq.	Coralgreens
红马蹄草	*Hydrocotyle nepalensis* Hook.	Neple Pennywort
天胡荽	*Hydrocotyle sibthorpioides* Lam.	Asiatic Pennywort
肾叶天胡	*Hydrocotyle wilfordii* Maxim.	Wilford's Pennywort
水芹(辣野菜)	*Oenanthe javanica* (Bl.) DC.	Water Celery
前胡(鸡脚前胡)	*Peucedanum praeruptorum* Dunn	Hogfennel

杜鹃花科 Ericaceae

红皮紫陵	*Craibiodendron scleranthum* (Dop) Judd. var. *kwangtungense* (S. Y. Hu) Judd. [*C. kwangtungense* S. Y. Hu]	Guangdong Goldleaf
吊钟花	*Enkianthus quingqueflorus* Lour.	Chinese New Year Flower
齿缘吊钟花	*Enkianthus serrulatus* (Wils.) Schneid.	Serrulate Pendent-bell
毛叶杜鹃(毛鹃)	*Rhododendron championae* Hook.	Champion's Rhododendron
丁香杜鹃(华丽杜鹃)	*Rhododendron farrerae* Tate ex Sweet	Farrer's Azalea

香港杜鹃(白马银花)	*Rhododendron hongkongense* Hutch.	Hongkong Azalea
*洋杜鹃(比利时杜鹃)	*Rhododendron indicum* (L.) Sweet	Indian Azalea
羊角杜鹃(毛棉杜鹃)	*Rhododendron moulmainense* Hook. [*R. westlandii* Hemsl.]	Maomian Azalea
*白花杜鹃(毛白杜鹃)	*Rhododendron mucronatum* (Bl.) G. Don	White Azalea
*粉花杜鹃	*Rhododendron mucronatum* (Bl.) G. Don var. *kemono* Hort.	Purple Doubled Snow Azalea
锦绣杜鹃(毛鹃)	*Rhododendron pulchrum* Sweet	Lovely Azalea
映山红(杜鹃花)	*Rhododendron simsii* Planch.	Red Azalea

越橘科 Vacciniaceae

乌饭树(牛筋树)	*Vaccinium bracteatum* Thunb.	Don Blue Berry

柿科 Ebenaceae

乌材(乌柿)	*Diospyros eriantha* Champ. ex Benth.	Woolly-flowered Persimmon
柿(朱柿)	*Diospyros kaki* Thunb.	Persimmon
罗浮柿(山柿)	*Diospyros morrisiana* Hance	Morris's Persimmon
毛柿(台湾黑檀)	*Diospyros strigosa* Hemsl.	Hairy Persimmon
怀德柿(曾氏柿)	*Diospyros tsangii* Merr.	Tsiang Persimmon
岭南柿(长叶柿)	*Diospyros tutcheri* Dunn	Tutcher's Persimmon
小果柿	*Diospyros vaccinioides* Lindl.	Small Persimmon

山榄科 Sapotaceae

金叶树	*Chrysopyllum lanceolatum* A. DC. var. *stellatocarpon* Van Royen	Golden-leaved Tree

紫荆木	*Madhuca pasquieri* (Dubard) Lam.	Pasquier Madhuca
*人心果	*Manilkara zapota* (L.) Van Royen	Sapodilla
*伊兰芷硬胶(牛乳树)	*Mimusops elengii* L.	Spanish-cherry
*蛋黄果(狮头果)	*Pouteria campechiana* (Kunth) Baehni	Egg Fruit
铁榄(革叶铁榄)	*Sinosideroxylon wightianum* (Hook. et Arn.) Aubr.	Iron Olive
*神秘果(奇迹果)	*Synsepalum dulcificum* (A. DC.) Daniell	Miracle Berry

肉实科 Sarcospermaceae

水石梓(肉实树)	*Sarcosperma laurinum* (Benth.) Hook. f.	Fleshy Nut Tree

紫金牛科 Myrsinaceae

蜡烛果(桐花树)	*Aegiceras corniculatum* (L.) Blanco	Corniculata Candleefruit
朱砂根(圆齿紫金牛)	*Ardisia crenata* Sims	Hilo Holly
郎伞木(小罗伞)	*Ardisia elegans* Andr.	Elegant Ardisia
大罗伞树(大罗伞)	*Ardisia hanceana* Mez	Hance's Ardisia
紫金牛	*Ardisia japonica* (Thunb.) Bl.	Japan Ardisia
虎舌红(红毡)	*Ardisia mamillata* Hance	Mamillate Ardisia
莲座紫金牛	*Ardisia primulifolia* Gardn. et Champ.	Rosula Ardisia
斑叶朱砂根(沿海紫金牛)	*Ardisia lindleyana* D. Dietr. [*A. punctata* Lindl.]	Spotted Ardisia
罗伞树(凉伞树)	*Ardisia quinquegona* Bl.	Asiatic Ardisia

酸藤子(酸藤果)	*Embelia laeta* (L.) Mez	Twig-hanging Embelia
多脉酸藤子(长圆叶酸藤子)	*Embelia vestita* Roxb. [*E. oblongifolia* Hemsl.]	Lenticel-bearing Embelia
当归藤	*Embelia parviflora* Wall. ex A. DC.	Small-leaved Embelia
白花酸藤子	*Embelia ribes* Burm. f.	White-flowered Embelia
杜茎山	*Maesa japonica* (Thunb.) Moritzi ex Zoll.	Japanese Maesa
鲫鱼胆	*Maesa perlaria* (Lour.) Merr.	Cruian Gall
柳叶杜茎山	*Maesa salcifolia* Walker	Willowleaf Maesa
打铁树	*Rapanea linearis* (Lour.) S. Moore	Forgeirontree
密花树(打铁树)	*Rapanea neriifolia* (Sieb. et Zucc.) Mez	Oleanderleaf Rapanea

安息香科 Styracaceae

赤杨叶(拟赤杨)	*Alniphyllum fortunei* (Hemsl.) Makino	Fortune's China-bell
银钟花(假杨桃)	*Halesia macgregorii* Chun	Silverbell
广东木瓜红(岭南木瓜红)	*Rehderodendron kwangtungensise* Chun	Kwangtung Rehdertree
白花笼	*Styrax faberi* Perk.	Faber Snowbell
大花安息香	*Styrax grandiflorus* Griff.	Bigflower Snowbell
芳香安息香(芬芳安息香)	*Styrax odoratissimus* Champ. et Benth.	Fragrant Snow-bell
齿叶安息香	*Styrax serrulatus* Roxb.	Toothleaf Snowbell
栓叶安息香(红皮树)	*Styrax suberifolius* Hook. et Arn.	Cork-leaved Snow-bell
越南安息香(白背安息香)	*Styrax tonkinensis* (Pierre) Craib ex Hartw.	Tonkin Snowbell

山矾科 Symplocaceae

腺叶山矾(腺叶灰木)	*Symplocos adenophylla* Wall.	Glandular-leaved Sweet-leaf
腺柄山矾(赤牙木)	*Symplocos adenopus* Hance	Glandular-stipe Sweet-leaf
华山矾(华灰木)	*Symplocos chinensis*(Lour.) Druce	Chinese Sweet-leaf
越南山矾(火灰树)	*Symplocos cochinchinensis*(Lour.) Moore	Vietnam Sweet-leaf
南岭山矾(南岭灰木)	*Symplocos confusa* Brand. [*S. pendula* Wight var. *hirtistylis*(C. B. Clarke) Noot.]	Asiatic Sweet-leaf
厚皮山矾(白布果)	*Symplocos crassifolia* Benth.	Thickbark Sweet-leaf
光亮山矾	*Symplocos lucida* Sieb. et Zucc.	Shining Sweet-leaf
美山矾	*Symplocos decora* Hance	Beautiful Sweet-leaf
长毛山矾	*Symplocos dolichotricha* Merr.	Longhair Sweet-leaf
灰山矾	*Symplocos dung* Eberm. et Dub.	Cinderlike Sweet-leaf
三裂山矾	*Symplocos fordii* Hance	Ford's Sweet-leaf
羊舌树(山羊耳)	*Symplocos glauca*(Thunb.) Koidz.	Glaucous Sweet-leaf
光叶山矾(剑叶灰木)	*Symplocos lancifolia* Sieb. et Zucc.	Smooth-leaved Sweet-leaf
黄牛奶树	*Symplocos laurina*(Retz.) Wall.	Laurel Sweet-leaf
白檀(檀香)	*Symplocos paniculata*(Thunb.) Miq.	Sapphire-berry Sweet-leaf
珠仔树	*Symplocos racemosa* Roxb.	Racemose Sweet-leaf

| 老鼠矢 | *Symplocos stellaris* Brand | Starshape Sweet-leaf |
| 山矾(山桂花) | *Symplocos sumuntia* Buch.-Ham. ex D. Don | Sumuntia Sweet-leaf |

马钱科 Loganiaceae

驳骨丹(小驳骨)	*Buddleja asiatica* Lour.	Asiatic Butterfly-bush
醉鱼草(闹鱼花)	*Buddleja lindleyana* Fort.	Lindley Butterflybush
*灰莉(华灰莉木)	*Fagraea ceilanica* Thunb.	Ceylon Fagraea
蓬莱葛(多花蓬莱葛)	*Gardneria multiflora* Makino	Many-flowered Gardneria
大茶药(胡蔓藤)	*Gelsemium elegans* (Gardn. et Champ.) Benth.	Gelsemium
水田白裸茎(小姬苗)	*Mitrasaceme pygmaea* R. Br.	Dwarf Mitrasacme
牛眼马钱(牛眼珠)	*Strychnos angustiflora* Benth.	Narrow-flowered Poisonnut
华马钱(三脉马钱)	*Strychnos cathayensis* Merr.	Cathay Poisonnut
伞花马钱	*Strychnos umbellata* (Lour.) Merr.	Umbel-flowered Poisonnut

木犀科 Oleaceae

白蜡树(青榔木)	*Fraxinus chinensis* Roxb.	Chinese Ash
苦枥木(台湾梣)	*Fraxinus insularis* Hemsl.	Retuse Ash
扭肚藤	*Jasminum elongatum* (Bergius) Willd. [*J. amplexicaule* Buch.-Ham.]	Mock Jasmine
北清香藤(光清香藤)	*Jasminum lanceolarium* Roxb.	HongKong Jasmine
云南黄素馨(云南迎春)	*Jasminum mesnyi* Hance	Yellow Jasmine
*茉莉	*Jasminum sambac* (L.) Ait.	Arabian Jasmine

华素馨(华清香藤)	*Jasminum sinensis* Hemsl.	China Jasmine
日本女贞(台湾女贞)	*Ligustrum amamianum* Koidz. [*L. japonicum* Thunb.]	Formosa Privet
李氏女贞(华女贞)	*Ligustrum lianum* Hsu	Li's Privet
女贞(大叶女贞)	*Ligustrum lucidum* Ait.	Glossy Privet
小蜡树(山指甲)	*Ligustrum sinense* Lour.	Chinese Privet
*金叶女贞	*Ligustrum vicaryi* Rehd.	Vicary Golden Privet
异叶木犀榄	*Olea dioica* Roxb.	Dioecious Olive
*尖叶木犀榄(锈鳞木犀榄)	*Olea cuspidate* Wall. et G. Don [*O. ferruginea* Royle]	Cuspidata Olive
桂花(木犀)	*Osmanthus fragrans* (Thunb.) Lour.	Kwai-Fah
*丹桂	*Osmanthus fragrans* (Thunb.) Lour. var. *aurantiacus* Makino	Orange Sweet Osmanthus
*金桂	*Osmanthus fragrans* Lour. (Thunb.) var. *thunbergii* Makino	Golden Sweet Osmanthus
牛矢果(长叶木犀)	*Osmanthus matsumuranus* Hayata	Taiwan Osmanthus

夹竹桃科 Apocynaceae

*沙漠玫瑰(天宝花)	*Adenium obesum* (Forssk.) Roem. et Schult.	Desert Rose
*软枝黄蝉(黄莺花)	*Allamanda cathartica* L.	Common Allamanda
*大花软枝黄蝉	*Allamanda cathartica* L. var. *hendersonii* (Bull ex Dombr.) Bailey et Raffill	Henderson Allamanda
*紫黄蝉(大紫蝉)	*Allamanda blanchetii* A. DC.	Blanchett Allamanda

被子植物 ANGIOSPERMAE

*黄蝉(黄兰蝉)	*Allamanda schottii* Pohl [*A. neriifolia* Hook.]	Small Allamanda
*糖胶树(面条树)	*Alstonia scholaris* (L.) R. Br.	Devil Tree
链珠藤(念珠藤)	*Alyxia sinensis* Champ. ex Benth.	Bead Vine
串珠子	*Alyxia vulgaris* Tsiang	Common Alyxia
鳝藤(神葛)	*Anodendron affine* (Hook. et Arn.) Druce	Asian Cable Creeper
保亭鳝藤	*Anodendron howii* Tsiang	How Eelvine
*长春花(雁来红)	*Catharanthus roseus* (L.) G. Don	Periwrinkle
海杧果	*Cerbera manghas* L.	Common Cerbera Tree
止泻木	*Holarrhena antidysenterica* Wall. ex A. DC.	Common Holarrhena
仔榄树	*Hunteria zelanica* (Retz.) Garden. ex Thwaites	Srilanka Hunteria
蕊木(假乌榄树)	*Kopsia lancibracteolata* Merr.	Lancebract Kopsia
*红蝉花(双腺藤)	*Mandevilla sanderi* (Hemsl.) Woodson	Brazilian Jasmine
尖山橙	*Melodinus fusiformis* Champ. ex Benth.	Fusiform Melodinus
山橙(猴子果)	*Melodinus suaveolens* Champ. ex Benth.	Mountain Orange
*夹竹桃(红花夹竹桃)	*Nerium oleander* L. [*N. indicum* Mill.]	Oleander
*玫瑰树	*Ochrosia borbonica* Gmel.	Bourbon Ochrosia
*红花鸡蛋花(红鸡蛋花)	*Plumeria rubra* L.	Red Frangipani
*鸡蛋花(缅栀子)	*Plumeria rubra* L. 'Acutifolia'	Frangipani

帘子藤	*Pottsia laxiflora* (Bl.) O. Kuntze	Loose-flowered Pottsia
四叶萝芙木	*Rauvolfia tetraphylla* L.	Fourleaf Devilpepper
萝芙木(萝芙藤)	*Rauvolfia verticillata* (Lour.) Baill	Devil-pepper
羊角坳(羊角扭)	*Strophanthus divaricatus* (Lour.) Hook. et Arn.	Goat Horns
*狗牙花	*Tabermaemontana divaricata* (L.)R. Br. ex Roem. et Schult. [*Ervatamia divaricata* (L.) Burk.]	Crepe Jasmine
*黄花夹竹桃(酒杯花)	*Thevetia peruviana* (Pers.) Schum.	Yellow Oleander
络石(扒墙虎)	*Trachelospermum jasminoides* (Lindl.) Lem.	Star Jasmine
石血	*Trachelospermum jasminoides* var. *heterophyllum* Tsiang	Stoneblood
杜仲藤(花皮胶藤)	*Urceola micrantha* (Wall. ex G. Don) A. DC. [*Parabarium micranthum* (A. DC.) Pierre]	Small-flowered Urceola
酸叶胶藤	*Urceola rosea* (Hook. et Arn.) D. J. Middleton [*Ecdysanthera rosea* Hook. et Arn.]	Sour Creeper
倒吊笔	*Wrightia pubescens* R. Br.	Common Wrightia
	萝藦科 Asclepiadaceae	
*马利筋(莲生桂子花)	*Asclepias curassavica* L.	Blood-flower Milkweed
*恋之蔓	*Ceropegia woodii* Schldl.	Wood Ceropegia

白叶藤	*Cryptolepis sinensis* (Lour.) Merr.	Chinese Cryptolepis
刺瓜(小刺瓜)	*Cynanchun corymbosum* Wight	Swallow Wort
眼树莲(瓜子藤)	*Dischidia chinensis* Champ. ex Benth.	Chinese Dischidia
*气球果(钉头果)	*Gomphocarpus fruticosus* (L.) W. T. Aiton	Fruricose Nailheadfruit
天星藤	*Graphistemma pictum* (Benth.) B. D. Jacks	Painted Graphistemma
匙羹藤(狗屎藤)	*Gymnema sylvestre* (Retz.) Schult.	Australian Cow-plant
牛皮消(白首乌)	*Cynanchum auriculatum* Royle ex Wight	Bunge Swallowwart
球兰	*Hoya carnosa* (L.f.) R. Br.	Wax Plant
折冠藤	*Lygisma inflexum* (Cost.) Kerr	Bent-corolla Lygisma
石萝藦	*Pentasacme caudatum* Wall. ex Wight [*P. championii* Benth.]	Caudate Pentasachme
*大花犀角(海星花)	*Stapelia grandiflora* Mass.	Largeflower Carrionflower
*犀角	*Stapelia hirsuta* L. [*S. pulchella* Mass.]	Hirsute Carrionflower
华南夜来香(中华夜来香)	*Telosma cathayensis* Merr.	South China Telosma
夜来香(夜香藤)	*Telosma cordata* (Burm. f.) Merr.	Night-fragrant Flower
平滑弓果藤	*Toxocarpus laevigatus* Tsiang	Laevigate Bowfruitvine
弓果藤	*Toxocarpus wightianus* Hook. et Arn.	Wight's Toxocarpus

三分丹(毛果娃儿藤)	*Tylophora atrofolliculata* Metc.	Sanfendan Childvine
七层楼	*Tylophora floribunda* Miq.	Manyflower Childvine
海南弓果藤	*Tylophora hainanensis* Tsiang	Hainan Childvine
娃儿藤(卵叶娃儿藤)	*Tylophora ovata* (Lindl.) Hook. ex Steud.	Ovate Tylophora
小叶蛙儿藤	*Tylophora tenuis* Bl.	Smallleaf Childvine

杠柳科 Periplocaceae

海岛藤(假络石)	*Gymnanthera oblonga* (Burm. f.) P. S. Green	Oblong Gymananthera

茜草科 Rubiaceae

水团花(水杨梅)	*Adina pilulifera* (Lam.) Franch. ex Drake	Chinese Buttonbush
*黄棉木	*Adina polycephala* Benth.	Polycephalous Adina
香楠(光叶山黄皮)	*Aidia canthioides* (Champ. ex Benth.) Masamune [*Randia canthioides* Champ. ex Benth.]	Moutain Wampi
茜树(越南香楠)	*Aidia cochinchinensis* Lour. [*Randia cochinchinensis* (Lour.) Merr.]	Cochinchina Randia
多毛茜草树(毛山黄皮)	*Aidia pycnantha* (Drake) Tirvenz [*Randia acuminatissima* Merr.]	Hairy Randia
白果香楠	*Alleizettella leucocarpa* (Champ. ex Benth.) Tirvenz	White-fruited Randia
毛茶	*Antirhea chinensis* (Champ. ex Benth.) Forbes et Hemsl.	Chinese Antirhea
糙叶丰花草	*Borreria articularia* (L. f.) F. N. Will.	Rough-leaved Borreria

阔叶丰花草(日本草)	*Borreria latifolia* (Aubl.) K. Schum.	Wide-leaved Borreria
丰花草	*Borreria stricta* (L. f.) G. E. W. Mey.	Narrow-leaved Borreria
鱼骨子(鱼骨木)	*Canthium dicoccum* (Gaertn.) Teysmenn et Binnedijk.	Butulang Canthium
猪肚木	*Canthium horridum* Bl.	Bristly Canthium
山石榴	*Catunaregam spinosa* (Thunb.) Tirv.	Moutain Pomegranate
*小粒咖啡树	*Coffea arabica* L.	Arabian Coffee
流苏子	*Coptosapelta diffusa* (Champ. ex Benth.) Van Steenis	Diffuse Coptosapelta
虎刺(绣花针)	*Damnacanthus indicus* (L.) Gaertn. f.	Indian Damnacanthus
狗骨柴	*Diplospora dubia* (Lindl.) Masamune [*D. viridiflora* DC.]	Common Tricalysia
多刺山黄皮	*Fagerlindia depauperata* (Drake) Tirv. [*Randia depauperata* Drake]	Manyspine Fagerlindia
栀子(黄栀子)	*Gardenia jasminoides* Ellis	Cape Jasmine
狭叶栀子	*Gardenia jasminoides* Ellis var. *angustifolia* Makino	Narrow-leaved Gardenia
白蟾	*Gardenia jasminoides* Ellis var. *fortuniana* (Lindl.) Hara	Fortune's Cape Jasmine
大花栀子(大叶栀子)	*Gardenia jasminoides* Ellis var. *grandiflora* Makino	Largeflower Cape Jasmine
水栀子(雀舌板子)	*Gardenia jasminoides* Ellis var. *radicans* (Thunb.) Makino	Water Cape Jasmine

爱地草	*Geophila herbacea* (Jacq.) K. Schum.	Herbaceous Geophila
*希茉莉(长隔木)	*Hamelia patens* Jacq.	Scarlet Bush
金草(方骨草)	*Hedyotis acutangula* Champ. ex Benth.	Angle-stemmed Hedyotis
广花耳草	*Hedyotis ampliflora* Hance	Ample-flowered Hedyotis
耳草	*Hedyotis auricularis* L.	Auriclar Hedyotis
双花耳草	*Hedyotis biflora* (L.) Lam.	Two-flowered Hedyotis
剑叶耳草	*Hedyotis caudatifolia* Merr. et Metcalf [*H. lancea* Thunb.]	Tailleaf Eargrass
金毛耳草(黄毛耳草)	*Hedyotis chrysotricha* (Palib.) Merr.	Goldhair Eargrass
拟金草	*Hedyotis consanguinea* Hance	Pink Bone-wort
伞房花耳草(水线草)	*Hedyotis corymbosa* (L.) Lam.	Corymbose Hedyotis
白花蛇舌草	*Hedyotis diffusa* Willd.	Diffuse Hedyotis
牛白藤(广花耳草)	*Hedyotis hedyotidea* (DC.) Merr.	White Ox Creeper
粤港耳草	*Hedyotis loganioides* Benth.	Similar-logania Eargrass
卵叶耳草	*Hedyotis ovata* Thunb. ex Maxim.	Ovate Eargrass
松叶耳草(鸟舌草)	*Hedyotis pinifolia* Wall. ex G. Don	Pineleaf Eargrass
纤花耳草	*Hedyotis tenelliflora* Bl.	Tender-flowered Hedyotis
方茎耳草	*Hedyotis tetrangularia* (Korth.) Walp.	Squarestem Eargrass
长节耳草	*Hedyotis uncinella* Hook. et Arn.	Small-hooked Hedyotis

被子植物 ANGIOSPERMAE

香港耳草	*Hedyotis vachellii* Hook.	Hong Kong Hedyotis
粗叶耳草	*Hedyotis verticillata* (L.) Lam. [*H. hispida* Retz.]	Coarseleaf Eargrass
黄叶耳草	*Hedyotis xanthochroa* Hance	Yellowleaf Eargrass
龙船花(仙丹花)	*Ixora chinensis* Lam.	Chinese Ixora
*红龙船花(红仙丹花)	*Ixora coccinea* L.	Dwarf Ixora
白龙船花(白龙船)	*Ixora henryi* Levl.	White Ixora
*黄龙船花(黄仙丹)	*Ixora lutea* (Veitch) Hutch.	Yellow Ixora
*小仙丹花(矮仙丹)	*Ixora parviflora* Vahl	Smallflower Ixora
粗叶木(白果鸡屎树)	*Lasianthus chinensis* (Champ.) Benth.	Chinese Lasianthus
西南粗叶木	*Lasianthus henryi* Hutch.	Herry Roughleaf
斜基粗叶木(斜脉鸡屎树)	*Lasianthus wallichii* (Wight et Arn.) Wight	Small-leaved Lasianthus
小叶巴戟天(百眼藤)	*Morinda parvifolia* Bartl. ex DC.	Littleleaf Indian-mulberry
鸡眼藤(羊角藤)	*Morinda umbellata* L.	Common Indian-mulberry
楠藤	*Mussaenda erosa* Champ.	Erose Mussaenda Wild Mussaenda
*红叶金花(血萼花)	*Mussaenda erythrophylla* Schum. et Thonn.	Red-leaved Mussaenda
大叶白纸扇	*Mussaenda esquirolii* Levl.	Esquirol Mussaenda
粗毛玉叶金花	*Mussaenda hirsutula* Miq.	Hirsute Mussaenda
广东玉叶金花(白纸扇)	*Mussaenda kwangtungensis* Li	Kwangtung Mussaenda

玉叶金花(野白纸扇)	*Mussaenda pubescens* Ait. f.	Splash-of-white
乌檀	*Nauclea officinalis* (Pierre ex Pitard) Merr. et Chun	Medicinal Fatheadtree
薄叶假耳草(落叶新耳草)	*Neanotis hirsuta* (L. f.) Lewis [*Anotis hirsuta* (L. f.) Boerl.]	Hirsuta Neanotis
广州蛇根草	*Ophiorrhiza cantoniensis* Hance	Canton Ophiorrhiza
日本蛇根草(蛇根草)	*Ophiorrhiza japonica* Bl.	Japanese Ophiorrhiza
短小蛇根草(小蛇根草)	*Ophiorrhiza pumila* Champ. ex Benth.	Dwarf Ophiorrhiza
鸡爪簕	*Oxyceros sinensis* Lour.	Chinese Randia
鸡屎藤	*Paederia scandens* (Lour.) Merr.	Chinese Feverine
毛鸡屎藤(白鸡屎藤)	*Paederia scandens* var. *tomentosa* (Bl.) Hand.-Mazz.	Tomentose Feverine
狭枝鸡矢藤(狭叶鸡矢藤)	*Paederia stenobotrya* Merr.	Narrow-leaf Fevervine
香港大沙叶(满天星)	*Pavetta hongkongensis* Bremek.	Hongkong Pavetta
*五星花(繁星花)	*Pentas lanceolata* (Forsk.) Schum.	Pentas
海南槽裂木	*Pertusadina hainanensis* (How) Ridsd.	Hainan Adina
九节(山大刀)	*Psychotria asiatica* L. [*P. rubra* (Lour.) Poir.]	Red Psychotria
蔓九节(穿根藤)	*Psychotria serpens* L.	Creeping Psychotria
假九节(小叶九节)	*Psychotria tutcheri* Dunn [*P. tutcherii* Dunn]	Tutcher Psychotria

六月雪(满天星)	*Serissa japonica* (Thunb.) Thunb.	Snow of June
白马骨(满天星)	*Serissa serissoides* (DC.) Druce	Serissa
乌口树(达仑木)	*Tarenna attenunata* (Voigt) Hutch.	Tapered Tarenna
密毛乌口树(白花苦灯笼)	*Tarenna mollissima* (Hook. et Arn.) Rob.	Whiteflower Tarenna
水锦树	*Wendlandia uvariifolia* Hance	Uvarialeaf Wendlandia

忍冬科 Caprifoliaceae

*糯米条(茶条树)	*Abelia chinensis* R. Br.	Chinese Abelia
*单花六道木	*Abelia uniflora* R. Br.	One-flowered Abelia
华南忍冬(山银花)	*Lonicera confusa* (Sweet) DC.	Wild Honeysuckle
金银花	*Lonicera japonica* Thunb.	Japanese Honeysuckle
*红白忍冬(红金银花)	*Lonicera japonica* Thunb. var. *chinensis* (Wats.) Baker	Chinese Honeysuckle
长花忍冬	*Lonicera longiflora* (Lindl.) DC.	Long-flowered Honeysuckle
大花忍冬	*Lonicera macrantha* (D. Don) Spreng.	Large-flowered Honeysuckle
皱叶忍冬(显脉忍冬)	*Lonicera rhytidophylla* Hand.-Mazz. [*L. reticulata* Champ.]	Grape Honeysuckle
接骨木	*Sambucus williamsii* Hance	Williams Elder
南方荚蒾(南方荚)	*Viburnum fordiae* Hance	Southern Viburnum
蝶花荚蒾	*Viburnum hanceanum* Maxim.	Hance's Viburnum
珊瑚树(极香荚蒾)	*Viburnum odoratissimum* Ker-Gawl.	Sweet Viburnum

坚荚蒾(常绿荚蒾)	*Viburnum sempervirens* K. Koch	Evergreen Viburnum

菊科 Compositae

紫花藿香蓟(熊耳草)	*Ageratum houstonianum* Mill.	Blue Billygoat-weed
下田菊	*Adenostemma lavenia* (L.) O. Kuntze	Common Adenostemma
藿香蓟(胜红蓟)	*Ageratum conyzoides* L.	Billygoat-weed
杏叶香兔儿风(吉香兔儿风)	*Ainsliaea fragrans* Champ.	Fragrant Ainsliaea
铁丁兔儿风(铁灯兔儿风)	*Ainsliaea macroclinidioides* Hayata	Integrifolious Ainsliaea
*红背兔儿风(红走马胎)	*Ainsliaea rubrifolia* Franch.	Redleaf Ainsliaea
山黄菊	*Anisopappus chinensis* (L.) Hook. et Arn.	Chinese Anisopappus
*牛蒡	*Arctium lappa* L.	Great Burdock
*白菊仔	*Argyranthemum frutescens* (L.) Sch.-Bip	Shrubby Argyranthemum
黄芩蒿(黄花蒿)	*Artemisia annua* L.	Sweet Wormwood
艾蒿(艾叶)	*Artemisia argyi* Levl. et Vant.	Argy's Wormwood
南毛蒿	*Artemisia chingii* Pamp.	Ching Wormwood
白苞蒿(甜菜子)	*Artemisia lactiflora* Wall. ex DC.	White Wormwood, Ghostplant Wormwood
山白菊(三脉紫菀)	*Aster ageratoides* Turcz.	Threevein Aster
白舌紫菀	*Aster baccharoides* (Benth.) Steetz.	Whiteligulate Aster
琴叶紫菀	*Aster panduratus* Nees ex Walp.	Fiddleleaf Aster

被子植物 ANGIOSPERMAE

*短舌紫菀	*Aster sampsonii* (Hance) Hemsl.	Sampson's Aster
*雏菊	*Bellis perennis* L.	English Daisy
*白花鬼针草	*Bidens alba* (L.) DC.	Shepherd's Needle
鬼针草(婆婆针)	*Bidens bipinnata* L.	Bur-Marigold
金盏银盘	*Bidens biternata* (Lour.) Merr. et Sherff.	Biternate Beggartick
三叶鬼针草(鬼针草)	*Bidens pilosa* L.	Hairy Bur-Marigold
异芒菊	*Blainivillea acmella* (L.) Phillipson [*B. latifolia* (L.) DC.	Broadleaf Blainvillea
艾纳香(大风艾)	*Blumea balsamifera* (L.) DC.	Balsamic Blumea
七里明	*Blumea clarkei* Hook. f.	Clark's Blumea
节节红	*Blumea fistulosa* (Roxb.) Kurz	Fistulose Blumea
大头艾纳香(东风草)	*Blumea megacephala* (Randeria) Chang et Tseng	Big-flowered Blumea
长圆叶艾纳香	*Blumea oblongifolia* Kitam.	Oblongleaf Blumea
无梗艾纳香	*Blumea sessiliflora* Decne.	Sessile Blumea
*金盏菊(金盏花)	*Calendula officinalis* L.	Potmarigold Calendula
*翠菊(八月菊)	*Callistephus chinensis* (L.) Nees	China Aster
*银叶菊(雪叶菊)	*Centaurea cineraria* L.	Dusty Miller
石胡荽(鹅不食草)	*Centipeda minima* (L.) A. Br. et Aschers	Small Centipeda
*白晶菊(晶晶菊)	*Chrysanthemum paludosum* Poir.	Daisy White Button
茼蒿(蒿子秆)	*Chrysanthemum segetum* L.	Corn Chrysanthemum

绿蓟	*Cirsium chinense* Gardn. et Champ.	Chinese Thistle
湖北蓟	*Cirsium hupehense* Pamp.	Hubei Thistle
小蓟(蓟)	*Cirsium japonicum* Fisch. ex DC.	Japanese Thistle
线叶蓟	*Cirsium lineare* (Thunb.) Sch. -Bip.	Linearleaf Thistle
香丝草(小加蓬)	*Conyza bonariensis* (L.) Cronq.	Bona Conyza
*加拿大蓬(小白酒草)	*Conyza canadensis* (L.) Cronq. [*Erigeron canadensis* L.]	Horseweed
白花白酒草	*Conyza leucantha* (D. Don) Ludlow er Raven	Whiteflower Conyza
*剑叶波斯菊(金鸡菊)	*Coreopsis lanceolata* L.	Lance Coreopsis
*蛇目菊(波斯菊)	*Coreopsis tinctoria* Nutt.	Plains Coreopsis
*大波斯菊(秋英)	*Cosmos bipinnatus* Cav.	Common Cosmos
*硫磺菊(黄波斯菊)	*Cosmos sulphureus* Cav.	Sulphur Cosmos
野茼蒿(革命菜)	*Crassocephalum crepidioides* (Benth.) S. Moore	Redflower Ragleaf
*芙蓉菊	*Crossostephium chinense* (L.) Makino	Chinese Crossostephium
*红大丽花	*Dahlia coccinea* Cav.	Fire Dahlia
*卷瓣大丽花	*Dahlia jaurezii* Hort. ex Sasaki	Jaurez Dahlia
*光滑大丽花	*Dahlia merckii* Lehm.	Bedding Dahlia
*大丽花(大丽菊)	*Dahlia pinnata* Cav.	Garden Dahlia

野黄菊(野菊)	*Dendranthema indicum* (L.) Des Moul.	Chinese Chrysanthemum
*菊花	*Dendranthema morifolium* (Ramat.) Tzvel.	Common Chrysanthemum
鱼眼草	*Dichrocephala integrifolia* (L. f.) O. Kuntze	Entireleaf Dichrocephala
短冠东风菜	*Doellingeria marchandii* (H. Lev.) Ling [*Aster marchandii* H. Lev.]	Marchand Doellingeria
东风菜	*Doelligeria scaber* (Thunb.) Nees [*Aster scaber* Thunb.]	Scabrous Doellingeria
鳢肠(旱莲草)	*Eclipta prostrata* (L.) L.	False Daisy
地胆草(地胆头)	*Elephantopus scaber* L.	Elephants-foot
白花地胆草(毛地胆草)	*Elephantopus tomentosus* L.	Tomentose Elephantfoot
一点红(叶下红)	*Emilia sonchifolia* (L.) DC.	Tassel-Flower
球菊(鹅不食草)	*Epaltes australis* Less.	Goose-no-eat
美洲菊芹(梁子菜)	*Erechtites hieracifolia* (L.) Raf. ex DC.	American Burnweed
*假臭草(猫腥草)	*Eupatorium catarium* Veldk. [*Praxelis clematideum* (Griseb.) R. M. King et H. Rob.]	Praxelis
华泽兰	*Eupatorium chinensis* L.	Chinensis Eupatorium
白头婆(泽兰)	*Eupatorium japonicum* Thunb.	Japanese Boneset
林泽兰(野马追)	*Eupatorium lindleyanum* DC. [*Eupatorium lindleyana* DC.]	Lindley's Boneset
*飞机草(香泽兰)	*Eupatorium odoratum* L.	Fragrant Eupatorium

大吴风菊(橐吾)	*Farfugium japonicum* (L. f.) Kitam.	Leopard Plant
*非洲菊(扶郎花)	*Gerbera jamesonii* Bolus [*Gerbera jamesonii* Bolus ex Hook. f.]	Flameray Gerbera
毛大丁草	*Gerberia piloselloides* (L.) Cass.	Hairy Gerbera
鹿角草	*Glossocardia bidens* (Retz.) Veldk.	Native Cobblers Pegs
鼠曲草(清明菜)	*Gnaphalium affine* D. Don	Cudweed
匙叶鼠曲草	*Gnaphalium pensylvanicum* Willd.	Pensylvanian Cudweed
多茎鼠曲草	*Gnaphalium polycaulon* Pers.	Manystem Cusweed
*凤凰菜(紫背菜)	*Gynura bicolor* (Roxb. ex Willd.) DC.	Twocolored Velvetplant
白籽菜(小毛花)	*Gynura divaricata* (L.) DC.	Divaricate Velvetplant
三七草(土三七)	*Gynura japonica* (L. f.) Juel	Japanese Velvetplant
*向日葵(葵花)	*Helianthus annuus* L.	Sunflower
*菊芋	*Helianthus tuberosus* L.	Girasole
泥湖菜	*Hemistepta lyrata* (Bunge) Bunge	Lyrata Hemistetea
羊耳菊(白牛胆)	*Inula cappa* (Buch.-Ham.) DC.	Elecampane
细叶小苦荬	*Ixeridium gracile* (DC.) Shih	Slender Ixeridium
纤细苦荬菜	*Ixeris gracilis* Stebb.	Slender Ixeris
剪刀股(假蒲公英)	*Ixeris japonica* (Burm. f.) Nakai	Japanese Ixeris
苦荬菜(匍匐苦荬菜)	*Ixeris repens* (L.) A. Gray	Creeping Ixeris

被子植物 ANGIOSPERMAE

马兰(鸡儿肠)	*Kalimeris indica* (L.) Sch.-Bip. [*Aster indicus* L.]	Indian Kalimeris
山莴苣	*Lactuca indica* L.	Indian Lettuce
莴苣(莴笋)	*Lactuca sativa* L.	Garden Lettuce
生菜(叶用莴苣)	*Lactuca sativa* L. var. *romana* L. H. Bailey [*L. sativa* L. var. *longifolia* Lam.]	Romana Salat
野莴苣	*Lactuca seriola* Torner (*L. serriola* Torner)	Wild Lettuce
六棱菊(六耳铃)	*Laggera alata* (D. Don) Sch.-Bip. ex Hochst.	Winged Laggera
匍茎栓果菊	*Launaea sarmentosa* (Willd.) Merr. et Chun	Creeping Launaea
滨菊(西洋滨菊)	*Leucanthemum vulgare* H. S. La (*Leucanthemum vulgare* Lam.)	General Leucanthemum
*蛇鞭菊(麒麟菊)	*Liatris spicata* Willd.	Spike Gayfeather
大头橐吾	*Ligularia japonica* (Thunb.) Less.	Japannese Goldenray
小舌菊(梨叶小舌菊)	*Microglossa pyrifolia* (Lam.) O. Kuntze	Pearleaf Microglossa
*假泽兰(蔓菊)	*Mikania cordata* (Burm. f.) B. L. Robinson	Heartshape Mikania
*薇甘菊(小花假泽兰)	*Mikania scandens* (L.) Willd. [*M. micrantha* H. B. K.]	Mile-a-minute Weed
栌菊木	*Nouelia insignis* Franch.	Insignis Nouelia
黄瓜菜	*Paraixeris denticulata* (Houtt.) Nakai [*Youngia denticulata* (Houtt.) Kitam.]	Denticulate Paraixeris

*瓜叶菊(千日莲、瓜叶莲、千里光)	*Pericallis hybrida* B. Nord. [*Senecio cruentus* DC.]	Common Ragwort
阔苞菊	*Pluchea indica* (L.) Less.	Marsh Fleabane
翅果菊(山莴苣)	*Pterocypsela indica* (L.) Shih	Common Pterocypsela
*除虫菊(白花除虫菊)	*Pyrethrum cinerariifolium* Trev.	Dalmatian Pyrethrum
千里光	*Senecio scandens* Buch.-Ham. ex D. Don	Climbing Groundsel
闽粤千里光	*Senecio stauntonii* DC.	Staunton's Ragwort
豨莶	*Siegesbeckia orientalis* L.	Shrimp Claw Plant
水飞蓟	*Silybum marianum* (L.) Gaertn.	Holy Thistle
*北美一枝黄花(加拿大一枝黄花)	*Solidago canadensis* L.	Canadian Goldenrod
一枝黄花(野黄菊)	*Solidago decurrens* Lour.	Goldenrod
裸柱菊(假吐金菊)	*Soliva anthemifolia* (Juss.) R. Br.	Camomileleaf Soliva
苣荬菜	*Sonchus arvensis* L.	Field Sowthistle
续断菊(圆耳苦苣菜)	*Sonchus asper* (L.) Hill.	Prickly Snowthistle
苦苣菜(苦芥菜)	*Sonchus oleraceus* L.	Common Snowthistle
戴星草(流星草)	*Sphaeranthus africanus* L.	African Sphaeranthus
金钮扣	*Spilanthes paniculata* Wall. ex DC.	Gold Button
*金腰箭	*Synedrella nodiflora* (L.) Gaertn.	Nodalflower Synedrella
*万寿菊(蜂窝菊)	*Tagetes erecta* L.	Big Marigold

*孔雀草(小万寿菊)	*Tagetes patula* L.	French Marigold
肿柄菊(墨西哥向日葵)	*Tithonia diversifolia* A. Gray	Maxican Sunflower
羽芒菊	*Tridax procumbens* L.	Tridax
款冬	*Tussilago farfara* L.	Common Coltsfoot
细脉斑鸠菊	*Vernonia andersonii* Clarke [*Vernonia andersonii* C. B. Clarke]	Anderson's Veronia
糙叶斑鸠菊	*Vernonia aspera* (Roxb.) Buch. -Ham.	Roughleaf Ironweed
夜香牛	*Vernonia cinerea* (L.) Less.	Iron-Weed
毒根斑鸠菊	*Vernonia cumingiana* Benth.	Poisonousroot Ironweed
咸虾花	*Vernonia patula* (Dryand.) Merr.	Iron-Weed
柳叶斑鸠菊	*Vernonia saligna* (Wall.) DC.	Willow-leaved Iron-weed
斑鸠菊(茄叶斑鸠菊)	*Vernonia solanifolia* Benth.	Large-leaved Iron-weed
双花蟛蜞菊	*Wedelia biflora* (L.) DC.	Twoflower Wedelia
蟛蜞菊	*Wedelia chinensis* (Osbeck) Merr.	Chinese Wedelia
卤地菊	*Wedelia prostrata* (Hook. et Arn.) Hemsl.	Beach Wedelia
*三裂蟛蜞菊(美洲蟛蜞菊)	*Wedelia trilobata* (L.) Hitchc.	Trilobate Wedelia
山蟛蜞菊(麻叶蟛蜞菊)	*Wedelia wallichii* Less.	Wallich's Wedelia
苍耳	*Xanthium sibiricum* Patrin ex Widder	Siberian Cocklebur

黄鹌菜(黄瓜菜)	*Youngia japonica* (L.) DC.	Hawk's Beard
*百日菊(鱼尾菊)	*Zinnia violacea* Cav.	Common Zinnia

龙胆科 Gentianaceae

华南龙胆(紫花地丁)	*Gentiana loureirii* (D. Don) Griseb.	Florists Gentian
香港双蝴蝶(肺形草)	*Tripterospermum nienkui* (C. Marq.) C. J. Wu [*Crawfurdia fasciculata* Wall.]	Creeping Gentian

睡菜科 Menyanthaceae

*荇菜	*Nymphoides peltatum* (Gmel.) O. Kuntze	Floating Heart

报春花科 Primulaceae

点地梅(铜钱草)	*Androsace umbellata* (Lour.) Merr.	Umbellate Rockjasmine
*仙客来(兔子花)	*Cyclamen persicum* Mill.	Florists Cyclamen
泽珍珠菜(泽星宿菜)	*Lysimachia candida* Lindl.	White Loosestrife
*过路黄	*Lysimachia christinae* Hance	Christina Loosestrife
珍珠菜	*Lysimachia clethroides* Duby	Clethra Loosestrife
红根草(星宿菜)	*Lysimachia fortunei* Maxim.	Fortune's Loosestrife
*多花报春(西洋樱草)	*Primula polyantha* Mill.	English Primrose

白花丹科 Plumbaginaceae

*阔叶补血草	*Limonium latifolium* O. Kuntze	Wideleaf Statice
补血草(勿忘我)	*Limonium sinense* (Girard) O. Kuntze	Sea-lavender
*深波叶补血草(不凋花)	*Limonium sinuatum* (L.) Mill.	Notchleaf Statice

*蓝花丹(蓝雪花)	*Plumbago auriculata* Lam.	Blueflower Leadword
白花丹(白雪花)	*Plumbago zeylanica* L.	White flower Leadword

车前草科 Plantaginaceae

车前草	*Plantago asiatica* L. [*P. major* L.]	Plantain

桔梗科 Campanulanceae

*杏叶沙参(三叶沙参)	*Adenophora hunanensis* Nannf.	Hunan Ladybell
金钱豹(土党参)	*Campanumoea javanica* Bl.	Java Campanumoea
羊乳(四叶参)	*Codonopsos lanceolata* (Sieb. et Zucc.) Trautv.	Lance Asiabell
桔梗(六角荷)	*Platycodon grandiflorus* (Jacq.) A. DC.	Balloon Flower
蓝花参(娃儿草)	*Wahlenbergia marginata* (Thunb.) A. DC.	Marginate Rockbell

半边莲科 Lobeliaceae

短柄半边莲	*Lobelia alsinoides* Lam. [*L. triangulata* Roxb.]	Shortstalk Lobelia
半边莲	*Lobelia chinensis* Lour.	Chinese Lobelia
卵叶半边莲(疏毛半边莲)	*Lobelia zeylanica* L. [*L. affinis* Wall.]	Bigflower Fleshy Lobelia
铜锤玉带草	*Pratia nummularia* (Lam.) A. Br. et Aschers. [*Lobelia nummularia* Lam.]	Common Pratia

草海桐科 Goodeniaceae

离根草	*Calogyne pilosa* R. Br. ssp. *chinensis* (Benth.) H. S. Kiu	Chinese Calogyne

小叶草海桐(海南草海桐)	*Scaevola hainanensis* Hance	Hainan Naupaka
草海桐	*Scaevola sericea* Vahl	Beach Naupaka

花柱草科 Stylidiaceae

花柱草(小叶花柱草)	*Stylidium uliginosum* Swartz	Small Stylidium

田基麻科 Hydrophyllaceae

田基麻(假芹菜)	*Hydrolea zeylanica* (L.) Vahl	Srilanka Hydrolea

紫草科 Boraginaceae

斑种草(柔弱斑种草)	*Bothriospermum tenellum* (Hornem.) Fisch. et Mey.	Tender Bothriospermum
*福建茶(基及树)	*Carmona microphylla* (Lam.) G. Don	Small-leaf Carmona
破布叶(破布木)	*Cordia dichotoma* Forst. f.	Dichotomous Cordia
长花厚壳树(鸡肉树)	*Ehretia longiflora* Champ. ex Benth.	Longflower Ehretia
厚壳树	*Ehretia thyrsiflora* (Sieb. et Zucc.) Nakai	Heliotrope Ehretia
大尾摇	*Heliotropium indicum* L.	Indian Heliotrope

茄科 Solanaceae

*颠茄(颠茄草)	*Atropa belladonna* L. [*A. acuminata* Royle ex Lindl.]	Common Atropa
*鸳鸯茉莉(二色茉莉)	*Brunfelsia latifolia* Benth.	Broadleaf Raintree
辣椒(牛角椒)	*Capsicum annuum* L.	Bell Pepper
*观赏辣椒	*Capsicum annuum* L. var. *cerasiformis* Irish	Ornamental Pepper
*小米椒	*Capsicum frutescens* L.	Bush Tedpepper

被子植物 ANGIOSPERMAE

*夜香树(丁香花)	*Cestrum nocturnum* L.	Night Jessamine
*洋金花(白花曼陀罗)	*Datura metel* L.	Hindu Datura
*曼陀罗(洋金花)	*Datura stramonium* L.	Jimson Weed
红丝线(十萼茄)	*Lycianthes biflora* (Lour.) Bitter	Twoflower Lycianthes
枸杞(枸杞子)	*Lycium chinensis* Mill.	Chinese Wolfberry
番茄(西红柿)	*Lycopersicon esculentum* Mill.	Tomato
*红花烟草(烟草花)	*Nicotiana sanderae* W. Watson	Garden Nicotiana
烟草(烟)	*Nicotiana tabacum* L.	Tobacco
*碧冬茄(矮牵牛)	*Petunia hybrida* (Hook.) Vilm.	Garden Petunia
酸浆(苦蘵)	*Physalis alkekengi* L.	Chinese Lantern
挂金灯	*Physalis alkekengi* L. var. *francheti* (Mast.) Makino	Franchet Groundcherry
小酸酱(小酸浆)	*Physalis minima* L. [*Ph. parviflora* R. Br.]	Little Groundcherry
*金杯花(金杯藤)	*Solandra nitida* Zucc.	Cup of Gold Vine
少花龙葵(白花菜)	*Solanum americanum* Mill. [*S. nigrum* L. var. *pauciflrum* Liou]	Shiningfruit Nightshade
刺茄(癫茄)	*Solanum capsicoides* Allioni	Sodaapple Nightshade
野茄	*Solanum coagulans* Forsk.	Wild Nightshade
假烟叶(假烟叶树)	*Solanum erianthum* D. Don [*S. verbascifolium* L.]	Tobacco Tree
毛茄(大叶毛刺茄)	*Solanum lasiocarpum* Dunal [*S. ferox* L.]	Hairy Nightshade

白英(山甜菜、蔓茄、北风藤)	*Solanum lyratum* Thunb. [*S. dulcamara* L. var. *pubescens* Bl.]"	Bittersweet
山茄(大丁茄子)	*Solanum macaonense* Dunal	Macao Nightshade
*树茄(生毛将军)	*Solanum macranthum* Sw.	Giant Star Potato Tree
*乳茄(五代同堂茄)	*Solanum mammosum* L.	Nipple Fruit
茄子(矮瓜)	*Solanum melongena* L.	Garden Eggplant
龙葵(野海椒)	*Solanum nigrum* L.	Black Nightshade
木龙葵(白花籽草)	*Solanum suffruticosum* Schousb.	Subshrub Nightshade
水茄(一面针)	*Solanum torvum* Swartz	Water Nightshade
马铃薯(洋芋、土豆)	*Solanum tuberosum* L.	Potato

旋花科 Convolvulaceae

心萼薯(毛牵牛)	*Aniseia biflora* (L.) Choisy	Hairy Morning-glory
白鹤藤	*Argyreia acuta* Lour.	Acute Asiaglory
毛白鹤藤(头花银被藤)	*Argyreia capitata* (Vahl) Choisy	Capitate-flower Asiaglory
黄毛白鹤藤(银背藤)	*Argyreia mollis* (Burm. f.) Choisy	Softhair Asiaglory
美丽银背藤	*Argyreia nervosa* (Burm. f.) Boj.	Hawaiian Baby Woodrose
银背藤	*Argyreia obtusifolia* Lour.	Blunt-leaf Asiaglory
菟丝子	*Cuscuta chinensis* Lam.	Chinese Dodder
日本菟丝子	*Cuscuta japonica* Choisy	Japanese Dodder
马蹄金	*Dichondra micrantha* Urban [*D. repens* auct. non Forst.]	Creeping Dichondra

丁公藤	*Erycibe obtusifolia* Benth.	Obtuse-leaf Erycibe
土丁桂	*Evolvulus alsinoides* (L.) L.	Alsine-like Evolvuls
猪菜藤(细圆藤)	*Hewittia malabarica* (L.) Suresh [*H. sublobata* (L. f.) O. Kuntze]	Sublobate Hewittia
蕹菜(通心菜)	*Ipomoea aquatica* Forsk.	Water Spinach
番薯(红薯)	*Ipomoea batatas* (L.) Lam.	Sweet Potato
*五爪金龙(番仔藤)	*Ipomoea cairica* (L.) Sweet	Gairo Morning Glory
七爪龙	*Ipomoea digitata* L.	Fingerleaf Morning Glory
*树牵牛(南美旋花)	*Ipomoea fistulosa* Mart. ex Choisy	Bush Morning-Glory
紫心牵牛(圆叶牵牛)	*Ipomoea obscura* (L.) Ker-Gawl.	Obscure Morning-glory
厚藤(马鞍藤)	*Ipomoea pes-caprae* (L.) Sweet	Tow-leaves Morning-glory
盘苞牵牛(帽苞薯藤)	*Ipomoea pileata* Roxb.	Pileate Morning-glory
三裂叶牵牛(三裂叶薯)	*Ipomoea triloba* L.	Three-lobed Morning-glory
小牵牛	*Jacquemontia paniculata* (Burm. f.) Hall. f.	Paniculate Jacquemonia
鱼黄草(小花山猪菜)	*Merremia hederacea* (Burm. f.) Hall. f.	Ivy-like Merrema
毛山猪菜(毛茉栾藤)	*Merremia hirta* (L.) Merr.	Hairy Merremia
尖萼山猪菜(尖萼鱼黄草)	*Merremia tridentata* (L.) Hall. f. ssp. *hastata* (Desr.) Oststr.	Narrow-leaved Merrema

伞花鱼黄草	*Merremia umbellata* (L.) Hall. f.	Umbellate Merremia
山猪菜(伞花茉栾藤)	*Merremia umbellata* Hall. f. ssp. *orientalis* (Hall. f.) Oststr.	Umbellate Merremia
盒果藤	*Operculina turpethum* (L.) Manso	Foully Operculina
*变色牵牛	*Pharbitis indica* (Burm.) R. C. Fang	Indian Pharbitis
*裂叶牵牛	*Pharbitis nil* (L.) Choisy	Imperial Japanese Morningglory
圆叶牵牛(紫色牵牛)	*Pharbitis purpurea* (L.) Voigt	Common Morning-Glory
*茑萝	*Quamoclit pennata* (Desr.) Bojer [*Q. pennata* (Lam.) Boj.]	Cypress-vine
腺叶藤(大萼旋花)	*Stictocardia tiliifolia* (Desr.) Hall. f.	Queensland Woodrose

玄参科 Scrophulariaceae

毛麝香(凉草)	*Adenosma glutinosum* (L.) Druce	Sticky Adenosma
球花毛麝香	*Adenosma indianum* (Lour.) Merr.	Headed-flowered Adenosma
*金鱼草(龙口花)	*Antirrhinum majus* L.	Common Snapdragon
假马齿苋(水马齿)	*Bacopa monnieri* (L.) Wettst.	Water Hyssop
鬼羽箭(黑草)	*Buchnera cruciata* Hamilt	Blue Hearts
*蒲包花(荷包花)	*Calceolaria crenatiflora* Cav.	Crenateflower Calceolaria
胡麻草	*Centranthera cochinchinensis* (Lour.) Merr.	Cochinchina Centranthera

虻眼	Dopatrium junceum (Roxb.) Buch.-Ham. ex Benth.	Rushlike Dopatrium
紫苏草(麻省草)	Limnophila aromatica (Lam.) Merr.	Sparraw Herb
长蒴母草(长果母草)	Lindernia anagallis (Burm. f.) Pennell	Longcapsuled Falsepimpernel
泥花草(鸭利草)	Lindernia antipoda (L.) Alston	Creeping Falsepimpernel
刺齿叶泥花草(齿叶泥花草)	Lindernia ciliata (Colsm.) Pennell	Ciliate Falsepimpernel
母草	Lindernia crustacea (L.) F. Muell.	Brittle Falsepimpernel
狭叶母草	Lindernia angustifolia (Benth.) Wettst. [L. micrantha D. Don]	Narrowleaf Falsepimpernel
陌上菜	Lindernia procumbens (Krock.) Philcox	Procumbent Falsepimpernel
旱田草	Lindernia ruellioides (Colsm.) Pennell	Dry Falsepimpernel
荨麻叶母草	Lindernia urticifolia (Hance) Bonati [Vandellia urticifolia Hance]	Nettleleaf Falsepimpernel
通泉草	Mazus pumilus (Burm. f.) Steenis [M. japonicus (Thunb.) O. Kuntze]	Japanese Mazus
小果草(微果草)	Microcarpaea minima (Koenig) Merr.	Small Microchloa
苦玄参	Picria fel-terrae Lour.	Common Picria
*炮仗竹(吉祥草)	Russelia equisetiformis Schlcht. et Cham.	Coral-plant
野甘草	Scoparia dulcis L.	Sweet Broomwort

独脚金(疳积草)	*Striga asiatica* (L.) O. Kuntze	Asiatic Striga
单色蝴蝶草(单色翼萼)	*Torenia concolor* Lindl.	Concolorous Torenia
黄花翼萼(黄花蝴蝶草)	*Torenia flava* Buch.-Ham. ex Benth.	Yellowflower Torenia
*夏堇(蓝猪耳)	*Torenia fournieri* Linden ex Fourn.	Blue Torenia
光叶蝴蝶草	*Torenia glabra* Osbeck	Glabrous Torenia
水苦荬(水莴苣)	*Veronica undulata* Wall.	Unulate Speedwell

列当科 Orobanchaceae

| 野菰(金牙齿、僧帽花) | *Aeginetia indica* L. | Indian Aeginetia |

狸藻科 Lentibulariaceae

黄花狸藻(黄花挖耳藻)	*Utricularia aurea* Lour.	Floating Bladderwort
挖耳草(耳挖草)	*Utricularia bifida* L.	Small Yellow Bladerwort
少花狸藻(丝叶狸藻)	*Utricularia exoleta* R. Br.	Fewflower Bladderwort
禾叶挖耳草	*Utricularia graminifolia* Vahl	Grassleaf Bladderwort
圆叶挖耳草(圆叶狸藻)	*Utricularia striatula* J. Sm.	Roundleaf Bladderwort
齿萼挖耳草(蓝花狸藻)	*Utricularia uliginosa* Vahl	Swamp Bladderwort

苦苣苔科 Gesneriaceae

| 中华芒毛苣苔(芒毛苣苔) | *Aeschynanthus acuminatus* Wall. ex A. DC. | Acuminate Basket Vine |
| *毛萼口红花(胭脂花) | *Aeschynanthus lobbianus* Hook. | Lipstick Vine |

*美丽口红花(美丽芒毛苣苔)	*Aeschynanthus speciosus* Hook.	Basket Vine
紫花短筒苣苔(佳氏苣苔)	*Boeica guileana* B. L. Burtt	Guile's Boeica
牛耳朵	*Chirita eburnea* Hance	Ivorywhite Chirita
蚂蝗七	*Chirita fimbrisepala* Hand.-Mazz.	Fimbriate-sepal Chirita
两广唇柱苣苔(长蒴苣苔)	*Chirita sinensis* Lindl.	Chinese Chirita
狭叶长蒴苣苔	*Chirita sinensis* Lindl. var. *angustifolia* Dunn	Narrowleaf Chirita
双片苣苔	*Didymostigma obtusum* (Clarke) W. T. Wang	Obtuse Didymostigma
广东半蒴苣苔	*Hemiboea subcapitata* Clarke var. *guangdongensis* (Z. Y. Li) Z. Y. Li	Guangdong Half-capitate Hemiboea
石吊苣苔(吊石苣苔)	*Lysionotus pauciflorus* Maxim.	Few-flower Lysionotus
大叶石上莲(石莲)	*Oreocharis benthamii* Clarke	Bentham Oreocharis
石上莲	*Oreocharis benthamii* Clarke var. *reticulata* Dunn	Reticulate Oreocharis
冠萼线柱苣苔	*Rhynchotechum formosanum* Hatusima	Taiwan Rhynchotechum
*非洲紫罗兰(非洲堇)	*Saintpaulia ionantha* H. Wendl.	African Violet
*大岩桐(落雪泥)	*Sinningia speciosa* Benth. et Hook.	Brazilian Gloxinia

紫葳科 Bignoniaceae

*凌霄	*Campsis grandiflora* (Thunb.) Schum.	Chinese Trumpet-creeper

*十字架树(叉叶木)	*Crescentia alata* H. B. K.	Gourd Tree
猫尾木	*Dolichandrone cauda-felina* (Hance) Benth. et Hook. f.	Cat-tail Tree
*蓝花楹	*Jacaranda mimosifolia* D. Don	Blue Jacaranda
*吊瓜树(吊灯树)	*Kigelia africana* (Lam.) Benth.	Sausage Tree
*猫爪藤(猫儿爪)	*Macfadyena unguis-cati* (L.) A. Gentry	Cat-claw Vine
*火烧花(火花树)	*Mayodendron igneum* (Kurz) Kurz	Brightred Mayodendron
*木蝴蝶(千张纸)	*Oroxylum indicum* (L.) Benth. ex Kurz	Indian Trumpetflower
*炮仗花(炮仗红)	*Pyrostegia venusta* (Ker-Gawl.) Miers	Fire-cracker Vine
*海南菜豆树	*Radermachera hainanensis* Merr.	Hainan Belltree
菜豆树	*Radermachera sinica* (Hance) Hemsl.	Asia Belltree
*蒜香藤	*Saritaea magnifica* (Sprague ex van Steenis) Dugand	Glow Vine
*火焰木(火焰树)	*Spathodea campanulata* Beauv.	African Tulip Tree
*黄风铃花	*Tabebuia chrysantha* (Jacq.) Nichols.	Yellow Pui
*蔷薇风铃花(蔷薇钟花)	*Tabebuia rosea* (Bertol.) DC.	Pink Trumpet-tree
*硬骨凌霄(洋凌霄)	*Tecoma capensis* (Thunb.) Lindl.	Cape-Honeysuckle
*黄钟花(金钟花)	*Tecoma stans* (L.) A. L. Juss. ex Kunth	Trumpet-flower

爵床科 Acanthaceae

老鼠簕	*Acanthus ilicifolius* L.	Spiny Bears Breech
穿心莲	*Andrographis paniculata* (Burm. f.) Nees	Kariyat
白接骨	*Asystasiella chinensis* (S. Moore) E. Hoss. [*A. chinensis* (S. Moore) E. Hossain]	Chinese Asystasiella
假杜鹃	*Barleria cristata* L.	Philippine Violet
*虾衣花(麒麟吐珠)	*Calliaspidia guttata* (T. S. Brand.) Bremek.	Lollypops Super Goldy
中华赛爵床(杜根藤)	*Calophanoides chinensis* (Champ.) C. Y. Wu et H. S. Lo [*Adhatoda chinensis* Champ.]	Chinese Calophanoides
海南赛爵床	*Calophanoides hainanensis* C. Y. Wu	Hainan Calophanoides
杜根藤	*Calophanoides quadrifaria* Ridl.	
黄琼草	*Championella tetrasperma* (Champ. ex Benth.) Brem.	Common Championella
针刺草	*Codonacanthus pauciflorus* Nees	Fewflower Codonacanthus
十字爵床	*Crossandra infundibuliformis* Nees	Funnel-shaped Crossandra
狗肝菜	*Dicliptera chinensis* (L.) Juss.	Chinese Dicliptera
*可爱花(蓝仔花)	*Eranthemum pulchellum* Andrews. [*Eranthemum nervosum* R. Br]	Blue Eranthemum

*白网纹草(白脉网纹草)	*Fittonia verschaffeltii* (Lemaire) Van Houtte var. *argyroneura* Nichols.	Snail Plant
*网纹花(银网草)	*Fittonia verschaffeltii* (Lemaire) Van Houtte	Nerve Plant
水蓑衣	*Hygrophila salicifolia* (Vahl) Nees	Willowleaf Hygrophila
红丝线(刀枪药)	*Hypoestes purpurea* (L.) R. Br. [*Justicia purpurea* L.]	Purple Hypoestes
小驳骨(驳骨丹)	*Justicia gendarussa* Burm. f. [*Gendarussa vulgaris* Nees]	White Justicia
爵床(鼠尾红)	*Justicia procumbens* L.	Purple Justicia
大驳骨(黑叶爵床)	*Justicia ventricosa* Wall. [*Gendarussa ventricosa* (Wall.) Nees]	Obliqueswollen Adhatoda
*鸭嘴花	*Justicia adhodata* L.	Malabar Nut
鳞花草	*Lepidagathis incurva* Buch.-Ham. ex D. Don	Common Lepidagathis
拟地皮消	*Leptosiphonium venustum* (Hance) E. Hossain	Beautiful Leptosiphonium
*金苞花(黄鸭咀花)	*Pachystachys lutea* Nees	Lollypops
曲枝假蓝	*Strobilanthus dalzielii* (W. W. Smith) R. Ben. [*Pteroptychia dalzielii* (Smith.) H. S. Lo]	Dalziel Strobilanthus
兰花草	*Ruellia brittoniana* Leonard	Britton Ruellia
孩儿草(中华孩儿草)	*Rungia pectinata* (L.) Nees [*Rungia pectinata* Bentn.]	Pectinate Rungia
*金脉爵床(黄脉爵床)	*Sanchezia nobilis* Hook. f.	Noble Sanchezia

*小苞黄脉爵床	*Sanchezia parvibracteata* Sprag. et Hutch.	Small-breact Sanchezia
板蓝	*Strobilanthes cusia* (Nees) O. Kuntze	Common Conehead
*波斯红草(红背马蓝)	*Strobilanthes dyerianus* Mast.	Burma Conehead
四子马蓝(狗肝菜)	*Strobilanthes tetraspermus* (Champ. ex Benth.) Druce [*Championella debilis* (Hemsl.) C. B. Clarke]	Fourseed Conehead
疏花马蓝(金鸡蜡)	*Strobilantheus divaricatus* (Nees) T. Anders. [*Diflugossa divaricata* (Nees) Bremek.]	Divaricate Conehead
*硬枝老鸦嘴(立鹤花)	*Thunbergia erecta* (Benth.) T. Anders.	Bush Thunbergia
老鸦嘴(碗花草)	*Thunbergia fragrans* Roxb.	White Thunbergia
大花老鸦嘴	*Thunbergia grandiflora* Roxb.	Large-flowered Thunbergis
*桂叶山牵牛(樟叶老鸦嘴)	*Thunbergia laurifolia* Lindl.	Laurel Clockvine

马鞭草科 Verbenaceae

海榄雌(白骨壤)	*Avicennia marina* (Forsk.) Vierh.	Black Mangrove
短柄紫珠(短序紫珠)	*Callicarpa brevipes* (Benth.) Hance [*C. brevipes* (Benth.) Hance f. *serrulata* Péi]	Shortstalk Beautyberry
白毛紫珠	*Callicarpa candicans* (Burm. f.) Hochr. [*C. americana* Lour.]	Whitehairy Beautyberry

华紫珠	*Callicarpa cathayana* H. T. Chang	Beauty Berry
白棠子树	*Callicarpa dichotoma* (Lour.) K. Koch	Purple Beauty-berry
杜虹花	*Callicarpa formosana* Rolfe [*C. pedunculata* Lam, *C. ningpoensis* Mats., *C. aspera* Hand.-Mazz.]	Tiwan Beautyberry
老鸦糊	*Callicarpa giraldii* Hesse ex Rehd. [*C. giraldiana* Hesse, *C. mairei* Lévl.]	Girald Beautyberry
全缘叶紫珠	*Callicarpa integerrima* Champ. [*C. integrifolia* Forbes et Hemsl.]	Entire Beautyberry
枇杷叶紫珠(劳来氏紫珠)	*Callicarpa kochiana* Makino [*C. loureiri* Hook. et Arn.]	Loquatleaf Beautyberry
广东紫珠	*Callicarpa kwangtuangensis* Chun [*C. japonica* Thunb. var. *angustata* Rehd.]	Kwangtung Beautyberry
大叶紫珠	*Callicarpa macrophylla* Vahl	Large-leaved Beautyberry
裸花紫珠(裸花紫珠叶)	*Callicarpa nudiflora* Hook. et Arn. [*C. macrophylla* Vahl. var. *sinensis* C. B. Clarke]	Nakedflower Beautyberry
狭叶红紫珠	*Callicarpa rubella* Lindl. f. *angustata* Péi (*C. rubella* Rehd.)	Narrowleaf Reddish Beautyberry
红紫珠(红叶紫珠)	*Callicarpa rubella* Lindl.	Red Beautyberry
钝齿红紫珠	*Callicarpa rubella* Lindl. f. *crenata* Péi	Crenate Reddish Beautyberry

被子植物 ANGIOSPERMAE

蓝香草(荗)	*Caryopteris incana* (Thunb.) Miq.	Common Bluebeard
臭牡丹	*Clerodendrum bungei* Steud. [*C. chinense* (Osbeck) Mabb.]	Fragrant Gloryberry
灰毛大青(毛赪桐)	*Clerodendrum canescens* Wall. [*C. haematocalyx* Hance]	Greyhair Glorybower
腺茉莉	*Clerodendrum colebrookianum* Walp. [*C. glandulosum* Colebr. ex Wall.]	Glandular Glorybower
大青	*Clerodendrum cyrtophyllum* Turcz.	Mayflower Gloryberry
白花灯笼草(鬼灯笼)	*Clerodendrum fortunatum* L. [*C. lividum* Lindl., *C. pumilum* (Lour.) Spreng.]	Redcalyx Glorybower
假茉莉(苦郎树)	*Clerodendrum inerme* (L.) Gaertn.	Unarmed Glorybower
*赪桐(状元红)	*Clerodendrum japonicum* (Thunb.) Sweet	Japanese Glorybower
广东赤桐(广东大青)	*Clerodendrum kwangtungense* Hand.-Mazz.	Kwangtung Glorybower
*重瓣臭茉莉	*Clerodendrum philippinum* Schauer [*C. chinense* (Osbeck) Mabb.]	Philippine Glorybower
臭茉莉	*Clerodendrum philippinum* Schauer var. *simplex* Moldenke [*C. chinense* (Osbeck) Mabb.]	Simplex Glorybower
美丽桢桐(爪哇桢桐)	*Clerodendrum speciosissimum* Van Geert	Java Glorybower
*红龙吐珠(龙吐珠藤)	*Clerodendrum splendens* G. Don	Flaming Glorybower

*龙吐珠(珍珠宝莲)	*Clerodendrum thomsonae* Balf.	Bleeding Heart Glorybower
*蝴蝶花(紫蝶花)	*Clerodendrum ugandense* Prain	Blue Butterfly
*假连翘(金露花)	*Duranta erecta* L. [*D. repens* L.]	Golden Dewdrops
苦梓(海南石梓)	*Gmelina hainanensis* Oliv.	Hainan Bushbeech
*冬红(帽子花)	*Holmskioldia sanguinea* Retz.	Chinese Hatplant
*马樱丹(五色梅)	*Lantana camara* L.	Common Lantana
*黄花马樱丹(黄马樱丹)	*Lantana camara* L. var. *flava* Mold.	Yellowflower Lanatan
*五彩马樱丹	*Lantana camara* L. var. *hybrida* Mold.	Hybrid Lantana
*橙红五色梅	*Lantana camara* L. var. *mista* L. H. Bailey	Orange Lantana
*蔓马缨丹(小叶马缨丹)	*Lantana montevidensis* (Spreng.) Briq.	Trailing Lantana
过江藤	*Phyla nodiflora* (L.) Greene [*Lippia nodiflora* (L.) Rich.]	Knottedflower Phyla
*豆腐柴(山麻糍)	*Premna microphylla* Turcz. [*P. microphylla* Turcz. var. *glabra* Nakai.]	Japanese Premna
钝叶臭黄荆	*Premna obtusifolia* R. Br. [*P. integrifolia* L. var. *obtusifolia* Péi]	Obtuseleaf Premna
假马鞭	*Stachytarpheta jamaicensis* (L.) Vahl	Jamaica Vervain
*柚木	*Tectona grandis* L. f.	Common Teak
*美女樱(美人樱)	*Verbena hybrida* Voss.	Hybrid Verbena

被子植物 ANGIOSPERMAE

马鞭草(铁马鞭)	*Verbena officinalis* L. [*V. officinalis* L. var. *ramosa* Lévl.]	European Verbena
细叶美女樱	*Verbena tenera* Spreng.	Common Verbena
黄荆(布荆)	*Vitex negundo* L.	Yellow Bramble
牡荆	*Vitex negundo* L. var. *cannabifolia* (Sieb. et Zucc.) Hand.-Mazz. [*V. cannabifolia* Sieb. et Zucc.]	Hempleaf Negundo Chastetree
拟黄荆	*Vitex negundo* var. *thyrsoides* Péi et S. L. Liou	Thyrselike negundo Chastetree
山牡荆(五叶牡荆)	*Vitex quinata* (Lour.) F. N. Will.	Wild Vitex
蔓荆	*Vitex trifolia* L. [*V. trifolia* L. var. *trifoliolata* Schauer]	Threeleaf Chastetree
单叶蔓荆(沙荆子)	*Vitex trifolia* L. var. *simplicifolia* Cham. [*V. rotundifolia* L. f.]	Simpleleaf Shrub Chastetree

唇形科 Labiatae

筋骨草(金疮小草)	*Ajuga decumbens* Thunb.	Decumbent Bugle
紫背金盘	*Ajuga nipponensis* Makino [*A. argyi* Lévl. ex Dunn]	Japanese Bugle
广防风(土防风)	*Anisomeles indica* (L.) O. Kuntze [*A. ovata* R. Br.]	Indian Epimerdei
*肾茶(猫须草)	*Clerodendranthus spicatus* (Thunb.) C. Y. Wu ex H. W. Li [*C. spicatum* Thunb.]	Spicate Clerodendranthus
风轮菜(野凉粉藤)	*Clinopodium chinensis* (Benth.) O. Kuntze [*Calamintha chinensis* Benth.]	Chinese Clinopodium

瘦风轮(细风轮菜)	*Clinopodium gracile* (Benth.) Matsum.	Slender Clinopodium
*假紫苏(小洋紫苏)	*Coleus scutellarioides* (L.) Benth. var. *crispipilus* (Merr.) H. Keng [*C. pumilus* Blanco]	Small Coleus
*洋紫苏(彩叶草)	*Coleus scutellarioides* (L.) Benth. [*C. blumei* Benth.]	Skullcaplike Coleus
紫花香薷(野薄荷)	*Elsholtzia argyi* Lévl. [*E. macrostemon* Hand.-Mazz.]	Purpleflower Elsholtzia
*退色香薷	*Elsholtzia ciliata* (Thunb.) Hyl.	Common Elsholtzia
海州香薷(香薷)	*Elsholtzia splendens* Nakai ex F. Maekawa [*Elsholtzia haichowensis* Sun ex C. H. Hu]	Haichow Elsholtzia
连钱草	*Glechoma longituba* (Nakai) Kupr. [*G. hederacea* L. var. *longituba* Nakai]	Longtube Ground Ivy
中华锥花	*Gomphostemma chinense* Oliv.	Chinese Gomphostemma
山香(山薄荷)	*Hyptis suaveolens* (L.) Poiteau [*Ballota suaveolens* L.]	Wild Spikenard
香茶菜	*Isodon amethystoides* (Benth.) Hara [*Plectranthus amethystoides* (Benth.) Hara]	Common Isodon
益母草(月母草)	*Leonurus japonicus* Houtt.	Wormwoodlike Mother-wout
疏毛白绒草(柔毛绣球防风)	*Leucas mollissima* Wall. var. *chinensis* Benth.	Loosehairy Leucas
绉面草	*Leucas zeylanica* (L.) R. Br.	Ceylon Leucas

薄荷(人丹草)	*Mentha haplocalyx* Briq. [*M. canadensis* L.]	Peppermint
凉粉草(仙草)	*Mesona chinensis* Benth.	Chinese Mesona
小鱼仙草	*Mosla dianthera* (Buch.-Ham. ex Roxb.) Maxim. [*Lycopus dianthera* Buch.-Ham.]	Twoanther Mosla
石荠苧	*Mosla scabra* (Thunb.) C. Y. Wu et H. W. Li [*M. lanceolata* (Benth.) Maxim.]	Scabrous Mosla
罗勒(金不换)	*Ocimum basilicum* L.	Sweet Basil
丁香罗勒	*Ocimum gratissimum* L. var. *suave* (Willd.) Hook. f.	Sweetscented Basil
短齿假糙苏	*Paraphlomis albida* Hand.-Mazz var. *brevidens* Hand.-Mazz.	Short-toothed White-hairy Paraphlomis
狭叶假糙苏	*Paraphlomis javanica* (Bl.) Prain var. *angustifolia* (C. Y. Wu) C. Y. Wu et H. W. Li	Narrowleaf Java Paraphlomis
*紫苏	*Perilla frutescens* (L.) Britt.	Common Perilla
*回回苏(鸡冠紫苏)	*Perilla frutescens* (L.) Britt. var. *crispa* (Thunb.) Hand.-Mazz. [*Dentidia nankinensis* (Lour.) Dence.]	Crisped Common Perilla
野生紫苏(野紫苏)	*Perilla frutescens* (L.) Britt. var. *purpurascens* (Hayata) H. W. Li	Wild Perilla
水珍珠菜(毛水珍珠菜)	*Pogostemon auricularius* (L.) Hassk. [*M. joetida* Burm. f.]	Auriculate Pogostemon

*广藿香(藿香)	*Pogostemon cablin* (Blanco) Benth. [*Pogostemon patchouly* Pellet.]	Cablin Patchouli
短穗刺蕊草(杉氏刺蕊草)	*Pogostemon championii* Prain	Shortspike Pogostemon
夏枯草	*Prunella vulgaris* L. [*P. japonica* Makino, *P. vulgaris* L. var. *japonica* Kudo]	Common Selfheal
华鼠尾草	*Salvia chinensis* Benth.	Chinese Sage
*红花鼠尾草	*Salvia coccinea* L.	Redlip Sage
*粉萼鼠尾草(蓝丝线)	*Salvia farinacea* Benth.	Blue Sage
荔枝草(雪见草)	*Salvia plebeia* R. Br.	Common Salvia
*一串红(绯衣草)	*Salvia splendens* Ker-Gawl.	Scarlet Sage
半枝莲	*Scutellaria barbata* D. Don [*S. rivularis* Wall., *S. minor* L., *S. adenophylla* Miq.]	Barbed Skullcap
韩信草(耳挖草)	*Scutellaria indica* L.	Indian Skullcap
血见愁(山藿香)	*Teucrium viscidum* Bl.	Sticky Germander

单子叶植物 MONOCOTYLEDONEAE

花蔺科 Butomaceae

*黄花蔺(黄天鹅绒叶)	*Limnocharis flava* (L.) Buch. [*L. plumieri* Rich, *L. emarginata* Humb. et Bonpl.]	Yellow Velvetleaf

水鳖科 Hydrocharitaceae

水筛(日本箦草)	*Blyxa japonica* (Miq.) Maxim.	Janpanese Blyxa

黑藻	*Hydrilla verticillata* (L. f.) Royle	Verticillate Hydrilla
软骨草(虾子草)	*Nechamandra alternifolia* (Roxb. ex Wight) Thw. [*Lagarosiphon alternifolia* (Roxb.) Druce]	Alternateleaf Nechamandra
水车前	*Ottelia alismoides* (L.) Pers.	Water Plantain
*苦草(扁草)	*Vallisneria natans* (Lour.) Hara	Eel Grass

泽泻科 Alismataceae

*东方泽泻(泽泻)	*Alisma orientale* (Samuel.) Juz.	Water-plantain Ottelia
*泽苔草(水兰)	*Echinodorus paleafolius* (Nees et Mart.) Macbr.	Paleleaf Echinodorus
*慈姑	*Sagittaria sagittifolia* L. ssp. *leucopetala* (Miq.) Hartog [*S. trifolia* L. var. *sinensis* (Sims) Makino]	Chinese Arrow-head
野慈姑(长瓣慈姑)	*Sagittaria trifolia* L.	Three-leaf Arrow-head

水蕹科 Aponogetonaceae

水蕹	*Aponogeton lakhonensis* A. Camus	Common Pond Weed

眼子菜科 Potamogetonaceae

竹叶眼子菜(箬叶藻)	*Potamogeton malaianus* Miq. [*P. gaudichaudii* Cham. et Schl.]	Bambooleaf Pondweed

鸭跖草科 Commelinaceae

穿鞘花	*Amischotype hispida* (Less et A. Rich.) Hong	Hispid Amischotype

耳苞鸭跖草(耳叶鸭跖草)	*Commelina auriculata* Bl.	Auriculata Dayflower
饭包草(火柴头)	*Commelina bengalensis* L.	Bengal Dayflower
鸭跖草(竹叶草)	*Commelina communis* L.	Common Dayflower
竹节草	*Commelina diffusa* Burm. f.	Diffuse Dayflower
大苞鸭跖草(大叶鸭跖草)	*Commelina paludosa* Bl.	Bigbract Dayflower
露水草	*Cyanotis arachnoidea* C. B. Clarke	Arachnoid Cyanotis
蓝耳草(鸡冠参)	*Cyanotis vaga* (Lour.) Roem. et Schult.	Common Cyanotis
聚花草(蔓襄荷)	*Floscopa scandens* Lour.	Climber Floscopa
大苞水竹叶	*Murdannia bracteata* (Clarke) J. K. Morton ex Hong	Largebract Murdannia
牛轭草(细竹蒿草)	*Murdannia loriformis* (Hassk.) R. Rao et Kamm.	Severalflower Dewflower
大果水竹叶	*Murdannia macrocarpa* Hong	Largefruit Murdannia
裸花水竹叶	*Murdannia nudiflora* (L.) Brenan.	Nakedflower Murdannia
田蒿草(细竹蒿草)	*Murdannia simplex* (Vahl) Brenan.	Simplex Murdannia
细柄水竹叶	*Murdannia vaginata* (L.) Bruckn	Sheathed Murdannia
毛果网子草(钩毛子草)	*Rhopalephora scaberrima* (Bl.) Faden [*Dictyospermum scaberrimum* (Bl.) J. K. Morton ex Hong]	Hairyfruit Netseedgrass
*紫鸭趾草(紫竹梅)	*Setcreacea purpurea* B. K. Boom	Purple Heart

被子植物 ANGIOSPERMAE

*紫叶水竹草(吊竹梅)	*Tradescantia fluminensis* Vell.	Wandering Jew
*紫万年青(蚌花)	*Tradescantia spathacea* Sw. [*Rhoeo discolor* L'Her.]	Oyster Plant
*吊竹梅(水竹草)	*Tradescantia zebrina* Hort. ex Bosse	Wandering Jew Zebrina

黄眼草科 Xyridaceae

黄眼草	*Xyris indica* L.	Yellow Eye Grass
葱草(少花黄眼草)	*Xyris pauciflora* Willd.	Onion Grass

谷精草科 Eriocaulaceae

谷精草	*Eriocaulon buergerianum* Koern.	Buerger Pipewort
白药谷精草(赛谷精草)	*Eriocaulon cinereum* R. Br.	Siebold Pipewort
小叶谷精草	*Eriocaulon luzulifolium* Mart.	Small Pipewort
华南谷精草(谷精珠)	*Eriocaulon sexangulare* L.	Sixangular Pipewort
瑶山谷精草	*Eriocaulon yaoshanense* Ruhl.	Yaoshan Cupgrass
粤港谷精草	*Eriocaulon yuegangensis* F. W. Xing et Y. X. Zhang	Hongkong-Canton Cupgrass

凤梨科 Bromeliaceae

*蜻蜓凤梨(银纹凤梨)	*Aechmea fasciata* (Lindl.) Baker	Fasciate Aechmea
*亮绿凤梨	*Aechmea lueddemanniana* (K. Koch) Brongn. ex Mez	Lueddemann Aechmea
*凤梨(菠萝花)	*Ananas comosus* (L.) Merr.	Pineapple
*水塔花(火焰凤梨)	*Billbergia pyramidalis* (Sims) Lindl.	Fool Proof Plant

*姬凤梨(紫锦凤梨)	*Cryptanthus acaulis* (Lindl.) Beer	Stemless Cryptanthus
*环带姬凤梨(虎纹小凤梨)	*Cryptanthus zonatus* Beer	Zonate Cryptanthus
*短叶雀舌兰(小雀舌兰)	*Dyckia brevifolia* Hort. ex Baker	Shortleaf Dyckia
*红杯凤梨	*Neoregelia carolinae* (Beer) L. B. Sm.	Blushing Bromeliad
*阔叶彩叶凤梨(同心彩叶凤梨)	*Neoregelia concentrica* (Vell.) L. B. Sm.	Concentrate Neoregelia
*端红彩叶凤梨(艳美彩叶凤梨)	*Neoregelia spectabilis* (T. Moore) L. B. Sm.	Beautiful Neoregelia
*白被穗花凤梨(巴西翠凤草)	*Pitcairnia muscosa* Mart. ex Schult. f.	Brazilian Pitcairnia
*紫花铁兰(铁兰、紫花木柄凤梨)	*Tillandsia cyanea* Linden ex K. Koch	Pink Quill
*莺哥凤梨(黄苞莺哥)	*Vriesea carinata* Wawra	Lobster Claws
*虎纹凤梨(火剑)	*Vriesea splendens* (Brongn.) Lem.	Flaming-sword

芭蕉科 Musaceae

*象腿蕉(象腿芭蕉)	*Ensete glaucum* (Roxb.) Cheesm. (*Musa glauca* Roxb.)	Greyblue Ensete
*小果野蕉(阿加蕉)	*Musa acuminata* Colla	Acuminate Banana
*芭蕉(甘蕉)	*Musa basjoo* Sieb. et Zucc.	Japanese Banana
野蕉(山芭蕉)	*Musa balbisiana* Colla	Wild Banana
*红蕉(红花蕉)	*Musa coccinea* Andr.	Flowering Banana

香蕉(香牙蕉)	Musa nana Lour. (Musa acuminata Colla 'Cavendishii')	Dwarf Banana
*紫花芭蕉(紫梦幻蕉)	Musa ornata Roxb.	Bronze Banana
*地涌金莲(地金莲)	Musella lasiocarpa (Franch.) C. Y. Wu ex W. W. Li [Musa lasiocarpa Fr.]	Hairyfruit Musella

旅人蕉科 Strelitziaceae

*火鸟蕉(比海蝎尾蕉)	Heliconia bihai L.	Carib Heliconia
*粉鸟蝎尾蕉	Heliconia collinsiana Griggs	Carib Heliconia
*艳红赫蕉(赫蕉)	Heliconia humilis Jacq.	Lobster Claw
*粉鸟赫蕉	Heliconia platystachys Baker	False Bird of Paradise
*金鸟赫蕉(垂花赫蕉)	Heliconia rostrata Ruiz et Pav.	Hanging Lobster Claw
*黄丽鸟赫蕉(黄丽鸟蕉)	Heliconia subulata Ruiz et Pav.	Guatemalan Bird of Paradise
*艳黄赫蕉(垂花赫蕉)	Heliconia wagneriana Petersen	Rainbow Heliconia
*旅人蕉(扇芭蕉)	Ravenala madagascariensis Sonn.	Traveller's Palm
*鹤望兰(天堂鸟花)	Strelitzia reginae Aiton	Bird of Paradise

姜科 Zingiberaceae

华山姜(华良姜)	Alpinia oblongifolia Hayata [A. chinensis (Retz.) Rosc.]	China Galangal
红豆蔻(大高良姜)	Alpinia galanga (L.) Willd.	Great Galangal

草豆蔻(海南山姜)	Alpinia hainanensis K. Schum. [A. katsumadai Hayata]	Hainan Galangal
山姜(箭秆风)	Alpinia japonica (Thunb.) Miq.	Japan Galangal
高良姜(小良姜)	Alpinia officinalis Hance	Lesser Galangal
益智(益智子)	Alpinia oxyphylla Miq.	Sharpleaf Galangal
花叶山姜(矮山姜)	Alpinia pumila Hook. f.	Dwarf Galangal
密苞山姜(杆枫)	Alpinia stachyoides Hance [A. densibracteata T. L. Wu et Senjen]	Dense-bract Galangal
艳山姜(月桃)	Alpinia zerumbet (Pers.) Burtt et Smith	Shell Ginge
闭鞘姜(水蕉花)	Costus speciosus (Koen.) Smith	Crape Ginger
郁金(玉金)	Curcuma aromatica Salisb.	Aromatic Turmeric
*毛莪术(姜七)	Curcuma kwangsiensis Lee et Liang	Guangxi Turmeric
姜黄(姜黄子)	Curcuma longa L.	Common Turmeric
莪术	Curcuma phaeocaulis Val.	Zedoary Turmeric
*姜花	Hedychium coronarium Koen.	Ginger Lily
土田七(姜田七)	Stahlianthus involucratus (King ex Baker) Craib	Involucrate Stahlianthus
珊瑚姜(阴姜)	Zingiber corallinum Hance	Coral Ginger
蘘荷(野姜)	Zingiber mioga (Thunb.) Rosc.	Mioga Ginger
姜(均姜)	Zingiber officinale Rosc.	Common Ginger
阳荷(白蘘荷)	Zingiber striolatum Diels	Striolate Ginger
红球姜(姜花)	Zingiber zerumbet (L.) Smith	Wild Ginger

美人蕉科 Cannaceae

*蕉芋(姜芋)	*Canna edulis* Ker	Edible Canna
*大花美人蕉	*Canna generalis* Bailey	Common Garden Canna
*美人蕉(凤尾花)	*Canna indica* L.	Indian Shot
*黄花美人蕉(黄兰蕉)	*Canna indica* L. var. *flava* Roxb.	Yellow Canna
*兰花美人蕉	*Canna orchioides* Bailey	Orchid Canna

竹芋科 Marantaceae

*孔雀肖竹芋(孔雀竹芋)	*Calathea makoyana* E. Morr. et Boom	Peacock Plant
*彩虹肖竹芋(玫瑰竹芋)	*Calathea roseopicta* (Linden) Regel	Rose-painted Calathea
*银羽栉花竹芋(箭羽竹芋)	*Ctenanthe oppenheimiana* (E. Morr.) K. Schum.	Oppenheim's Ctemanthe
*竹芋(紫背)	*Maranta arundinacea* L.	Bermuda Arrowroot
*花叶竹芋(孔雀草)	*Maranta bicolor* Ker-Gawl.	Twocolor Arrowroot
柊叶(棕叶)	*Phrynium rheedei* Suresh et Nichols. [*P. capitatum* Willd.]	Capitate Phrynium
*红背卧花竹芋(红背竹芋)	*Stromanthe sanguinea* (Hook.) Sonder	Tricolored Stromanthe
*水竹芋(载力花)	*Thalia dealbata* J. Fraser	Powdery Alligator-flag

百合科 Liliaceae

*不夜城芦荟(高尚芦荟)	*Aloe mitriformis* Mill.	Mitre-flower Aloe
*芦荟	*Aloe vera* L. var. *chinensis* (Haw.) Berger	Barbados Aloe
天门冬	*Asparagus cochinchinensis* (Lour.) Merr.	Wild Asparagus

*狐尾武竹(狐尾天门冬)	*Asparagus densiflorus* (Kunth) Jessop	Africa Asparagus
*松叶武竹	*Asparagus macowanii* Baker	Ming Fern
*武竹(蓬莱松)	*Asparagus myriocladus* Baker	Zigzag Asparagus
*石刁柏(小叶天冬)	*Asparagus officinalis* L.	Garden Asparagus
文竹(云片竹)	*Asparagus plumosus* Baker	Asparagus Fern
蜘蛛抱蛋(一叶兰)	*Aspidistra elatior* Bl.	Common Aspidistra
小花蜘蛛抱蛋	*Aspidistra minutiflora* Stapf	Smallflower Aspidistra
绵枣儿(地枣儿)	*Scilla scilloides* (Lindl.) Druce [*Barnardia japonica* (Thunb.) Schult. et Schult. f.]	Common Squill
*白纹草	*Chlorophytum bichetii* (Karrer) Backer	Siam Lily
*银边吊兰	*Chlorophytum capense* (L.) O. Kuntze var. *variegatum* Hort.	Variegated Spider Plant
*吊兰	*Chlorophytum comosum* Baker	Bracket plant
三角草	*Chlorophytum laxum* R. Br.	Small-flower Bracket Plant
山菅兰(桔梗兰)	*Dianella ensifolia* (L.) DC.	Dianella
万寿竹(富贵竹)	*Disporum cantoniense* (Lour.) Merr.	Cantonese Fairy Bells
宝铎草	*Disporum nantouense* S. S. Ying [*D. sessile* D. Don]	Fairy Bells
*嘉兰(嘉兰百合)	*Gloriosa superba* L.	Climbing Lily
*水晶掌(玻璃爪)	*Haworthia cymbiformis* (Haw.) Duval var. *translucens* Triebner et Poelln.	Translucent Haworthia

被子植物 ANGIOSPERMAE

*条纹十二卷(锦鸡尾)	Haworthia fasciata (Willd.) Haw.	Zebra Haworthia
*点纹十二卷(龙之爪)	Haworthia margaritifera Haw.	Pearl Haworthia
*萱草(金针花)	Hemerocallis fulva (L.) L.	Day-Lily
*玉簪	Hosta plantaginea (Lam.) Aschers.	Fragrant Plantain-lily
*风信子(洋水仙)	Hyacinthus orientalis L.	Hyacinth
百合	Lilium brownii F. E. Brown ex Miellez	Chinese lily
*卷丹(虎皮百合)	Lilium tigrinum Ker-Gawl. [L. lancifolium Thunb.]	Tiger Lily
禾叶土麦冬(长叶麦冬)	Liriope graminifolia (L.) Baker	Grassleaf Liriope
*阔叶山麦冬(阔叶土麦冬)	Liriope muscari (Decne.) Bailey [L. platyphylla Wang et Tang]	Broadleaf Liriope
山麦冬(土麦冬)	Liriope spicata (Thunb.) Lour.	Lily Turf
中型沿阶草	Ophiopogon intermedius D. Don	Intermediate Lilyturf
沿阶草	Ophiopogon japonicus (L. f.) Ker-Gawl.	Blue Grass
广东沿阶草	Ophiopogon reversus Hwang	Guangdong Lily-turf
大盖球子草	Peliosanthes macrostegia Hance	Bigcover Peliosanthes
*多花黄精(玉竹)	Polygonatum cyrtonema Hua [P. odoratum (Mill.) Druce]	Manyflower Landpick
*吉祥草	Reineckia carnea (Andr.) Kunth	Pink Reineckia

*万年青	*Rohdea japonica* (Thunb.) Roth	Omoto Nipponlily
异蕊草	*Thysanotus chinensis* Benth.	Chinese Fringelily
油点草	*Tricyrtis macropoda* Miq.	Speckled Toadlily
*郁金香	*Tulipa gesneriana* L.	Tulip
黑紫藜芦	*Veratrum japonicum* (Baker) Loes. f.	Japanese False Hellebore
*藜芦(黑藜芦)	*Veratrum nigrum* L.	Black Falsehellebore

延龄草科 Trilliaceae

七叶一枝花(七叶莲)	*Paris polyphylla* Smith [*P. chinensis* Fr.]	Manyflower Paris

雨久花科 Pontederiaceae

*凤眼莲(水葫芦)	*Eichhornia crassipes* (Mart.) Solms	Water Hyacinth
*雨久花(蓝鸟花)	*Monochoria korsakowii* Regel et Maack	Korsakow Monochoria
鸭舌草	*Monochoria vaginalis* (Burm. f.) Kunth	Duck's Tongue Grass

菝葜科 Smilacaceae

华肖菝葜(肖菝葜)	*Heterosmilax chinensis* Wang	China Heterosmilax
合丝肖菝葜	*Heterosmilax japonica* var. *gaudichaudiana* (Kunth) Maxim.	Gaudichaud Heterosmilax
灰叶菝葜	*Smilax astrosperma* Wang et Tang	Grey-leaved Greenbrier
菝葜(金刚头)	*Smilax china* L.	Green brier
粉叶菝葜(里白菝葜)	*Smilax corbularia* Kunth	Long-leaved Greenbrier
小果菝葜	*Smilax davidiana* A. DC.	Smallfruit Greenbrier

被子植物 ANGIOSPERMAE

土茯苓	*Smilax glabra* Roxb.	Glabrous Greenbrier
*粉背菝葜	*Smilax hypoglauca* Benth.	Hypoglaucous Greenbrier
剑叶菝葜(马甲菝葜)	*Smilax lanceifolia* Roxb.	Vest Greenbrier
暗色菝葜(白茯苓)	*Smilax lanceifolia* Roxb. var. *opaca* A. DC.	Opaque Greenbrier
软叶菝葜(牛尾菜)	*Smilax riparia* A. DC.	Oxtail Greenbrier

假叶树科 Ruscaceae

*假叶树(百劳金霍花)	*Ruscus aculeatus* L.	Butchers Broom

天南星科 Araceae

金钱蒲(山菖蒲)	*Acorus gramineus* Soland.	Grassleaf Sweetflag
石菖莆(苦菖蒲、)	*Acorus tatarinowii* Schott	Sweet-Flag
*白斑亮丝草	*Aglaonema commutatum* Schott	Aglaonema
*白雪亮丝草	*Aglaonema crispum* (Pitcher et Manda) Nicolson	Painted Drop-tongue
*广东万年青(亮丝草)	*Aglaonema modestum* Schott ex Engl.	Chinese Evergreen
*黑叶观音莲(黑叶芋)	*Alocasia amazonica* (Lour.) Schott	African Mask
尖尾芋	*Alocasia cucullata* (Lour.) Schott	Chinese Taro
海芋	*Alocasia macrorrhiza* (L.) Schott	Giant Alocasia
南蛇棒(野生大头芋)	*Amorphophallus dunnii* Tutcher	Voodoo Lily
香港魔芋	*Amorphophallus oncophyllus* Prain ex Book. f.	Hongkong Giantarum

魔芋	*Amorphophallus rivieri* Durieu ex Riviere	Rivier Giantarum
野魔芋(半夏)	*Amorphophallus variabilis* Bl.	Snake Aroid
*花烛(红掌)	*Anthurium andraeanum* Linden	Palette Flower
*圆叶花烛(克拉利安祖花)	*Anthurium clarinervium* Matuda	White-veined Anthurium
*水晶花烛(美叶花烛)	*Anthurium crystallinum* Linden ex Andre	Crystal Anthurium
*掌叶花烛(趾叶花烛)	*Anthurium pedato-radiatum* Schott	Footleaf Anthurium
*红鹤芋(火鹤花)	*Anthurium scherzerianum* Schott	Common Anthurium
心檐天南星	*Arisaema cordatum* N. E. Br.	Heart-leaved Arisaema
天南星(掌叶半夏)	*Arisaema erubeacens* (Wall.) Schott	Dragon Arum
*花叶芋	*Caladium bicolor* Vent.	Common Caladium
*彩叶芋(五彩芋)	*Caladium hortulanum* Birdsey	Fancy-leaved Caladium
野芋	*Colocasia antiquorum* Schott	Dashen
芋(芋头)	*Colocasia eaculenta* (L.) Schott	Taro
*大王黛粉叶(大王万年青)	*Dieffenbachia amoena* Nichols.	Giant Dumbcane
*黛粉叶(斑叶万年青)	*Dieffenbachia maculata* (Lodd.) G. Don	Potted Dumbcane
*花叶万年青(银斑万年青)	*Dieffenbachia seguine* (Jacq.) Schott [*D. picta* (Lodd.) Schott]	Dumbcane
绿萝(麒麟叶)	*Epipremnum pinnatum* (L.) Engl. [*E. aureum* (Linden et Andre) Bunting]	Ivy-arum

被子植物 ANGIOSPERMAE

千年健	*Homalomena occulta* (Lour.) Schott	Hidden Homalomena
* 龟背竹(蓬莱蕉)	*Monstera deliciosa* Liebm.	Ceriman
* 绿宝石喜林芋(长心叶蔓绿绒)	*Philodendron emerald* Duke	Emerald Philodendron
* 红叶蔓绿绒	*Philodendron erubescens* C. Koch et Aug.	Blushing Philodendron
* 大叶蔓绿绒(箭叶蔓绿绒)	*Philodendron grandifolium* Schott	Largeleaf Philodendron
* 红柄蔓绿绒(红背蔓绿绒)	*Philodendron imbe* Schott ex Engl.	Weak Philodendron
* 姬喜林芋(心叶喜林芋)	*Philodendron oxycardium* Schott	Heart-leaved Philodendron
* 琴叶喜林芋(琴叶蔓绿绒)	*Philodendron panduraeforme* Kunth	Fiddle-leaved Philodendron
* 羽叶喜林芋(春羽)	*Philodendron pittieri* Engl.	Pittier Philodendron
* 春羽(羽裂喜林芋)	*Philodendron selloum* C. Koch	Lacy Tree Philodendron
* 大藻(水浮莲)	*Pistia stratiotes* L.	Water Lettuce
石柑子(石蒲藤)	*Pothos chinensis* (Raf.) Merr.	Rock Vine
蜈蚣藤(百足藤)	*Pothos repens* (Lour.) Druce	Creeping Pothos
崖角藤(狮子尾)	*Rhaphidophora hongkongensis* Schott	Hong Kong Taro-vine
* 白鹤芋(银苞芋)	*Spathiphyllum floribundum* (Linden et Andre) N. E. Br.	Snow Flower
* 白掌(白鹤芋)	*Spathiphyllum kochii* Engl. et Krause	White Flag
* 合果芋(白蝴蝶)	*Syngonium podophyllum* Schott	African Evergreen

犁头尖(土半夏)	*Typhonium blumei* Nicols. et Sivadasan [*T. divaricatum* (L.) Decne.]	Divaricate Typhonium
马蹄犁头尖(山半夏)	*Typhonium trilobatum* (L.) Schott	Trilobed Typhonium
*雪铁芋(金钱树)	*Zamioculcas zamifolia* (Lodd.) Engl.	Money Tree
*马蹄莲	*Zantedeschia aethiopica* (L.) Spreng.	Arum Lily
*黄花马蹄莲(马蹄金)	*Zantedeschia elliottiana* (H. Knight) Engl.	Golden Calla

浮萍科 Lemnaceae

浮萍(青萍)	*Lemna minor* L.	Lesser Duck-weed
紫萍(浮萍)	*Spirodela polyrrhiza* (L.) Schleid.	Greater Duck-weed
无根萍(微萍)	*Wolffia arrhiza* (L.) Wimm.	Water-meal

香蒲科 Typhaceae

水烛(长苞香蒲)	*Typha angustifolia* L.	Narrow-leaved Cat-tail
*香蒲(水蜡烛)	*Typha orientalis* Presl	Oriental Cattail

石蒜科 Amaryllidaceae

*百子莲(紫君子兰)	*Agapanthus africanus* (L.) Hoffmanns	African Lily
荞头(藠荞)	*Allium chinense* G. Don	China Onion
葱(大葱)	*Allium fistulosum* L.	Welsh Onion
蒜(大蒜)	*Allium sativum* L.	Garlic
*球序韭(大花葱)	*Allium thunbergii* G. Don	Thunberg Leek
*韭菜	*Allium tuberosum* Rottl. ex Spreng.	Chinese Chives

被子植物 ANGIOSPERMAE

*大花君子兰(剑叶石蒜)	Clivia miniata Regel	Kaffir Lily
*红花文殊兰(苏门答腊文殊兰)	Crinum amabile Donn.	Giant Spider Lily
文殊兰(十八学士)	Crinum asiaticum L. var. sinicum (Roxb. ex Herb.) Baker	Chinese Crinum
*网球花(网球石蒜)	Haemanthus multiflorus Martyn	Blood Lily
*朱顶红(百子莲)	Hippeastrum vittatum Herb.	Common Knight's Star
蜘蛛兰(水鬼蕉)	Hymenocallis americana M. Roem. [H. littoralis (Jacq.) Salisb.]	American Spider Lily
忽地笑(黄花石蒜)	Lycoris aurea (L'Her.) Herb.	Chinese Amaryllis
石蒜(龙爪花)	Lycoris radiata (L'Her.) Herb.	Stonegarlic
*中国水仙	Narcissus tazetta L. var. chinensis Roem.	Chinese Sacred Lily
*葱兰	Zephyranthes candida (Lindl.) Herb.	White Zephyr Flower
*红花葱兰(风雨花)	Zephyranthes carinata Herb.	Autumn Zephyr-lily
*小韭兰(花韭)	Zephyranthes rosea Lindl.	Cuban Zephyr-lily
	鸢尾科 Iridaceae	
射干(乌扇)	Belamcanda chinensis (L.) DC.	Black-berry Lily
*香雪兰(小苍兰)	Freesia refracta Klatt	Common Freesia
*唐菖蒲(剑兰)	Gladiolus gandavensis Van Houtte.	Bredders Gladiolus

*蝴蝶花(日本鸢尾)	*Iris japonica* Thunb.	Fringed Iris
*鸢尾(蓝蝴蝶)	*Iris tectorum* Maxim.	Grested Iris
*巴西鸢尾(美丽鸢尾)	*Neomarica gracilis* Spragne	New Nymph Flower

百部科 Stemonaceae

对叶百部(山百部)	*Stemona tuberosa* Lour.	Tuber Stemona

薯蓣科 Dioscoreaceae

*参薯(白薯)	*Dioscorea alata* L.	Greater Yam
大青薯(山药薯)	*Dioscorea benthamii* Prain et Burkill	Bentham's Yam
黄独(零余薯)	*Dioscorea bulbifera* L.	Air Potato
薯莨	*Dioscorea cirrhosa* Lour.	Shoulang Yam
山薯(山芋)	*Dioscorea fordii* Prain et Burkill	Ford's Yam
白薯莨	*Dioscorea hispida* Dennst.	White Yam
五叶薯蓣(五叶薯)	*Dioscorea pentaphylla* L.	Five-leaved Yam
褐苞薯蓣(土淮山)	*Dioscorea persimilis* Prain et Burkill	Mountain Yam
薯蓣	*Dioscorea polystachya* Turcz. [*D. batatas* Decne.]	Chinese Yam

龙舌兰科 Agavaceae

*龙舌兰(龙舌掌)	*Agave americana* L.	Century Plant
*狐尾龙舌兰(无刺龙舌兰)	*Agave attenuata* Salm-Dyck	Fox-tail Agave
*雷神(棱叶龙舌兰)	*Agave potatorum* Zucc. var. *verschaffeltii* (Lem.) Berger	Verschaffelt Agave
*白丝龙舌兰(泷之白丝)	*Agave schidigera* Lemm.	Maguey

*剑麻(万年麻)	*Agave sisalana* Perr. ex Engelm	Sisal Agave
*锦叶龙舌兰	*Agave victoriae-reginae* T. Moore	Queen Victoria Agave
*朱蕉(红竹)	*Cordyline fruticosa* (L.) A. Cheval.	Iron Plant
*小花龙血树(不老松)	*Dracaena cambodiana* Pierre ex Gagnep.	Cambodia Dragonblood
*香龙血树(巴西木)	*Dracaena fragrans* (L.) Ker-Grawl.	Fragrant Dracaena
*星点木(星斑千年木)	*Dracaena godseffiana* Hort. ex Baker	Gold-dust Dracaena
*百合竹(反折龙血树)	*Dracaena reflexa* Lam.	Song of India
*金边富贵竹(镶边竹蕉)	*Dracaena sanderiana* Sander ex M. T. Mast.	Sander's Dracaena
*万年麻(万年兰)	*Furcraea foetida* (L.) Haw.	Giant Cabuga
*酒瓶兰(象腿树)	*Nolina recurvata* (Lem.) Hemsl.	Bottle Palm
*圆叶虎尾兰(棒叶虎尾兰)	*Sansevieria cylindrica* Bojer	Spear Sansevieria
*虎尾兰(虎皮兰)	*Sansevieria trifasciata* Prain	Snake Plant
*金边虎尾兰	*Sansevieria trifasciata* Prain var. *laurentii* N. E. Br.	Veriegated Snake Plant
*金边千手兰	*Yucca aloifolia* L. var. *marginata* Bommer	Marginate Spanish Bayonet
*象腿丝兰(巨丝兰)	*Yucca elephantipes* Hort. ex Regel	Spineless Yucca
*丝兰(凤尾丝兰)	*Yucca filamentosa* L.	Weakleaf Yucca

*凤尾丝兰(刺叶王兰)	*Yucca gloriosa* L.	Spanish Dagger
*鸟喙丝兰	*Yucca rostrata* Engelm. ex Trelease	Beaked Yucca

棕榈科 Arecaceae

*假槟榔(亚历山大椰子)	*Archontophoenix alexandrae* (F. Muell.) Wendl. et Drude	King Palm
*槟榔	*Areca catechu* L.	Betel Palm
*三药槟榔(丛生槟榔)	*Areca triandra* Roxb. ex Buch.-Ham.	Triandra Palm
*散尾棕(矮桄榔)	*Arenga engleri* Becc.	Arenga Palm
*桄榔(砂糖椰子)	*Arenga pinnata* (Wurmb.) Merr.	Sugar Palm
*南椰	*Arenga westerhoutii* Griffith	Westerhout's Sugar Palm
*霸王棕(霸王榈)	*Bismarckia nobilis* Hildebr. et H. Wendl.	Bismarck Palm
*糖棕(扇椰子)	*Borassus flabellifer* L.	Sweetpalm
*弓葵(冻子椰子)	*Butia capitata* (Mart.) Becc.	Pindo Palm
华南省藤	*Calamus rhabdocladus* Burret	Stick Rattanpalm
白藤(鸡藤)	*Calamus tetradactylus* Hance	Four-finger Rattan Palm
毛鳞省藤	*Calamus thysanolepis* Hance	Hairy-scale Rattan Palm
*短穗鱼尾葵(酒椰子)	*Caryota mitis* Lour.	Small Fishtail Palm
*单穗鱼尾葵	*Caryota monostachya* Becc.	Singlespike Fishtailpalm
鱼尾葵(假桄榔)	*Caryota maxima* Bl. [*C. ochlandra* Hance]	Fishtail Palm
*董棕(孔雀椰)	*Caryota urens* L.	Wine Palm

被子植物 ANGIOSPERMAE

*袖珍椰子(矮生椰子)	*Chamaedorea elegans* Mart.	Parlor Palm
*鱼尾椰子(燕尾葵)	*Chamaedorea metallica* O. F. Cook ex H. E. Moore	Metal Palm
*散尾葵(黄椰子)	*Chrysalidocarpus lutescens* Wendl.	Bamboo Palm
*琼棕(桄榔木)	*Chuniophoenix hainanensis* Burret	Qiongpalm
*矮琼棕(小琼棕)	*Chuniophoenix nana* Burret	Dwalf Qiongpalm
*老人葵(华盛顿葵)	*Coccothrinax crinita* Becc.	Old Man Palm
*椰子(椰树)	*Cocos nucifera* L.	Coconut Palm
*贝叶棕(团扇葵)	*Corypha umbraculifera* L.	Cowryleafpalm
*吕宋糖棕(金丝棕)	*Corypha utan* Lam.	Talipot Palm
*红柄椰(红槟椰)	*Cyrtostachys renda* Bl.	Lipstick Palm
黄藤(省藤)	*Daemonorops margaritae* (Hance) Becc.	Rattan Palm
*三角椰子(三角棕)	*Dypsis decaryi* (H. Jumelle) Beentje et J. Dransf.	Triangle Palm
*油棕(油椰子)	*Elaeis guineensis* N. J. Jacq.	African Oilpalm
*荷威棕(富贵椰子)	*Howea belmoreana* Becc.	Sentry Palm
*酒瓶椰子(酒瓶棕)	*Hyophorbe lagenicaulis* (L. H. Bailey) H. E. Moore	Bottle Palm
*红脉葵(红脉榈)	*Latania lontaroides* (J. Gaertn.) H. E. Moore	Red Latan Palm
轴榈(穗花轴榈)	*Licuala fordiana* Becc.	Spikeflower Axispalm
*刺轴榈(裂叶棕枚)	*Licuala spinosa* Thunb.	Spine Axispalm

*圆叶轴榈(团圆椰子)	*Licuala grandis* (Bull.) H. A. Wendl.	Licuala Palm
*蒲葵(葵树)	*Livistona chinensis* (Jacq.) R. Br.	Chinese Fan-palm
*裂叶蒲葵	*Livistona decipiens* Becc.	Ribbon Fan Palm
*长叶刺葵(加纳利海枣)	*Phoenix canariensis* Hort. ex Chabaud	Canary Island Date Palm
*枣椰(海枣)	*Phoenix dactylifera* L.	Date Palm
刺葵(猪嵫怕)	*Phoenix hanceana* Naudin	Spiny Date-palm
*泰国刺葵(大刺葵)	*Phoenix paludosa* Roxb.	Mangrove Date Palm
*软叶刺葵(美丽针葵)	*Phoenix roebelenii* O. Brien	Dwarf Date-palm
*燕尾山槟榔	*Pinanga sinii* Burret	Swallowtail Pinangapalm
*青棕	*Ptychosperma macarthurii* Nichols.	MacArthur Palm
*国王椰子	*Ravenea rivularis* Jum. et H. Perrier	Majesty Palm
棕竹(棕榈竹)	*Rhapis excelsa* (Thunb.) Henry ex Rehd.	Lady Palm
*细棕竹(细叶棕竹)	*Rhapis gracilis* Burret	Slender Palm
*多裂棕竹(金山棕)	*Rhapis multifida* Burret	Dissected Ladypalm
*菜王椰(菜棕)	*Roystonea oleracea* (Jacq.) O. F. Cook	South American Royal Palm
*大王椰子(王棕)	*Roystonea regia* (Kunth) O. F. Cook	Royal Palm
*金山葵(皇后葵)	*Syagrus romanzoffiana* (Cham.) Glassm.	Queen Palm

被子植物 ANGIOSPERMAE

棕榈(棕树、山棕)	*Trachycarpus fortunei* (Hook.) Wendl.	Windmill Palm
*龙棕	*Trachycarpus nana* Becc.	Dragon palm
*马尼拉椰子(圣诞椰)	*Veitchia merrillii* (Becc.) H. E. Moore	Manila Palm
*丝葵(老人葵)	*Washingtonia filifera* (Linden) H. Wendl.	Petticoat Palm
*狐尾棕	*Wodyetia bifurcata* A. K. Irvine	Foxtail Palm

露兜树科 Pandanaceae

露兜草(野菠萝)	*Pandanus austrosinensis* T. L. Wu	South China Screwpine
簕古子(露兜簕)	*Pandanus forceps* Martelli	False Pineapple
分叉露兜簕角(分叉露兜)	*Pandanus furcatus* Roxb.	Furcate Screwpine
露兜树(露兜簕)	*Pandanus tectorius* Sol.	Tatch Screwpine
*红刺露兜树	*Pandanus utilis* Bory	Common Screw Pine
*斑叶露兜树	*Pandanus veitchi* (Dall.) Hort.	Veitch Screwpine

仙茅科 Hypoxidaceae

短葶仙茅(大莎草)	*Curculigo breviscapa* S. C. Chen	Shortscape Curculigo
大叶仙茅	*Curculigo capitulata* (Lour.) O. Kuntze	Large-leaved Curculigo
光叶仙茅	*Curculigo glabrescens* (Ridl.) Merr.	Glabrous Curculigo
仙茅	*Curculigo orchioides* Gaertn.	Golden-eyed-grass
小金梅草	*Hypoxis aurea* Lour.	Gold Stargrass

田葱科 Philydraceae

田葱(中葱)	*Philydrum lanuginosum* Banks et Sol. ex Gaertn.	Common Philydrum

水玉簪科 Burmanniaceae

纤草	*Burmannia itoana* Makino	Ito Burmannia

兰科 Orchidaceae

多花脆兰(蕉兰)	*Acampe rigida* (Buch.-Ham. ex J. E. Smith) P. F. Hunt	Banana Orchid
香港安兰(香港带唇兰)	*Tainia hongkongensis* (Rolfe) Tang et Wang [*Ania hongkongensis* (Rolfe) Tang et Wang]	Hongkong Tainia
*绿花安兰(绿花带唇兰)	*Tainia hookeriana* Ring et Prantl [*Ania hookeriana* (King et Pantl.) Tang et Wang]	Greenflower Tainia
花叶开唇兰(筒瓣兰)	*Anoectochilus roxburghii* (Wall.) Lindl.	Roxburgh Goldlineorchis
牛齿兰(台湾牛齿兰)	*Appendicula cornuta* Bl.	Ladder Orchid
竹叶兰(笔竹)	*Arundina graminifolia* (D. Don) Hochr.	Bamboo Orchid
短距苞叶兰	*Brachycorythis galeandra* (Rchb. f.) Summ.	Short Spur Brachycorythis
赤唇石豆兰(高士佛豆兰)	*Bulbophyllum affine* Lindl.	Redlip Stonebean-orchis
芳香石豆兰(梳帽卷瓣兰)	*Bulbophyllum ambrosia* (Hance) Schltr.	Ambrosia Orchid
广东石豆兰(乐东石豆兰)	*Bulbophyllum kwangtungense* Schltr.	Antennae Orchid

密花石豆兰(小果上叶)	*Bulbophyllum odoratissimum* (J. E. Smith) Lindl.	Fragrant Bullbophyllum
银线虾脊兰	*Calanthe argenteo-striata* C. Z. Tang et S. J. Cheng	Silverline Calanthe
虾脊兰(钩距虾脊兰)	*Calanthe discolor* Lindl.	Commom Calanthe
二列虾脊兰(台湾根节兰)	*Calanthe formosana* Rolfe	Yellow-flowered Calanthe
反瓣虾脊兰(反卷根节兰)	*Calanthe reflexa* (O. Kuntze) Maxim.	Reflexpetal Calanthe
三褶虾脊兰(白花虾脊兰)	*Calanthe triplicata* (Willem.) Ames	Triplicata Calanthe
*两色卡特兰	*Cattleya bicolor* Lindl.	Two-color Cattleya
*橙黄卡特兰	*Cattleya citrina* Lindl.	Lemon-yellow Cattleya
*玫红卡特兰	*Cattleya labiata* Lindl.	Common Cattleya
*特丽安娜卡特兰	*Cattleya labiata* Lindl. var. *trianae* (Linden et Reichb. f.) Regel	Trian Cattleya
鞭花卷瓣兰(直唇卷瓣兰)	*Cirrhopetalum delitescens* (Hance) Rolfe	Straight-lip Cirrhopetalum
福氏隔距兰(尖喙隔距兰)	*Cleisostoma fordii* Hance [*C. nostiata* (Lindl.) Garay]	Rostrate Cleisostoma
尖啄隔距兰	*Cleisostoma rostratum* (Lodd.) Seidenf. ex Averyanov	Acute Closedspurorchis
*毛柱隔距兰(蜜蜂兰)	*Cleisostoma simondii* (Gagnep.) Seidenf.	Simond Closedspurorchis
广东隔距兰	*Cleisostoma simondii* (Gagnep.) Seidenf. var. *guangdongense* Z. H. Tsi	Terete Cleisostoma
流苏贝母兰(贝母兰)	*Coelogyne fimbrata* Lindl.	Brown Rock-orchid

*冬凤兰(凤兰)	*Cymbidium dayanum* Rchb. f.	Pendulous Orchis
建兰(四季兰)	*Cymbidium ensifolium* (L.) Sw.	Swordleaf Cymbidium
阔叶建兰	*Cymbidium ensifolium* (L.) Sw. var. *munronianum* (King et Prantl) Tang et Wang	Wideleaf Cymbidium
*大花蕙兰(土百部)	*Cymbidium faberi* Rolfe	Faber Orchis
*春兰(朵朵香)	*Cymbidium goeringii* (Rchb. f.) Rchb. f.	Spring Orchis
寒兰	*Cymbidium kanran* Makino	Smooth-lipped Cymbidium
兔耳兰(竹柏兰)	*Cymbidium lancifolium* Hook.	Lanceolate Cymibidum
*硬叶吊兰(壁兰、信兰)	*Cymbidium simulans* Rolfe	Imitating Cymbidium
墨兰(报岁兰)	*Cymbidium sinense* (Jackson ex Andr.) Willd.	Chinese Cymbidium
剑叶石斛	*Dendrobium acinaciforme* Roxb.	Sword-leaved Dendrobium
*鼓槌石斛(晶帽石斛)	*Dendrobium chrysotoxum* Lindl.	Yellow-bow Dendrobium
美花石斛(粉花石斛)	*Dendrobium loddigesii* Rolfe	Loddiges's Dendrobium
*石斛	*Dendrobium nobilis* Lindl.	Noble Dendrobium
蛇舌兰(船唇兰)	*Diploprora championii* (Lindl.) Hook. f.	Champion's Orchid
半柱毛兰	*Eria corneri* Rchb. f.	Corner's Eria
足茎毛兰(半柱花兰)	*Eria coronaria* (Lindl.) Rchb. f.	Corolla Hairorchis

被子植物 ANGIOSPERMAE

白绵毛兰(龙陵毛兰)	*Eria lasiopetala* (Willd.) Ormerod	Wolly Eria
对茎毛兰(五脊毛兰)	*Eria pusilla* (Griff.) Lindl.	Mini Hairorchis
钳喙兰(钳唇兰)	*Erythrodes chinensis* (Rolfe) Schltr.	China Entireliporchis
美冠兰(紫花美冠兰)	*Eulophia graminea* Lindl.	Pale Purple Eulophia
无叶美冠兰(毛唇美冠兰)	*Eulophia zollingerii* (Rchb. f.) J. J. Smith	Hong Kong Eulophia
高斑叶兰(大斑叶兰)	*Goodyera procera* (Ker-Gewl.) Hook.	Bottlebrush Orchid
鹅毛玉凤花	*Habenaria dentata* (Sw.) Schltr.	Phantom Orchid
细裂玉凤花(仙子兰)	*Habenaria leptoloba* Benth.	Narrow Habeuaria
坡参(玲姬玉凤花)	*Habenaria linguella* Lindl.	Tongue Habenaria
橙黄玉凤花(红唇玉凤花)	*Habenaria rhodocheila* Hance	Red-man Orchid
镰翅羊耳蒜(镰翅羊耳兰)	*Liparis bootanensis* Griff.	Bhutan Twayblade
见血青(脉羊耳蒜)	*Liparis nervosa* (Thunb. ex A. Murray) Lindl.	Purple Star Liparis
紫花羊耳蒜(香花羊耳蒜)	*Liparis nigra* Seidenf.	Giant Purple Liparis
绿花羊耳蒜(长茎羊耳蒜)	*Liparis viridiflora* (Bl.) Lindl.	Longstem Liparis
血叶兰(金线莲)	*Ludisia discolor* (Ker-Gawl.) A. Rich.	Discolor Ludisia
二脊沼兰	*Malaxis finetii* (Gagnap.) T. Tang et F. T. Wang	Finet Bogorchis

阔叶沼兰	*Malaxis latifolia* J. E. Smith	Broad-leaved Addermouth Orchid
云叶兰	*Nephelaphyllum tenuiflorum* Bl.	Cloud-leaf Orchid
紫纹兜兰(香港兜兰)	*Paphiopedilum purpuratum* (Lindl.) Stein	Hong Kong Lady's-slipper Orchid
旋药阔蕊兰(长穗阔蕊兰)	*Peristylus lacertiferus* (Lindl.) J. J. Sm. [*P. spiranthes* (Schauer) S. Y. Hu]	Splitlip Peristyle
鹤顶兰(大花鹤顶兰)	*Phaius tankervilleae* (Banks ex L'Her.) Bl.	Common Phaius
*蝴蝶兰(蝶兰)	*Phalaenopsis aphrodite* Rchb. f. [*P. amabilis* (L.) Bl.]	Moth Orchid
细叶石仙桃(广东石仙桃)	*Pholidota cantonensis* Rolfe	Slender-leaved Pholidota
石仙桃	*Pholidota chinensis* Lindl.	Rattlesnake Orchid
小叶寄树兰	*Robiquetia succisa* (Lindl.) Seidenf. et Garay	Big Ladder Orchid
苞舌兰(牛油杯)	*Spathoglottis pubescens* Lindl.	Buttercup Orchid
香港绶草	*Spiranthes hongkongensis* S. Y. Hu et Barretto	Hong Kong Spiranthes
绶草(盘龙参)	*Spiranthes sinensis* (Pers.) Ames	Ladies Tresses
绿锭带唇兰	*Tainia viridifusa* Hook.	Green-brown Tainia
仙茅竹茎兰	*Tropidia curculigoides* Lindl.	Bamboo-stemmed Orchid
宽叶线柱兰(黄唇线柱兰)	*Zeuxine affinis* (Lindl.) Benth. ex Hook. f.	Broadleaf Zeuxine

灯心草科 Juncaceae

小花灯心草	*Juncus articulatus* L.	Smallflower Rush

灯心草(灯草)	*Juncus effusus* L.	Commom Rush
江南灯心草(野灯芯草)	*Juncus leschenaultii* Gay	Leschenault Rush

莎草科 Cyperaceae

球柱草(畦莎)	*Bulbostylis barbata* (Rottb.) Kunth	Flag Grass
丝叶球柱草	*Bulbostylis densa* (Wall.) Hand.-Mazz.	Silk Grass
山稗子(芭茅草)	*Carex baccans* Nees	Berry-fruited Sedge
中华苔草(茅叶苔草)	*Carex chinensis* Retz.	Chinese Sedge
十字苔草(黄牛草)	*Carex cruciata* Wahlenb.	Cruciate Sedge
隐穗苔草(茅叶苔草)	*Carex cryptostachys* Brongn.	Lesser Sedge
蕨叶苔草(蕨状苔草)	*Carex filicina* Nees	Fern-like Sedge
弯柄苔草	*Carex manca* Boott	Red Bract Sedge
条穗苔草	*Carex nemostachys* Steud.	Linear-spike Sedge
大叶苔草(花莛苔草)	*Carex scaposa* C. B. Clarke	Scapose Sedge
细梗苔草(长柱头苔草)	*Carex teinogyna* Boott	Slender-pedicel Sedge
截穗苔草	*Carex truncatigluma* C. B. Clarke	Truncate-glume Sedge
一本芒	*Cladium mariscus* Pohl ssp. *jamaicense* (Crantz.) Kukenthal [*C. chinensis* Nees]	Chinese Cladium
*风车草(伞草)	*Cyperus flabelliformis* Rottb.	Umbrella Plant

扁穗莎草(莎田草)	*Cyperus compressus* L.	Compressed Galingale
异型莎草(球花莎草)	*Cyperus difformis* L.	Galingale
绿穗莎草(多脉莎草)	*Cyperus diffusus* Vahl	Diffuse Galingale
高秆莎草(无翅莎草)	*Cyperus exaltatus* Retz.	Tall-culm Galingale
畦畔莎草(埃及莎草)	*Cyperus haspan* L.	Paddy Galingale
叠穗莎草(覆瓦状莎草)	*Cyperus imbricatus* Retz.	Imbricate Galingale
碎米莎草	*Cyperus iria* L.	Rice Galingale
咸水草(茳芏)	*Cyperus malaccensis* Lam.	Malacca Galingale
短叶茳芏(咸水草)	*Cyperus malaccensis* Lam. var. *brevifolius* Bocklr.	Short-leaved Malacca Galingale
具芒碎米莎草(阿穆尔莎草)	*Cyperus microiria* Steud.	Awned Rice Galingale
白鳞莎草	*Cyperus nipponicus* Franch. et Sav.	Whitescale Galingale
*纸莎草(纸草)	*Cyperus papyrus* L.	Paper Galingale
毛轴莎草(三棱草)	*Cyperus pilosus* Vahl	Pilosus Galingale
多穗莎草	*Cyperus polystachyus* (Rottb.) P. Beauv.	Manyspike Galingale
矮莎草	*Cyperus pygmaeus* Rottb.	Low Galingale
香附子(香头草)	*Cyperus rotundus* L.	Nutgrass Galingale
紫果蔺(黑果蔺)	*Eleocharis atropurpurea* (Retz.) Presl	Purple-fruited Spikesedge
马蹄(荸荠)	*Eleocharis dulcis* (Burm. f.) Trin. ex Hensch.	Water Chestnut

黑籽荸荠(弯形蔺)	*Eleocharis geniculata* (L.) Roem. et Schult.	Black-seeded Spikesedge
透明荸荠	*Eleocharis congesta* D. Don [*E. pellucida* Presl]	Dense-flowered Spikesedge
牛毛毡(油麻毡)	*Eleocharis acicularis* (L.) Roem. et Schult.	Needle Spikesedge
夏飘拂草(小畦畔飘拂草)	*Fimbristylis aestivalis* (Retz.) Vahl	Summer Fimberstylis
复序飘拂草	*Fimbristylis bisumbellata* (Forst.) Bubani	Bisumbellate Fluttergrass
扁鞘飘拂草(两歧飘拂草)	*Fimbristylis complanata* (Retz.) Link	Flat-sheath Fimbristylis
红血飘拂草	*Fimbristylis consanguinea* Kunth	Consanguine Fimbristylis
两歧飘拂草	*Fimbristylis dichotoma* (L.) Vahl	Dichotomous Fimbristylis
柔毛飘拂草	*Fimbristylis dichotoma* (L.) Vahl *f. tomentosa* (Vahl) Ohwi	Woolly Fimbristylis
锈穗飘拂草	*Fimbristylis ferruginea* (L.) Vahl	Feruginous-scaled Fimbristylis
硬穗飘拂草	*Fimbristylis insignis* Thw.	Stiff-spike Fimbristylis
纤茎飘拂草	*Fimbristylis leptoclada* Benth.	Tender-clavate Fimbristylis
长柄果飘拂草	*Fimbristylis longistipitata* Tang et Wang	Longstalk Fimbristylis
水虱草(日照飘拂草)	*Fimbristylis miliacea* (L.) Vahl [*F. littoralis* Gamdich]	Sunshing Fimbristylis
少穗飘拂草(嘉义飘拂草)	*Fimbristylis schoenoides* (Retz.) Vahl	Few-spikelet Fimbristylis

绢毛飘拂草	*Fimbristylis sericea* (Poir.) R. Br.	Silky-haired Fimbristylis
叶状苞飘拂草	*Fimbristylis cymosa* (Lam.) R. Br. [*F. spathacea* Roth]	Cymose Fimbristylis
畦畔飘拂草(曲芒飘拂草)	*Fimbristylis velata* R. Br. [*F. squarrosa* Vahl]	Curved-awn Fimbristylis
双穗飘拂草	*Fimbristylis subbispicata* Nees et Meyen	Two-spikelet Fimbristylis
四棱飘拂草(四穗飘拂草)	*Fimbristylis tetragona* R. Br.	Tetragonal Fimbristylis
西南飘拂草(鬼野飘拂草)	*Fimbristylis thomsonii* Bocklr.	Thomson's Fimbristylis
毛瓣莎(毛异花草)	*Fuirena ciliaris* (L.) Roxb.	Ciliate Fuirena
芙兰草(黑珠蒿)	*Fuirena umbellata* Rottb.	Umbrella Grass
黑莎草	*Gahnia tristis* Nees	Gahnia
海南割鸡芒(海南大风子)	*Hypolytrum hainanense* (Merr.) Tang et Wang	Hainan Hypolytrum
割鸡芒(海薄利)	*Hypolytrum nemorum* (Vahl) Spreng.	Wooded Hypolytrum
短叶水蜈蚣(水蜈蚣)	*Kyllinga brevifolia* Rottb.	Short-leaved Killinga
黑籽水蜈蚣	*Kyllinga melanosperma* Nees	Blackseed Water-centipede
单穗水蜈蚣	*Kyllinga monocephala* Rottb.	Uni-spike Killinga
鳞子莎(鳞籽莎)	*Lepidosperma chinense* Nees	Chinese Scaly Seed
蒲草	*Lepironia articulata* (Retz.) Dom.	Cattail-leaved Lipocarpha
华湖瓜草	*Lipocarpha chinensis* (Osb.) Kern	Chinese Lipocarpha
*银穗湖瓜草	*Lipocarpha sengalensis* (Lam.) Dandy	Silverspike Lakemelongrass

被子植物 ANGIOSPERMAE

羽穗砖子苗	*Mariscus javanicus* (Houtt.) Merr. et Metc.	Feathery Spike Sawgrass
辐射砖子苗	*Mariscus radians* (Nees et Meyen) Tang et Wang	Radiate Sawgrass
砖子苗	*Mariscus sumatrensis* (Retz.) Koyama [*M. umbellatus* Vahl]	Sawgrass
球穗扁莎	*Pycreus flavidus* (Retz.) Koyama [*P. globosus* (All.) Reichb.]	Globular Spike Pycreus
多穗扁莎(多柱扁莎)	*Pycreus polystachyos* (Rottb.) P. Beauv.	Branch Pycreus
矮扁莎	*Pycreus pumilus* (L.) Domin	Dwarf Pycreus
红鳞扁莎	*Pycreus sanguinolentus* (Vahl) Nees	Red-scaled Pycreus
华刺子莞	*Rhynchospora rugosa* (Vahl) Gale [*R. chinensis* Nees et Mey.]	Chinese Beakrush
刺子莞(大叶谷精草)	*Rhynchospora rubra* (Lour.) Makino	Bristle Beakrush
萤蔺(庐山藨草)	*Scirpus juncoides* Roxb.	Rush-like Bulrush
水葱(管子草)	*Scirpus lacustris* L. ssp. *validus* (Vahl) Koyama [*S. validus* Vahl]	Softstem Bulrush
藨草(野荸荠)	*Scirpus triqueter* L.	Common Bulrush
二花珍珠茅	*Scleria biflora* Roxb.	Twoflower Pearlsedge
缘毛珍珠茅	*Scleria ciliaris* Nees	Chinese Nut Rush
宽叶珍珠茅	*Scleria elata* Thw. var. *latior* C. B. Clarke	Wideleaf Razorsedge
圆秆珍珠茅	*Scleria harlandii* Hance	Harland's Nut Rush

毛果珍珠茅(割鸡刀)	*Scleria levis* Retz.	Hairyfruit Pearlsedge
石果珍珠茅	*Scleria lithosperma* (L.) Sw.	Rock Seed Nut Rush
网果珍珠茅	*Scleria parvula* Steud.	Small Razorsedge
皱果珍珠茅	*Scleria rugosa* R. Br.	Rugose-fruited Razorsedge
高秆珍珠茅	*Scleria terrestris* (L.) Foss.	Tall Nut Rush

禾本科 Gramineae

凤头黍(门氏凤头黍)	*Acroceras munroanum* (Balansa) Henrard	Colorado-grass
毛颖草	*Alloteropsis semialata* (R. Br.) Hitchc.	Jet-glumed Grass
看麦娘	*Alopecurus aequalis* Sobol.	Shot-awn Foxtail
水蔗草(假雀麦)	*Apluda mutica* L.	Glutene-rice Grass
楔颖草(瑞氏楔颖草)	*Apocopis wrightii* Munro	Apocopis
华三芒草	*Aristida chinensis* Munro	Chinese Three-awn
*银边草(玉带草)	*Arrhenatherum elatius* (L.) Presl var. *bulbosum* (Willd.) Hyland f. *variegatum* Hitchc.	Tuber Oatgrass
荩草(马耳草)	*Arthraxon hispidus* (Thunb.) Makino	White Heart-leaf
矛叶荩草(竹叶草)	*Arthraxon lanceolatus* (Roxb.) Hochst.	Pikeleaf Ungeargrass
毛节野古草	*Arundinella barbinodis* Keng ex B. S. Sun et Z. H. Hu	Hairrode Arundinella
越南野古草(大花野古草)	*Arundinella cochinchinensis* Keng	Bigspike Arundinella

丈野古草	*Arundinella decenmppedalis* (O. Kuntze) Janow.	Zhang Arundinella
毛野古草	*Arundinella hirta* (Thunb.) Tanaka	Hirsute Arundinella
石芒草(野古草)	*Arundinella nepalensis* Trin.	Minireed
刺芒野古草(三芒野古草)	*Arundinella setosa* Trin.	Tri-awned Minireed
芦竹(荻芦竹)	*Arundo donax* L.	Giant Reed
*类地毯草	*Axonopus affinis* Chase	Carpetgrass
地毯草	*Axonopus compressus* (Swartz) Beauv.	Wide-leaved carpetgrass
臭根子草	*Bothriochloa bladhii* (Retz.) S. T. Blake [*B. intermedia* (R. Br.) A. Camus]	Australian Bluestem
*臂形草	*Brachiaria eruciformis* (J. E. Smith) Griseb.	Armgrass
多枝臂形草	*Brachiaria ramosa* (L.) Stapf	Manybranched Armgrass
四生臂形草(疏穗臂形草)	*Brachiaria subquadripara* (Trin.) Hitchc.	Two-spiked Signal-grass
毛臂形草	*Brachiaria villosa* (L.) A. Camus	Villous Armgrass
拂子草	*Calamagrostis epigejos* (L.) Roth	Chee Woodreed
竹枝细柄草	*Capillipedium assimile* (Steud.) A. Camus	Hard-stemmed
细柄草(吊丝草)	*Capillipedium parviflorum* (R. Br.) Stapf	Small-flowered Capillipedium
*蒺藜草	*Cenchrus echinatus* L.	Bur Grass
假淡竹叶(酸模芒)	*Centotheca lappacea* (L.) Desv.	Common Centotheca

孟仁草(红拂草)	*Chloris barbata* Sw. [*C. inflata* Link]	Peacock-plume Grass
台湾虎尾草	*Chloris formosana* (Honda) Keng	Taiwan Chloris
竹节草(粘人草)	*Chrysopogon aciculatus* (Retz.) Trin.	Wild Oat Grass
小丽草	*Coelachne simpliciuscula* (Wight et Arn.) Munro ex Benth.	Simple Coelachne
薏苡	*Coix lacryma-jobi* L.	Job's Tears
青香茅(橘香草)	*Cymbopogon caesius* (Nees) Stapf	Grey Lemongrass
*香茅(柠檬茅)	*Cymbopogon nardus* (L.) Rendle	Citronella Grass
芸香草(小香茅草)	*Cymbopogon distans* (Nees) A. Camus	Rue Lemongrass
扭鞘香茅(野香茅)	*Cymbopogon hamatulus* (Nees ex Hook. et Arn.) A. Camus	Tweaksheath Lemongrass
狗牙根(绊根草)	*Cynodon dactylon* (L.) Pers.	Bermuda Grass
散穗弓果黍	*Cyrtococcum accrascens* (Trin.) Stapf	Broad-leaved Bowgrass
弓果黍	*Cyrtococcum patens* (L.) A. Camus	Spreading Cyrtococcum
龙爪茅	*Dactyloctenium aegyptium* (L.) Beauv.	Egyptian Grass
*拌根草	*Digitaria adscendens* (H. B. K.) Henrard	Ascendent Crabgrass
毛马唐	*Digitaria chrysoblephara* Fig.	Hairy Crabgrass

升马唐	*Digitaria ciliaris* (Retz.) Koel.	Ciliate Crabgrass
异型马唐(二型马唐)	*Digitaria heterantha* (Hook. f.) Merr.	Dimorphic Crabgrass
长花马唐	*Digitaria longiflora* (Retz.) Pers.	Long-flower Crabgrass
短颖马唐(大马唐)	*Digitaria microbachne* (Presl.) Hitchc.	Short Glume Crabgrass
红尾翎(小马唐)	*Digitaria radicosa* (Presl) Miq.	Root Crabgrass
马唐	*Digitaria sanquinalis* (L.) Scop.	Hairy Crabgrass
五指草(紫马唐)	*Digitaria violascens* Link	Violet Crabgrass
镰形觹茅	*Dimeria flacata* Hack.	Falcate Dimeria
光头稗(旱稗)	*Echinochloa colona* (L.) Link	Jungle Rice
稗(稗草)	*Echinochloa crusgalli* (L.) Beauv.	Barn-yard Grass
孔雀稗	*Echinochloa crusgalli* (L.) Beauv. var. *cruspavonis* (H. B. K.) Hitchc.	Gulf Barnyard Grass
*无芒雀稗(无芒稗)	*Echinochloa crusgalli* (L.) Beauv. var. *mitis* (Pursh) Peterm.	Beardless Barnyardgrass
旱稗	*Echinochloa glabrescens* Munro ex Hook. f. *hispidula* (Retz.) Nees	Dryland Barnyardgrass
牛筋草(蟋蟀草)	*Eleusine indica* (L.) Gaertn.	Yard Grass
鲫鱼草(星星草)	*Eragrostis amabilis* (L.) Wight et Arn.	Feather Lovegrass
鼠妇草	*Eragrostis atrovirens* (Desf.) Trin. ex Steud.	Thalia Lovegrass

卡氏画眉草	*Eragrostis chariis* (Schult.) Hitchc.	Thalia Lovegrass
短穗画眉草	*Eragrostis cylindrica* (Roxb.) Nees ex Hook. et Arn.	Knotted Lovegrass
知风草(知风画眉草)	*Eragrostis ferruginea* (Thunb.) P. Beauv.	Korean Lovegrass
乱草(旱田草)	*Eragrostis japonica* (Thunb.) Trin.	Japanese Lovegrass
长穗画眉草	*Eragrostis longispicula* S. C. Sun et H. Q. Wang	Longspike Lovegrass
*小画眉草	*Eragrostis minor* Host. [*E. poaeoides* Beauv.]	Small Lovegrass
*华南画眉草	*Eragrostis nevinii* Hance	Nevin Lovegrass
宿根画眉草	*Eragrostis perennans* Keng	Perennial Lovegrass
疏穗画眉草	*Eragrostis perlaxa* Keng	Loosespike Lovegrass
画眉草	*Eragrostis pilosa* (L.) Beauv.	India Lovegrass
多毛画眉草(多毛知风草)	*Eragrostis pilosissima* Link	Pilose Lovegrass
扭枝画眉草	*Eragrostis reflexa* Hack.	Flex Lovegrass
鲫鱼草	*Eragrostis tenella* (L.) P. Beauv. ex Roem. et Schult.	Japanese Lovegrass
灰穗画眉草	*Eragrostis tephrosanthus* Schult.	Ashygreyflower Lovergrass
牛虱草	*Eragrostis unioloides* (Retz.) Nees ex Steud.	Chinese Lovegrass
长画眉草(长穗画眉草)	*Eragrostis zeylanica* Nees et Mey.	Elongated Lovegrass
蜈蚣草(娱蛤草)	*Eremochloa ciliaris* (L.) Merr.	Ciliate Centipede-grass

被子植物 ANGIOSPERMAE

假俭草(小牛鞭草)	*Eremochloa ophiuroides* (Munro) Hack.	Smooth Lawn Grass
高野黍	*Eremochloa procera* (Retz) Hubb.	Cupgrass
鹧鸪草	*Eriachne pallescens* R. Br.	Common Eriachine
台湾蔗茅(紫台蔗茅)	*Erianthus formosanus* Stapf	Taiwan Plumegrass
四脉金茅(羊茅)	*Eulalia quadrinervis* (Hack.) O. Kuntze	Four-veined Eulalia
金茅(小颖羊茅)	*Eulalia speciosa* (Debeaux) O. Kuntze	Golden-sheathed Eulalia
假蛇尾草	*Heteropholis cochinchinensis* (Lour.) C. E. Hubb.	Vietnam Fakesnaketailgrass
黄茅(扭黄茅)	*Heteropogon contortus* (L.) Beauv. ex Roem. et Schult.	Yellow Grass
*大麦	*Hordeum vulgare* L.	Barley
弊草	*Hymenachne assamica* (Hook. f.) Hitchc.	Assam Hymenachne
长耳膜稃草(硬骨草)	*Hymenachne aurita* (Presl) Balansa	Longear Water Hymenachne
距花黍	*Ichnanthus vicinus* (F. M. Bailey) Merr.	Scargrass
白茅(茅草)	*Imperata cylindrica* (L.) Beauv. var. *major* (Nees) C. E. Hubb.	Wooly Grass
白花柳叶箬(中华淡竹叶)	*Isachne albens* Trin.	Nepal Twinballgrass
二型柳叶箬(粉绿竹)	*Isachne dispar* Trin.	Menten Twinballgrass
柳叶箬	*Isachne globosa* (Thunb.) O. Kuntze	Globose Twinball Grass

有芒鸭嘴草(芒穗鸭嘴草)	*Ischaemum aristatum* L.	Awned Duck-beak
粗毛鸭嘴草	*Ischaemum barbatum* Retz.	Bearded Duck-beak
细毛鸭嘴草(印度鸭嘴草)	*Ischaemum indicum* (Houtt.) Merr. [*I. ciliare* Retz.]	India Duck-beak
田间鸭嘴草	*Ischaemum rugosum* Salisb.	Wrinkled Duck Beak
六蕊假稻	*Leersia hexandra* Swartz	Bareet Grass
千金子	*Leptochloa chinensis* (L.) Nees	Field Grass
矶子草	*Leptochloa panicea* (Retz.) Ohwi	Thread Sprangletop
淡竹叶	*Lophatherum gracile* Brongn.	Sasagrass
刚莠竹	*Microstegium ciliatum* (Trin.) A. Camus	Ciliate Sasagrass
莠竹	*Microstegium nodosum* (Kom.) Trvel.	Beardless Microstegium
蔓生莠竹(蔓生莠草)	*Microstegium vagans* (Nees ex Steud.) A. Camus	Vagabondage Microstegium
柔枝莠竹	*Microstegium vimineum* (Trin.) A. Camus	Vimineous Microstegium
五节芒	*Miscanthus floridulus* (Labill.) Warb. ex Suhum. et Lauterb.	Japanese Silbergrass
*荻芦	*Miscanthus sacchariflorus* (Maxim.) Benth.	Amur Silvergrass
芒	*Miscanthus sinensis* Anderss.	Chinese Silbergrass
山鸡谷草	*Neohusnotia tonkinensis* A. Camus	Tonkin Neohusnotia
望冬草	*Neyraudia arundinacea* (L.) Henrard	Madagascar Grass

被子植物 ANGIOSPERMAE

山类芦	*Neyraudia montana* Keng	Montain Burmareed
类芦(假芦)	*Neyraudia reynaudiana* (Kunth) Keng ex A. S. Hitchc.	Burmareed
蛇尾草	*Ophiuros exaltatus* (L.) O. Kuntze	Exaltated Snaketailgrass
竹叶草(大缩箬草)	*Oplismenus compositus* (L.) Beauv.	Composite Oplismenus
中间型竹叶草	*Oplismenus compositus* (L.) Beauv. var. *intermedius* (Honda) Ohwi	Running Montaingrass
求米草	*Oplismenus undulatifolius* (Ard.) Roem. et Schult.	Undulateleaf Oplismenus
水稻	*Oryza sativa* L.	Rice
小花露籽草	*Ottochloa malabarica* (L.) Dandy	Malabar Ottochloa
露子草(奥图草)	*Ottochloa nodosa* (Kunth) Dandy	Common Ottochloa
短叶黍(狼尾草)	*Panicum brevifolium* L.	Panic Grass
藤竹草	*Panicum incomtum* Trin.	Vinebamboo Panicgrass
大黍(羊草)	*Panicum maximum* Jacq.	Guinea Grass
心叶稷(山黍)	*Panicum notatum* Retz.	Vine Panic Grass
水生黍	*Panicum dichotomiforum* Michaux [*P. paludosum* Roxb.]	Aquatic Panic Grass
铺地黍	*Panicum repens* L.	Panic Grass
两耳草	*Paspalum conjugatum* Berg.	Hilo Grass
毛花雀稗	*Paspalum dilatatum* Poir.	Hairflower Dallisgrass

台湾雀稗	*Paspalum hirsutum* Retz. [*P. fomosanum* Honda]	Taiwan Dallisgrass
*雀稗(鲫鱼草)	*Paspalum thunbergii* Kunth ex Steud.	India Paspalum
圆果雀稗	*Paspalum scrobiculatum* L. var. *orbiculare* (G. Forst.) Hack. [*P. orbiculare* G. Forst.]	Ditch Millet
双穗雀稗	*Paspalum distichum* L. [*P. paspaloides* (Michx.) Scribn.]	Knotgrass
海滨雀稗(夏威夷草)	*Paspalum vaginatum* Sw.	Ditch Millet
狼尾草	*Pennisetum alopecuroides* (L.) Spreng.	Plume Grass
多穗狼尾草	*Pennisetum polystachyon* (L.) Schult.	Mission Grass
*象草(紫狼尾草)	*Pennisetum purpureum* Schum.	Napier Grass
茅根	*Perotis indica* (L.) O. Kuntze	India Perotis
芦苇(芦根)	*Phragmites australis* Trin. ex Steud.	Common Reedgrass
卡开芦(过江卢荻)	*Phragmites karka* (Retz.) Trin. ex Steud.	Tall Reed
金丝草(金丝茅)	*Pogonatherum crinitum* (Thunb.) Kunth	Golden-hair Grass
*红毛草	*Rhynchelytrum repens* (Willd.) Hubb.	Rose Natal Grass
筒轴草(筒轴茅)	*Rottboellia exaltata* (L.) L. f.	Rottboell's Grass
斑茅(大密)	*Saccharum arundinaceum* Retz.	Reed-like Sugarcane

甘蔗(果蔗)	*Saccharum officinarum* L.	Sugar Cane
甜根子草	*Saccharum spontaneum* L.	Wild Cane
囊颖草	*Sacciolep indica* (L.) A. Chase	Panic Grass
鼠尾囊颖草	*Sacciolep myosuroides* (R. Br.) A. Gamus	Mousetail Cupscale
短叶裂稃草(黄背草)	*Schizachyrium brevifolium* (Sw.) Nees ex Buse	Poverty Grass
红裂稃草(红稃草)	*Schizachyrium sanguineum* (Retz.) Alston	Red Poverty Grass
莠狗尾草(狗尾草)	*Setaria geniculata* (Lam.) P. Beauv.	Perennial Fox-tail
金色狗尾草	*Setaria glauca* (L.) Beauv. [*S. pumila* (Poir.) Roem. et Schult.]	Yellow Bristlegrass
褐穗狗尾草	*Setaria pallidifusca* (Schum.) Stapf et C. E. Hubb.	Brownhair Bristlegrass
棕叶狗尾草(台风草)	*Setaria palmifolia* (Koen.) Stapf	Palm-grass
皱叶狗尾草	*Setaria plicta* (Lam.) T. Cooke	Wrinkledleaf Bristlegrass
狗尾草	*Setaria viridis* (L.) P. Beauv.	Green Fox-tail
光高粱(野高粱)	*Sorghum nitidum* (Vahl) Pers.	Glossy Wild Sorghum
稗荩(稃荩)	*Sphaerocaryum malaccense* (Trin.) Pilg.	Water Ball-fruit
鬣刺(滨刺草)	*Spinifex littoreus* (Burm. f.) Merr.	Littoral Spinegrass
鼠尾粟	*Sporobolus fertilis* (Steud.) W. D. Clayton	Australian Smut-grass

盐地鼠尾粟	*Sporobolus virginicus* (L.) Kunth	Seashore Dropseed
钝叶草	*Stenotaphrum helferi* Munro ex Hook. f.	Centipede Grass
苞子草(老虎须)	*Themeda caudata* (Nees) A. Camus	Tail Themeda
大菅	*Themeda gigantea* (Cav.) Hack.	Big Themeda
西南菅草(小菅草)	*Themeda hookeri* (Griseb.) A. Camus	Hooker Themeda
菅(野菅)	*Themeda villosa* (Poir.) A. Camus	Silky Kangaroo Grass
棕叶芦(棕叶芦)	*Thysanolaena maxima* (Roxb.) O. Kuntze	Tiger-grass
*香根草(岩兰草)	*Vetiveria zizanioides* (L.) Nash	Aromaticroot
玉米王(包米)	*Zea mays* L.	Corn
*结缕草(日本结缕草)	*Zoysia japonica* Steud.	Korean Lawngrass
马尼拉草(半细叶结缕草)	*Zoysia matrella* (L.) Merr.	Manila Grass
中华结缕草(青岛结缕草)	*Zoysia sinica* Hance	Chinese Grass
*台湾草(细叶结缕草)	*Zoysia tenuifolia* Willd.	Mascarene Grass
*粉单竹(白粉单竹)	*Bambusa chungii* McClure	White Powdery Bamboo
*小簕竹	*Bambusa flexuosa* Munro	Lesser Thorny Bamboo
*坭竹(水黄竹)	*Bambusa gibba* McClure	Nai Bamboo
油簕竹(烂眼竹)	*Bambusa lapidea* McClure	Bambusa Lapidea

*凤尾竹(观音竹)	*Bambusa multiplex* (Lour.) Raeusch. ex J. A. et J. H. Schult. 'Fernleaf'	Fernleaf Hedge Bamboo
*观音竹(凤尾竹)	*Bambusa multiplex* (Lour.) Raeusch. ex J. A. et J. H. Schult. var. riviereorum R. Maire	Chinese Goddess Bamboo
水竹(大眼竹)	*Bambusa eutuldoides* McClure	Da Yan Bamboo
*绿竹(甜竹)	*Bambusa oldhami* Munro	Green Bamboo
撑篙竹(泥竹)	*Bambusa pervariablilis* McClure	Punting Pole Bamboo
*车筒竹(水簕竹)	*Bambusa sinospinosa* McClure	Chinese Thorny Bamboo
青皮竹(篾竹)	*Bambusa textilis* McClure	Weaver's Bamboo
青秆竹(水竹)	*Bambusa tuldoides* Munro	Verdant Bamboo
*佛肚竹(龙头竹)	*Bambusa vulgaris* Schrad. ex Wendl. 'Wamin'	Buddha Bamboo
*黄金间碧竹(挂绿竹)	*Bambusa vulgaris* Schrad. ex Wendl. 'Vittata'	Stripe Bamboo
大头典竹(大头竹)	*Dendrocalamopsis beecheyana* (Munro) Keng f. var. *pubescens* (P. F. Li) Keng f.	Common Bamboo
麻竹(甜竹)	*Dendrocalamus latiflorus* Munro	Broadflower Dragonbamboo
棕巴箬竹(光箨箬竹)	*Indocalamus herklotsii* McClure	Herklots Cane
箬叶竹(箬竹)	*Indocalamus longiauritus* Hand.-Mazz.	Long-ear Cane
*箬竹(楣竹)	*Indocalamus tessSelatus* (Munro) Keng f.	Chequer-shape Indocalamus
人面竹(罗汉竹)	*Phyllostachys aurea* Carr. ex Cam. et Riv.	Fishpole Bamboo

篌竹	*Phyllostachys nidularia* Munro	Broom Bamboo
*紫竹(黑竹)	*Phyllostachys nigra* (Lodd. ex Lindl.) Munro	Black Bamboo
*苦竹	*Pleioblastus amarus* (Keng) Keng f. [*P. simonii* (Carr.) Nakai]	Bitterbamboo
*茶秆竹(青篱竹)	*Pseudosasa amabilis* (McClure) Keng f.	Lovable Pseudosasa
托竹(篱竹)	*Pseudosasa cantorii* (Munro) Keng f. [*Arundinaria cantori* (Munro) Chia]	Cantor Pseudosasa
箸竹(寒山竹)	*Pseudosasa hindsii* (McClure) C. D. Chu et C. S. Chao	Hinds Pseudosasa
苗笋竹(薄竹)	*Schizostachyum pseudolima* McClure	Si-lao Bamboo

索　引

中文名称

一画

一本芒　173
一串红　146
一枝黄花　124
一品红　63
一点红　121

二画

丁公藤　131
丁癸草　80
丁香杜鹃　102
丁香罗勒　145
丁香蓼　44
七爪龙　131
七叶一枝花　156
七层楼　112
七里明　119
九丁树　86
九节　116
九里香　95
了哥王　44
二乔木兰　21
二列叶柃　51
二列虾脊兰　169
二色波罗蜜　84
二形卷柏　1
二花珍珠茅　177
二歧鹿角蕨　15
二型柳叶箬　183
二脊沼兰　172

人心果　104
人面子　99
人面竹　189
八仙花　66
八角枫　100
八角枫科　100
八角金盘　100
八角科　23
八角莲　29
十字花科　33
十字苔草　173
十字架树　136
十字爵床　137
十蕊大参　101
十蕊槭　97

三画

万年青　156
万年麻　163
万寿竹　154
万寿菊　124
万点星　36
丈野古草　179
三七草　122
三叉苦　94
三叉蕨　13
三叉蕨科　12
三分丹　112
三叶木蓝　77
三叶爬墙虎　93
三叶鬼针草　119

三叶崖爬藤　93
三白草　32
三白草科　32
三尖杉　19
三尖杉科　19
三羽新月蕨　9
三色苋　41
三花冬青　89
三角草　154
三角椰子　165
三角槭　97
三点金　76
三相蕨　12
三药槟榔　164
三裂山矾　106
三裂叶牵牛　131
三裂叶野葛　79
三裂蟛蜞菊　125
三蕊沟繁缕　37
三褶虾脊兰　169
下田菊　118
下延三叉蕨　12
下延沙皮蕨　12
千日红　41
千年健　159
千里光　124
千屈菜　43
千屈菜科　42
千果榄仁　55
千金子　184
千金藤　30

三 画

千根草　63
卫矛科　89
卫矛蒲桃　53
土丁桂　131
土人参　38
土牛膝　40
土田七　152
土沉香　44
土茯苓　157
土荆芥　40
土密树　62
土密藤　61
大马蓼　39
大王椰子　167
大王黛粉叶　158
大风子科　46
大叶千斤拔　77
大叶山蚂蝗　75
大叶山棟　96
大叶马兜铃　31
大叶木莲　22
大叶仙茅　167
大叶田繁缕　37
大叶白纸扇　115
大叶石上莲　135
大叶华南云实　72
大叶肉托果　99
大叶罗汉松　19
大叶苔草　173
大叶南洋杉　19
大叶相思　69
大叶骨碎补　13
大叶拿身草　76
大叶桂樱　67
大叶桃花心木　96
大叶桉　53
大叶臭花椒　95

大叶蛇葡萄　92
大叶紫珠　140
大叶落地生根　36
大叶榕　87
大叶算盘子　63
大叶蔓绿绒　159
大头艾纳香　119
大头典竹　189
大头茶　51
大头菜　34
大头橐吾　123
大羽铁角蕨　10
大芒萁　3
大血藤　30
大血藤科　30
大丽花　120
大吴风菊　122
大尾摇　128
大花三色堇　35
大花五桠果　45
大花安息香　105
大花老鸦嘴　139
大花君子兰　161
大花忍冬　117
大花软枝黄蝉　108
大花栀子　113
大花美人蕉　153
大花惠兰　170
大花犀角　111
大花紫薇　43
大花酢浆草　42
大豆　77
大驳骨　138
大麦　183
大岩桐　135
大果木莲　22

大果水竹叶　148
大果油麻藤　78
大果榕　85
大波斯菊　120
大罗伞树　104
大苞水竹叶　148
大苞白山茶　50
大苞赤瓟　48
大苞鸭跖草　148
大青　141
大青薯　162
大型双子铁　17
大茶药　107
大盖球子草　155
大菅　188
大戟科　60
大琴叶榕　86
大黍　185
大黑桫椤　5
大藻　159
女贞　108
小二仙草　44
小二仙草科　44
小刀豆　74
小毛蓼　39
小仙丹花　115
小叶三点金　76
小叶山龙眼　45
小叶乌药　25
小叶云实　72
小叶五月茶　61
小叶巴戟天　115
小叶买麻藤　20
小叶冷水花　88
小叶谷精草　149
小叶爬崖香　32
小叶青冈　82

三　画

小叶草海桐　128
小叶海金沙　4
小叶寄树兰　172
小叶黄花稔　59
小叶蛙儿藤　112
小叶黑面神　61
小叶榄仁　55
小叶榕　86
小回回蒜　28
小米椒　128
小红栲　82
小舌菊　123
小丽草　180
小花龙血树　163
小花灯心草　173
小花扁担杆　56
小花柏拉木　54
小花蜘蛛抱蛋　154
小花露籽草　185
小驳骨　138
小果十大功劳　29
小果叶下珠　65
小果石笔木　52
小果皂荚　73
小果柿　103
小果草　133
小果香椿　96
小果铁冬青　89
小果菝葜　157
小果野蕉　150
小果葡萄　93
小果蕗蕨　4
小画眉草　182
小苞黄脉爵床　139
小金梅草　168
小鱼仙草　145
小型月季　68

小牵牛　131
小茴香　102
小韭兰　161
小盘木　66
小盘木科　66
小粒咖啡树　113
小槐花山蚂蟥　75
小蓟　120
小翠云　2
小蓼花　40
小蜡树　108
小酸浆　129
小檗科　29
小籁竹　188
小藜　40
山小橘　95
山乌桕　65
山木通　27
山玉兰　21
山白菊　118
山石榴　113
山龙眼科　45
山合欢　70
山芝麻　58
山杜英　57
山牡荆　143
山苍子　25
山鸡谷草　184
山麦冬　155
山枇杷　67
山油麻　84
山矾　107
山矾科　106
山茄　130
山姜　152
山柑子科　90
山柑藤　90

山类芦　185
山胡椒　25
山茱萸科　99
山茶花　51
山香　144
山香圆　98
山莴苣　123
山猪菜　132
山绿豆　80
山绿柴　91
山菅兰　154
山麻杆　60
山黄菊　118
山黄麻　84
山椒子　24
山羡叶泡花树　98
山葛藤　79
山楝　96
山榄科　103
山稗子　173
山蒟　32
山蒲桃　54
山蓼　39
山橘　94
山橙　109
山薯　162
山蟛蜞菊　125
川鄂栲　82
广东万年青　157
广东山龙眼　45
广东书带蕨　8
广东木瓜红　105
广东水马齿　44
广东半蒴苣苔　135
广东玉叶金花　115
广东石豆兰　169
广东羊蹄甲　71

三画 — 四画

广东赤桐　141
广东沿阶草　155
广东金钱草　76
广东相思子　73
广东绣线菊　69
广东茘柊　46
广东蛇菰　91
广东蛇葡萄　92
广东黄杞　99
广东紫珠　140
广东隔距兰　169
广东蓧菜　34
广东蔷薇　68
广州假卫矛　89
广州蛇根草　116
广州槌果藤　33
广西木姜子　26
广西新木姜　27
广防己　31
广防风　143
广花耳草　114
广寄生　90
广藿香　146
弓果黍　180
弓果藤　111
弓葵　164
飞龙掌血　95
飞扬草　63
飞机草　121
飞燕草　28
马兰　123
马占相思　69
马尼拉草　188
马尼拉椰子　167
马尼拉榄仁　55
马甲子　91
马关木莲　22

马利筋　110
马尾松　18
马拉巴栗　58
马松子　58
马齿苋　38
马齿苋树　38
马齿苋科　38
马峦梅　97
马唐　181
马钱科　107
马铃薯　130
马兜铃科　31
马樱丹　142
马蹄　175
马蹄参　100
马蹄金　130
马蹄莲　160
马蹄犁头尖　160
马鞭草　143
马鞭草科　139

四画

不夜城芦荟　153
中华三宝木　66
中华卫矛　89
中华双扇蕨　14
中华地桃花　60
中华芒毛苣苔　134
中华杜英　57
中华里白　3
中华刺蕨　13
中华苔草　173
中华复叶耳蕨　11
中华结缕草　188
中华胡枝子　77
中华润楠　26
中华绣线菊　69

中华旌节花　80
中华黄花稔　60
中华锥　82
中华锥花　144
中华赛爵床　137
中华鳞毛蕨　11
中间型竹叶草　185
中国无忧花　73
中国水仙　161
中国蕨科　7
中南鱼藤　75
中型沿阶草　155
丰花草　113
丰城崖豆藤　77
丹桂　108
乌口树　117
乌毛蕨　10
乌毛蕨科　10
乌材　103
乌饭树　103
乌药　25
乌桕　65
乌榄　96
乌蔹梅　93
乌墨　53
乌蕨　6
乌檀　116
书带蕨科　8
云叶兰　172
云实　72
云南含笑　23
云南拟单性木兰　23
云南黄素馨　107
五月茶　61
五爪金龙　131
五加　100
五加科　100

四画

五叶薯蓣 162	双片苣苔 135	心檐天南星 158
五节芒 184	双花耳草 114	文竹 154
五列木 52	双花假卫矛 89	文定果 56
五列木科 52	双花蟛蜞菊 125	文殊兰 161
五味子科 23	双荚决明 72	方叶五月茶 61
五指草 181	双扇蕨科 14	方茎耳草 114
五星花 116	双盖蕨 8	无毛小果叶下珠 65
五桠果 45	双穗雀稗 186	无叶美冠兰 171
五桠果科 45	双穗飘拂草 176	无芒雀稗 181
五彩马缨丹 142	反瓣虾脊兰 169	无花果 85
五蕊寄生 90	天门冬 153	无根萍 160
井栏边草 6	天仙果 85	无根藤 24
介蕨 8	天竺桂 24	无患子 97
元江苏铁 17	天竺葵 41	无患子科 96
六月雪 117	天轮柱 49	无梗艾纳香 119
六角莲 29	天南星 158	无盖鳞毛蕨 12
六棱菊 123	天南星科 157	无瓣蔊菜 34
六蕊假稻 184	天星藤 111	日本女贞 108
内伶丁双盖蕨 8	天胡荽 102	日本五月茶 61
内伶仃秤钩风 30	天香藤 70	日本石竹 37
凤仙花 42	天料木 46	日本杜英 57
凤仙花科 42	天料木科 46	日本扁柏 18
凤头黍 178	孔雀木 101	日本菟丝子 130
凤尾丝兰 164	孔雀肖竹芋 153	日本蛇根草 116
凤尾竹 189	孔雀草 125	月季 68
凤尾鸡冠 41	孔雀稗 181	木兰科 21
凤尾蕨 7	少花龙葵 129	木瓜 49
凤尾蕨科 6	少花狸藻 134	木龙葵 130
凤梨 149	少穗飘拂草 176	木防己 30
凤梨科 149	巴西含羞草 71	木芙蓉 59
凤凰木 72	巴西鸢尾 162	木豆 74
凤凰菜 122	巴西橡胶树 64	木油桐 66
凤眼莲 156	巴西蟹爪 50	木波罗 84
分叉露兜簕角 167	巴豆 62	木姜叶柯 83
切边铁角蕨 10	心叶黄花稔 60	木荷 51
升马唐 181	心叶稷 185	木莓 69
双子叶植物 21	心萼薯 130	木莲 21

四 画

木贼科 2	毛茄 129	水马齿 44
木通科 29	毛鱼藤 75	水马齿科 44
木麻黄 83	毛柄短肠蕨 8	水车前 147
木麻黄科 83	毛柱铁线莲 27	水东哥 52
木麻槿 59	毛柱隔距兰 169	水冬哥科 52
木棉 58	毛柿 103	水玉簪科 168
木棉科 58	毛相思子 73	水瓜 47
木犀科 107	毛茛 28	水生黍 185
木榄 55	毛茛科 27	水田白裸茎 107
木槿 59	毛茶 112	水石梓 104
木蝴蝶 136	毛草龙 43	水石榕 57
木薯 65	毛药花 54	水龙骨科 14
木鳖 47	毛轴芽蕨 12	水同木 85
木鳖子 48	毛轴莎草 174	水团花 112
木麒麟 50	毛轴铁角蕨 10	水竹 189
止泻木 109	毛轴假蹄盖蕨 8	水竹芋 153
毛八角枫 100	毛轴碎米蕨 7	水杉 18
毛大丁草 122	毛桂 24	水芹 102
毛山猪菜 131	毛桃木莲 22	水苎麻 87
毛马唐 181	毛莪术 152	水松 18
毛冬青 88	毛堇菜 35	水苦荬 134
毛叶杜鹃 102	毛排钱草 79	水茄 130
毛叶肾蕨 13	毛脚金星蕨 9	水虱草 175
毛叶轮环藤 30	毛野古草 179	水栀子 113
毛叶嘉赐树 46	毛麻楝 96	水珍珠菜 145
毛白鹤藤 130	毛萼口红花 134	水烛 160
毛节野古草 178	毛萼清风藤 98	水翁 53
毛杨桐 50	毛稔 54	水蛇麻 85
毛花猕猴桃 52	毛颖草 178	水黄皮 79
毛花雀稗 186	毛蓼 39	水塔花 149
毛鸡屎藤 116	毛蕨 9	水晶花烛 158
毛果巴豆 62	毛臂形草 179	水晶掌 155
毛果网子草 148	毛瓣莎 176	水筛 146
毛果青冈 82	毛鳞省藤 164	水葱 177
毛果枧 51	毛麝香 132	水蒲桃 53
毛果珍珠茅 178	气球果 111	水蓑衣 138
毛果算盘子 63	水飞蓟 124	水锦树 117

水蓼 39
水蔗草 178
水稻 185
水蕨 7
水蕨科 7
水蕹 147
水蕹科 147
水鳖科 146
火鸟蕉 151
火炭母 39
火烧花 136
火棘 68
火焰木 136
牛大力藤 78
牛毛毡 175
牛白藤 114
牛皮消 111
牛矢果 108
牛耳朵 135
牛耳枫 66
牛虱草 182
牛轭草 148
牛齿兰 168
牛栓藤 99
牛栓藤科 99
牛眼马钱 107
牛筋草 181
牛筋藤 87
牛蒡 118
牛膝 40
牛蹄豆 71
牛繁缕 37
牛藤果 29
王瓜 48
瓦韦 14
见血青 171
贝叶棕 165

贝壳杉 19
车轮梅 68
车前草 127
车前草科 127
车桑子坡柳 97
车筒竹 189
长毛山矾 106
长叶木兰 21
长叶竹柏 19
长叶刺葵 166
长叶实蕨 13
长叶肾蕨 13
长叶变叶木 62
长叶柞木 46
长叶瓶子草 37
长叶铁角蕨 10
长叶巢蕨 10
长叶鹿角蕨 16
长节耳草 114
长羽萝卜 34
长耳膜稃草 183
长尾毛蕊茶 50
长花马唐 181
长花忍冬 117
长花厚壳树 128
长画眉草 182
长剑叶蓼 39
长春花 109
长柄杜英 57
长柄果飘拂草 175
长柄野扁豆 76
长柄银叶树 58
长钩刺蒴麻 56
长圆叶卫矛 89
长圆叶艾纳香 119
长梗黄花稔 60
长萼堇菜 35

长萼猪屎豆 74
长蒴母草 133
长穗画眉草 182
风车子 55
风车草 174
风车藤 60
风轮菜 143
风信子 155

五画

东方水丝梨 81
东方泽泻 147
东方狗脊蕨 11
东方荚果蕨 10
东风菜 121
东京油楠 73
丝兰 164
丝叶球柱草 173
丝瓜 47
丝葵 167
乐东拟单性木兰 23
乐昌含笑 22
仔榄树 109
仙人球 49
仙人掌 49
仙人掌科 49
仙茅 167
仙茅竹茎兰 172
仙茅科 167
仙客来 126
仙湖苏铁 17
令箭荷花 49
仪花 73
兰花美人蕉 153
兰花草 138
兰科 168
冬凤兰 170

五　画

冬瓜　47
冬红　142
冬青科　88
冬樱花　67
凸尖野百合　74
凹叶红豆　78
凹叶厚朴　21
加拿大蓬　120
北江荛花　44
北美一枝黄花　124
北美鹅掌楸　21
北清香藤　107
半支莲　38
半月铁线蕨　7
半边莲　127
半边莲科　127
半边旗　7
半枝莲　146
半柱毛兰　170
半荷枫　81
卡开芦　186
卡氏画眉草　182
可爱花　137
台湾火筒树　93
台湾苏铁　17
台湾虎尾草　180
台湾青枣　92
台湾相思　69
台湾草　188
台湾栾树　97
台湾雀稗　186
台湾榕　85
台湾蔗茅　183
叶下珠　65
叶子花　45
叶状苞飘拂草　176
四子马蓝　139

四叶萝芙木　110
四生臂形草　179
四块瓦　32
四季米仔兰　96
四季秋海棠　49
四脉金茅　183
四棱豆　79
四棱飘拂草　176
对叶百部　162
对叶榕　85
对茎毛兰　171
尼泊尔鼠李　92
平滑弓果藤　111
打铁树　105
母草　133
玄参科　132
玉兰　21
玉叶金花　116
玉米王　188
玉树　36
玉簪　155
瓜叶菊　124
瓜馥木　24
甘木通　27
甘葛藤　79
甘蓝　34
甘蔗　187
生菜　123
田字草　16
田间鸭嘴草　184
田基麻　128
田基麻科　128
田基黄　56
田菁　79
田葱　168
田葱科　168
田蒿草　148

田繁缕　37
白千层　53
白马骨　117
白木通　29
白毛紫珠　139
白丝龙舌兰　163
白兰　22
白叶瓜馥木　23
白叶藤　111
白头婆　121
白瓜　47
白龙船花　115
白网纹草　138
白肉黄果榕　87
白舌紫菀　118
白纹草　154
白花丹　127
白花丹科　126
白花白酒草　120
白花地胆草　121
白花异木棉　58
白花灯笼草　141
白花羊蹄甲　71
白花杜鹃　103
白花油麻藤　78
白花鱼藤　75
白花柳叶箬　183
白花鬼针草　119
白花悬钩子　68
白花笼　105
白花菜科　33
白花蛇舌草　114
白花酸藤子　105
白饭树　66
白果香楠　112
白苞蒿　118
白英　130

五画 — 六画

白茅 183
白籽菜 122
白背叶 64
白背盐肤木 99
白背黄花稔 60
白背算盘子 64
白药谷精草 149
白桂木 84
白桐树 62
白粉藤 93
白被穗花凤梨 150
白接骨 137
白绵毛兰 171
白菊仔 118
白菜绍菜 34
白雪亮丝草 157
白掌 160
白斑亮丝草 157
白晶菊 119
白棠子树 140
白楸 64
白瑞香 44
白睡莲 28
白鼓钉 38
白蜡树 107
白颜树 83
白鹤芋 159
白鹤藤 130
白薯莨 162
白檀 106
白簕花 100
白藤 164
白蟾 113
白鳞莎草 174
矛叶荩草 178
石刁柏 154
石上莲 135

石子藤 1
石山苏铁 17
石韦 15
石仙桃 172
石龙芮 28
石吊兰苣苔 135
石竹 37
石竹科 37
石芒草 179
石血 110
石杉科 1
石岩枫 65
石松 1
石果珍珠茅 178
石柑子 159
石胡荽 119
石茅荸 145
石栗 61
石海椒 41
石笔木 52
石莲花 36
石斛 170
石菖莆 157
石萝藦 111
石碌含笑 22
石蒜 161
石蒜科 160
石榴 43
石蝉草 31
禾叶土麦冬 155
禾叶挖耳草 134
禾叶蕨科 16
禾本科 178
禾串树 61
艾纳香 119
艾堇 65
艾蒿 118

节瓜 47
节节红 119
节节草 2
节节菜 43
边缘鳞盖蕨 5
鸟喙丝兰 164
龙爪茅 180
龙爪槐 79
龙牙花 76
龙吐珠 142
龙舌兰 162
龙舌兰科 162
龙柏 19
龙胆科 126
龙须藤 71
龙珠果 46
龙脑香科 52
龙眼 97
龙脷叶 65
龙船花 115
龙棕 167
龙葵 130

六画

买麻藤科 20
亚马孙王莲 29
亚麻科 41
交让木 66
交让木科 66
伊兰芷硬胶 104
伏石蕨 14
伞形花科 101
伞花马钱 107
伞花木 97
伞花鱼黄草 132
伞房花耳草 114
光山香圆 98

六　画

光叶子花　45
光叶山矾　106
光叶山黄麻　84
光叶火筒树　93
光叶仙茅　167
光叶白颜树　83
光叶羊蹄甲　71
光叶海桐　45
光叶紫玉盘　24
光叶蔷薇　68
光叶蝴蝶草　134
光头稗　181
光里白　3
光枝勾儿茶　91
光亮山矾　106
光高粱　187
光棍树　63
光滑大丽花　120
光滑悬钩子　69
光萼猪屎豆　75
光蓼　39
全叶芥　33
全叶美丽芙蓉　59
全缘叶紫珠　140
全缘叶澳洲坚果　45
全缘贯众　11
全缘榕　86
全缘樱桃　67
决明　72
列当科　134
刚毛木蓝　77
刚莠竹　184
华三芒草　178
华山矾　106
华山姜　151
华马钱　107
华中瘤足蕨　3

华凤仙　42
华东膜蕨　4
华肖菝葜　156
华刺子莞　177
华泽兰　121
华青牛胆　30
华南十大功劳　29
华南山胡椒　25
华南马尾杉　1
华南云实　71
华南五针松　18
华南天料木　46
华南毛蕨　51
华南毛蕨　9
华南长筒蕨　4
华南龙胆　126
华南朴　83
华南舌蕨　13
华南忍冬　117
华南条蕨　13
华南皂荚　73
华南谷精草　149
华南赤车　87
华南夜来香　111
华南实蕨　13
华南画眉草　182
华南青皮木　90
华南省藤　164
华南胡椒　32
华南骨碎补　13
华南铁角蕨　10
华南假脉蕨　4
华南紫萁　3
华南鳞毛蕨　12
华南麟盖蕨　5
华素馨　108
华湖瓜草　176

华紫珠　140
华鼠尾草　146
印度田菁　79
印度红睡莲　28
合丝肖菝葜　156
合欢　70
合果木　23
合果芋　160
吉贝　58
吉祥草　156
吊兰　154
吊瓜树　136
吊皮锥　82
吊灯扶桑　59
吊竹梅　149
吊钟花　102
吊裙草　74
吐烟花　87
向日葵　122
吕宋糖棕　165
回回苏　145
团叶麟始蕨　5
团扇蕨　4
地阳桃　65
地果　86
地胆草　121
地桃花　60
地涌金莲　151
地毯草　179
地稔　54
地锦　63
地锦　93
多毛马齿苋　38
多毛画眉草　182
多毛茜草树　112
多羽瘤蕨　15
多序楼梯草　87

六 画

多花山竹子　56
多花山蚂蝗　76
多花勾儿茶　91
多花瓜馥木　24
多花报春　126
多花脆兰　168
多花野牡丹　54
多花黄精　155
多花蔷薇　68
多刺山黄皮　113
多枝臂形草　179
多茎鼠曲草　122
多型栝楼　48
多脉润楠　26
多脉假蕨　4
多脉酸藤子　105
多荚草　38
多裂棕竹　166
多穗金粟兰　32
多穗扁莎　177
多穗狼尾草　186
多穗莎草　174
夹竹桃　109
夹竹桃科　108
安石榴科　43
安息香科　105
尖山橙　109
尖叶木犀榄　108
尖叶杜英　57
尖叶唐松草　28
尖叶清风藤　98
尖头瓶尔小草　2
尖尾芋　157
尖脉木姜子　25
尖啄隔距兰　169
尖萼山猪菜　131
延龄草科　156

异叶山蚂蝗　76
异叶双唇蕨　5
异叶木犀榄　108
异叶爬墙虎　93
异叶南洋杉　19
异叶榕　85
异芒菊　119
异果毛蕨　9
异型马唐　181
异型莎草　174
异蕊草　156
异穗卷柏　1
当归藤　105
托竹　190
曲枝假蓝　138
曲轴海金沙　4
有芒鸭嘴草　184
有翅星蕨　15
朱顶红　161
朱砂根　104
朱蕉　163
朴树　83
汝蕨　12
江西悬钩子　68
江南灯心草　173
江南卷柏　2
江南星蕨　15
池杉　18
灯心草　173
灯心草科　173
灰山矾　106
灰木莲　21
灰毛大青　141
灰毛豆　80
灰毛猕猴桃　52
灰叶菝葜　156
灰莉　107

灰绿椒草　31
灰绿藜　40
灰缘耳蕨　12
灰穗画眉草　182
百子莲　160
百日菊　126
百合　155
百合竹　163
百合科　153
百部科　162
竹叶兰　168
竹叶青冈　82
竹叶草　185
竹叶眼子菜　147
竹叶楠　27
竹叶榕　86
竹节树　55
竹节秋海棠　49
竹节草　148
竹节草　180
竹节蓼　39
竹芋　153
竹芋科　153
竹枝细柄草　179
竹柏　19
米仔兰　96
米碎花　51
红千层　52
红大丽花　120
红马蹄草　102
红木　46
红木科　46
红毛丹　97
红毛禾叶蕨　16
红毛草　186
红车　54
红丝线　129

六 画

红丝线 138
红叶金花 115
红叶蔓绿绒 159
红叶藤 99
红白忍冬 117
红皮紫陵 102
红皮糙果茶 50
红龙吐珠 141
红龙船花 115
红色新月蕨 9
红血飘拂草 175
红尾翎 181
红花千日红 41
红花天料木 46
红花文殊兰 161
红花木莲 22
红花鸡蛋花 109
红花油茶 50
红花烟草 129
红花荷 81
红花寄生 90
红花葱兰 161
红花酢浆草 42
红花鼠尾草 146
红花檵木 81
红豆杉科 20
红豆蔻 151
红刺露兜树 167
红杯凤梨 150
红枝蒲桃 54
红泡刺 69
红柄椰 165
红柄蔓绿绒 159
红树科 55
红秋葵 59
红绒球 70
红背山麻杆 61

红背兔儿风 118
红背卧花竹芋 153
红背桂 63
红脉悬钩子 69
红脉葵 165
红草 40
红根草 126
红桑 60
红胶木 54
红淡比 51
红球姜 152
红绿草 41
红雀珊瑚 65
红紫珠 140
红裂稃草 187
红楠 26
红睡莲 28
红锥 82
红雾水葛 88
红蓼 40
红蝉花 109
红辣蓼 39
红蕉 150
红鹤芋 158
红麒麟 62
红鳞扁莎 177
纤花耳草 114
纤细苦荬菜 122
纤茎飘拂草 175
纤草 168
纤弱木贼 2
网纹花 138
网果珍珠茅 178
网脉山龙眼 45
网脉假卫矛 89
网脉琼楠 24
网球花 161

羊耳菊 122
羊舌树 106
羊角杜鹃 103
羊角坳 110
羊乳 127
羊乳榕 86
羊蹄甲 71
羽叶喜林芋 159
羽芒菊 125
羽衣甘蓝 34
羽裂圣蕨 9
羽穗砖子苗 177
老人葵 165
老鸦糊 140
老鸦嘴 139
老鼠矢 107
老鼠拉冬瓜 48
老鼠簕 137
耳叶柃 51
耳形瘤足蕨 3
耳苞鸭跖草 148
耳草 114
耳基水苋 42
耳基卷柏 2
肉实科 104
肉桂 24
舌柱麻 87
舌蕨科 13
芋 158
芒 184
芒萁 3
血见愁 146
血叶兰 171
血桐 64
西瓜 47
西瓜皮椒草 31
西南木荷 52

六画 — 七画

西南粗叶木　115
西南菅草　188
西南飘拂草　176
西洋菜　34
西番莲科　46
观光木　23
观音竹　189
观音座莲科　2
观赏瓜　47
观赏辣椒　128
过山枫　89
过江藤　142
过路黄　126
闭鞘姜　152
问荆　2
防己科　30
阳桃　42
阳荷　152
阴石蕨　13
阴地唐松草　28
阴香　24

七画

两广禾叶蕨　16
两广唇柱苣苔　135
两广梭罗　58
两广黄檀　75
两耳草　185
两色卡特兰　169
两歧飘拂草　175
两面针　95
串珠子　109
串钱柳　53
丽格秋海棠　48
乱草　182
伽蓝菜　36
何首乌　40

余甘子　65
佛手　94
佛手瓜　48
佛甲草　36
佛肚竹　189
佛肚花　64
克鲁兹王莲　29
冷水花　88
卤地菊　125
卤蕨　7
卤蕨科　7
卵叶半边莲　127
卵叶耳草　114
君达菜　40
含笑　22
含羞草　71
含羞草决明　72
含羞草科　69
含羞草黄檀　75
坚荚蒾　118
壳斗科　81
壳菜果　81
岗松　52
岗松　51
希茉莉　114
忍冬科　117
怀德柿　103
扭肚藤　107
扭枝画眉草　182
扭鞘香茅　180
折冠藤　111
报春花科　126
拟地皮消　138
拟杜英　57
拟金草　114
拟黄荆　143
旱田草　133

旱金莲　42
旱金莲科　42
旱稗　181
杉木　18
杉科　18
李　68
李氏女贞　108
杏叶沙参　127
杏叶香兔儿风　118
杜仲藤　110
杜英　57
杜英科　57
杜茎山　105
杜虹花　140
杜根藤　137
杜鹃花科　102
杠板归　40
杠柳科　112
条纹十二卷　155
条纹凤尾蕨　6
条裂三叉蕨　12
条裂伽蓝菜　36
条蕨科　13
条穗苔草　173
杧果　99
杨叶肖槿　60
杨柳科　81
杨梅　81
杨梅科　81
求米草　185
沉水樟　25
沙梨　68
沙漠玫瑰　108
沟繁缕科　37
灶地乌骨木　53
牡荆　143
皂荚　73

矾子草 184
秀柱花 80
纸莎草 174
肖黄栌 62
肖蒲桃 52
芙兰草 176
芙蓉菊 120
芥菜 33
芥蓝 33
芦竹 179
芦苇 186
芦荟 153
芫荽 102
芭蕉 150
芭蕉科 150
花叶万年青 158
花叶山姜 152
花叶开唇兰 168
花叶竹芋 153
花叶芋 158
花叶冷水花 88
花叶垂榕 85
花柱草 128
花柱草科 128
花烛 158
花椒簕 95
花蔺科 146
芳香石豆兰 168
芳香安息香 105
芸香 95
芸香科 94
芸香草 180
芹菜 102
苋科 40
苍白秤钩风 30
苍耳 125
苎麻 87

苏木 72
苏木科 71
苏铁 17
苏铁科 17
苏铁蕨 11
苣荬菜 124
补血草 126
角花乌蔹梅 93
角花胡颓子 92
谷木 55
谷木叶冬青 88
谷精草 149
谷精草科 149
豆角 80
豆梨 68
豆腐柴 142
豆薯沙葛 78
豆瓣绿 32
赤车 87
赤杨叶 105
赤唇石豆兰 168
赤桉 53
赤楠 53
足茎毛兰 171
还魂草 1
远志科 35
连钱草 144
里白 3
里白科 3
针刺草 137
阿江榄仁 55
韧荚红豆 78
饭包草 148
饭甑青冈 82
驳骨丹 107
鸡爪槭 97
鸡爪簕 116

鸡血藤 78
鸡屎藤 116
鸡柏胡颓子 92
鸡骨香 62
鸡眼草 77
鸡眼梅花草 37
鸡眼藤 115
鸡蛋花 109
鸡蛋果 46
鸡嘴簕 72
龟背竹 159

八画

乳茄 130
乳源木莲 22
使君子 55
使君子科 55
侧柏 18
兔耳兰 170
具芒碎米莎草 174
刺子莞 177
刺毛黧豆 78
刺叶非洲铁 17
刺头复叶耳蕨 11
刺瓜 111
刺田菁 79
刺芒野古草 179
刺芫荽 102
刺苋 41
刺果血桐 64
刺果苏木 71
刺果番荔枝 23
刺果藤 57
刺茄 129
刺齿凤尾蕨 6
刺齿叶泥花草 133
刺齿贯众 11

八　画

刺轴桐　166
刺桐　76
刺葵　166
刺蒴麻　56
刺蕨　13
单子叶植物　146
单牙狗脊蕨　11
单叶双盖蕨　8
单叶铁线莲　27
单叶新月蕨　9
单叶蔓荆　143
单色蝴蝶草　134
单花六道木　117
单室鱼木　33
单穗水蜈蚣　176
单穗金粟兰　32
单穗鱼尾葵　165
卷丹　155
卷柏　2
卷柏科　1
卷瓣大丽花　120
参薯　162
国王椰子　166
坡参　171
坭竹　188
垂叶榕　85
垂柳　81
垂盆草　36
夜来香　111
夜花藤　30
夜香木兰　21
夜香牛　125
夜香树　129
孟仁草　180
定心藤　90
宝莲花　54
宝铎草　154

宝绿　38
实蕨科　13
岭南山竹子　56
岭南柿　103
岭南臭椿　95
岭南械　97
岭南酸枣　99
帘子藤　110
建兰　170
忽地笑　161
披针骨牌蕨　14
抱石莲　14
抱树莲　14
拂子草　179
拌根草　180
昆士兰瓶干树　57
明镜　35
昙花　49
杯苋　41
松叶红千层　52
松叶耳草　114
松叶武竹　154
松叶景天　36
松叶蕨　1
松叶蕨科　1
松科　17
板栗　81
板蓝　139
构树　84
枇杷　67
枇杷叶紫珠　140
林泽兰　121
枣树　92
枣椰　166
枫香树　80
枫香槲寄生　91
欧洲蕨　6

武竹　154
油杉　17
油松　18
油点草　156
油茶　51
油茶离瓣寄生　90
油桐　66
油棕　165
油渣果　47
油楠　73
油簕竹　188
沿阶草　155
泡叶冷水花　88
波斯红草　139
泥花草　133
泥湖菜　122
泽水苋　42
泽泻科　147
泽苔草　147
泽珍珠菜　126
浅裂绣毛莓　69
爬藤榕　86
狐尾龙舌兰　162
狐尾武竹　154
狐尾棕　167
狗牙花　110
狗牙根　180
狗尾红　60
狗尾草　187
狗肝菜　137
狗骨柴　113
狗脊蕨　11
玫红卡特兰　169
玫瑰　68
玫瑰树　109
环带姬凤梨　150
画眉草　182

知风草 182
空心泡 69
空心莲子草 41
线叶蓟 120
线羽凤尾蕨 6
线条楼梯草 87
线蕨 14
细毛鸭嘴草 184
细叶丁香蓼 44
细叶小苦荬 122
细叶白千层 53
细叶石仙桃 172
细叶石斑木 68
细叶美女樱 143
细叶桉 53
细叶野牡丹 54
细叶黄杨 81
细叶萼距花 43
细花丁香蓼 44
细花冬青 88
细枝木麻黄 83
细枝柃 51
细齿叶柃 51
细齿桫椤 5
细柄水竹叶 148
细柄草 179
细脉斑鸠菊 125
细轴荛花 44
细圆藤 30
细梗苔草 173
细棕竹 166
细裂玉凤花 171
绉面草 144
罗汉松 19
罗汉松科 19
罗伞树 104
罗浮买麻藤 20

罗浮柿 103
罗浮栲 82
罗浮槭 97
罗蓟 145
肾叶天胡 102
肾茶 143
肾蕨 13
肾蕨科 13
肿柄菊 125
苘麻 59
苞子草 188
苞舌兰 172
苦木科 95
苦玄参 133
苦瓜 48
苦竹 190
苦苣苔科 134
苦苣菜 124
苦枥木 107
苦树 95
苦草 147
苦荬菜 122
苦梓 142
苦梓含笑 22
苦楝 96
苹科 16
苹婆 58
茄子 130
茄科 128
茅瓜 48
茅根 186
茅莓 69
茅膏菜 37
茅膏菜科 37
茉莉 107
茑萝 132
虎皮楠 66

虎耳草科 36
虎舌红 104
虎克鳞盖蕨 5
虎尾兰 163
虎杖 39
虎纹凤梨 150
虎刺 113
虎刺梅 63
虎刺楤木 100
虎颜花 55
贯众 11
轮叶木姜子 26
轮叶蒲桃 54
轮环藤 30
软叶刺葵 166
软叶菠萝 157
软枝黄蝉 108
软荚红豆 78
软骨草 147
软鳞苏铁 17
郁金 152
郁金香 156
郎伞木 104
金山葵 167
金不换 35
金凤花 72
金毛耳草 114
金毛狗 4
金丝草 186
金丝桃 56
金丝桃科 56
金叶女贞 108
金叶含笑 22
金叶树 103
金瓜 47
金边千手兰 163
金边虎尾兰 163

八画 — 九画

金边富贵竹　163
金鸟赫蕉　151
金合欢　69
金色狗尾草　187
金花茶　51
金杯花　129
金线吊乌龟　30
金线草　39
金苞花　138
金英　60
金茅　183
金虎尾科　60
金钗凤尾蕨　6
金鱼草　132
金鱼藻　28
金鱼藻科　28
金星蕨　9
金星蕨科　8
金玲花　59
金脉爵床　138
金草　114
金钮扣　124
金桂　108
金盏菊　119
金盏银盘　119
金钱豹　127
金钱蒲　157
金银花　117
金粟兰　32
金粟兰科　32
金缕梅科　80
金腰箭　124
金锦香　55
金樱子　68
金橘　94
陌上菜　133
降香黄檀　75

降真香　94
雨久花　156
雨久花科　156
雨树　71
雨蕨　14
雨蕨科　14
青牛胆　30
青叶苎麻　87
青皮竹　189
青江藤　89
青果榕　86
青竿竹　189
青香茅　180
青梅　52
青菜　33
青棕　166
青葙　41
青锁龙　36
青檀　84
青藤　27
青藤科　27
非洲凤仙花　42
非洲菊　122
非洲紫罗兰　135
非洲棟　96
鱼木　33
鱼尾椰子　165
鱼尾葵　165
鱼骨子　113
鱼眼草　121
鱼黄草　131
鱼腥草　32
鱼藤　75
鸢尾　162
鸢尾科　161
齿牙毛蕨　8
齿叶安息香　105

齿叶睡莲　28
齿果草　35
齿缘吊钟花　102
齿萼挖耳草　134

九画

亮叶冬青　89
亮叶崖豆藤　77
亮叶雀梅藤　92
亮叶猴耳环　70
亮叶槭　97
亮叶霍氏鱼藤　75
亮绿凤梨　149
保亭鳝藤　109
冠盖藤　66
冠萼线柱苣苔　135
前胡　102
剑叶凤尾蕨　6
剑叶木姜子　26
剑叶石斛　170
剑叶耳草　114
剑叶卷柏　2
剑叶波斯菊　120
剑叶菝葜　157
剑叶鳞始蕨　5
剑麻　163
剑蕨科　16
匍茎栓果菊　123
南山茶　51
南五味子　23
南天竹　29
南方荚蒾　117
南方碱蓬　40
南毛蒿　118
南瓜　47
南亚松　18
南亚新木姜　27

九 画

南岭山矾 106	孩儿草 138	柄果槲寄生 91
南岭黄檀 75	李叶豆 73	柊叶 153
南洋杉 19	宫粉羊蹄甲 71	柏木 18
南洋杉科 19	帝王秋海棠 49	柏拉木 54
南洋参 101	带状瓶尔小草 2	柏科 18
南洋假脉蕨 4	庭藤 77	柑橘 94
南洋楹 70	弯曲碎米荠 34	柔毛齿叶睡莲 28
南蛇棒 157	弯枝黄檀 75	柔毛堇菜 35
南蛇簕 72	弯柄苔草 173	柔毛飘拂草 175
南椰 164	箪笋竹 190	柔枝莠竹 184
南紫薇 43	扁豆 77	柘树 85
南酸枣 98	扁担杆 56	柚 94
厚斗柯 82	扁担藤 93	柚木 142
厚叶铁线莲 27	扁枝槲寄生 90	柠檬 94
厚叶算盘子 63	扁桃 99	柠檬桉 53
厚皮山矾 106	扁蓄 39	柯 82
厚皮香 52	扁鞘飘拂草 175	柱果铁线莲 27
厚皮香八角 23	扁穗莎草 174	柳叶五月茶 61
厚壳树 128	挂金灯 129	柳叶毛蕊茶 51
厚壳桂 25	挖耳草 134	柳叶石斑木 68
厚果崖豆藤 78	星毛冠盖藤 66	柳叶杜茎山 105
厚藤 131	星花睡莲 29	柳叶剑蕨 16
变叶树参 100	星果滕 60	柳叶润楠 26
变叶榕 86	星点木 163	柳叶菜科 43
变异鳞毛蕨 12	星蕨 15	柳叶斑鸠菊 125
变色牵牛 132	映山红 103	柳叶箬 184
咸水草 174	春云实 72	柳叶蓼 39
咸虾花 125	春兰朵朵香 170	柿 103
响铃豆 74	春羽 159	柿科 103
复羽叶栾树 97	显脉山蚂蝗 76	栀子 113
复序飘拂草 175	显脉星蕨 15	栌菊木 123
姜 152	显脉新木姜 27	栎叶柯 83
姜花 152	枳椇 91	树参 100
姜科 151	枸杞 129	树茄 130
姜黄 152	枸骨 88	树牵牛 131
威灵仙 27	柃叶茶 50	毒根斑鸠菊 125
娃儿藤 112	柄叶鳞毛蕨 12	洋吊钟 36

九 画

洋杜鹃　103
洋金花　129
洋常春藤　100
洋紫苏　144
洋紫荆　71
洋蒲桃　54
洒金榕　62
炮仗竹　133
炮仗花　136
点地梅　126
点纹十二卷　155
独子藤　89
独活叶秋海棠　49
独脚金　134
狭叶山黄麻　84
狭叶长萌苣苔　135
狭叶母草　133
狭叶红紫珠　140
狭叶栀子　113
狭叶桃叶珊瑚　100
狭叶海金沙　4
狭叶海桐　45
狭叶假糙苏　145
狭叶紫萁　2
狭枝鸡矢藤　116
狭基巢蕨　10
狮子头　50
珊瑚姜　152
珊瑚树　117
珊瑚菜　102
珊瑚藤　39
珍珠菜　126
相思子　73
盾叶冷水花　88
盾柱木　73
省沽油科　98
看麦娘　178

矩形叶鼠刺　66
砖子苗　177
神秘果　104
秋枫　61
秋茄　55
秋海棠科　48
穿心莲　137
穿破石　84
穿鞘花　147
类地毯草　179
类芦　185
绒毛山蚂蝗　76
绒毛润楠　26
结缕草　188
络石　110
绞股蓝　47
美人蕉　153
美人蕉科　153
美女樱　142
美山矾　106
美丽口红花　135
美丽刺桐　76
美丽胡枝子　77
美丽桢桐　141
美丽银背藤　130
美花石斛　170
美花崖豆藤　78
美国黄莲　28
美冠兰　171
美洲商陆　40
美洲菊芹　121
美洲榄仁　55
美脉琼楠　24
胎生狗脊　11
胡桃科　99
胡萝卜　102
胡麻草　132

胡椒　32
胡椒木　95
胡椒科　31
胡颓子　92
胡颓子科　92
茜树　112
茜草科　112
茴茴蒜　28
茶　51
茶秆竹　190
茶科　50
茶荣萸科　90
茶梅　51
茼蒿　119
荇菜　126
草龙　43
草豆蔻　152
草珊瑚　33
草胡椒　31
草海桐　128
草海桐科　127
草莓　67
荔枝　97
荔枝草　146
荜拔　32
荞头　160
荠菜　34
荨麻叶母草　133
荨麻科　87
苎草　178
虹鳞肋毛蕨　12
虹眼　133
虾衣花　137
虾脊兰　169
蚂蟥七　135
蚤缀　37
贴生石韦　15

费氏秋海棠 48
轴榈 166
迷人鳞毛蕨 12
退色香薷 144
重阳木 61
重瓣安石榴 43
重瓣臭茉莉 141
钝叶草 188
钝叶核果木 62
钝叶臭黄荆 142
钝叶椒草 31
钝齿红紫珠 140
钮子瓜 48
闽粤千里光 124
闽粤石楠 67
除虫菊 124
面包树 84
韭菜 161
饶平石楠 67
首冠藤 71
香水月季 68
香丝草 120
香叶天竺葵 41
香叶树 25
香皮树 98
香龙血树 163
香花枇杷 67
香花崖豆藤 77
香附子 174
香茅 180
香籽含笑 22
香茶菜 144
香根草 188
香堇 35
香雪兰 161
香雪球 34
香港大沙叶 116

香港马兜铃 31
香港凤仙花 42
香港双蝴蝶 126
香港木兰 21
香港毛蕊茶 50
香港四照花 100
香港瓜馥木 24
香港安兰 168
香港耳草 115
香港坚木 96
香港杜鹃 103
香港远志 35
香港油麻藤 78
香港胡颓子 92
香港崖豆藤 78
香港绶草 172
香港蛇菰 91
香港黄檀 75
香港新木姜 27
香港蒟 32
香港算盘子 64
香港鹰爪花 23
香港魔芋 158
香椿 96
香楠 112
香睡莲 28
香蒲 160
香蒲科 160
香蒲桃 54
香膏萼距花 42
香蓼 40
香蕉 151
香橼 94
骨牌蕨 14
骨碎补科 13
鬼羽箭 132
鬼针草 119

鸦胆子 95

十画

倒吊笔 110
倒地铃 96
倒卵叶野木瓜 29
倒挂金钟 43
倒挂铁角蕨 10
凉粉草 145
凌霄 135
唇形科 143
唐印 36
唐菖蒲 162
圆叶节节菜 43
圆叶红苋 41
圆叶花烛 158
圆叶细辛 31
圆叶虎尾兰 163
圆叶南蛇藤 89
圆叶挖耳草 134
圆叶牵牛 132
圆叶轴榈 166
圆叶豺皮樟 26
圆叶野百合 75
圆叶野扁豆 76
圆叶野桐 64
圆叶福禄桐 101
圆果雀稗 186
圆秆珍珠茅 178
圆齿碎米荠 34
圆柏 19
圆绒鸡冠 41
圆盖阴石蕨 14
圆锥柯 83
夏枯草 146
夏堇 134
夏飘拂草 175

十 画

姬凤梨 150
姬喜林芋 159
姬蕨 6
姬蕨科 6
宽叶线柱兰 172
宽叶变叶木 62
宽叶珍珠茅 178
宽羽毛蕨 9
宽羽线蕨 14
宽苞茅膏菜 37
射干 161
展毛野牡丹 54
恋之蔓 110
扇叶铁线蕨 7
捕蝇草 37
旅人蕉 151
旅人蕉科 151
栓叶安息香 105
栗豆树 74
栗寄生 90
栗蕨 6
栝楼 48
根用芥菜 34
格木 73
格药柃 51
桂木 84
桂北木姜子 26
桂叶山牵牛 139
桂花 108
桂南木莲 21
桃
桃叶珊瑚 99
桃叶黄杨 81
桃叶猩猩草 63
桃花心木 96
桃金娘 53
桃金娘科 52

桃榔 164
桑 87
桑科 84
桑寄生 90
桑寄生科 90
桔梗 127
桔梗科 127
梠叶黄花稔 59
梧桐 58
梧桐科 57
梨叶悬钩子 69
泰国刺葵 166
流苏子 113
流苏贝母兰 170
浙江润楠 26
浮萍 160
浮萍科 160
海刀豆 74
海马齿 38
海风藤 23
海州香薷 144
海红豆 69
海芋 157
海岛十大功劳 29
海岛藤 112
海杧果 109
海金沙 4
海金沙科 3
海南大风子 46
海南弓果藤 112
海南木莲 22
海南红豆 78
海南杨桐 50
海南苏铁 17
海南草珊瑚 33
海南梧桐 58
海南菜豆树 136

海南割鸡芒 176
海南赛爵床 137
海南槽裂木 116
海南黧豆 78
海桐 45
海桐花科 45
海榄雌 139
海滨木槿 59
海滨雀稗 186
海漆树 63
烟斗石栎 82
烟草 129
爱地草 114
特丽安娜卡特兰 169
狸尾草 80
狸藻科 134
狼尾草 186
珠子草 65
珠仔树 106
珠兰 32
瓶子草科 37
瓶尔小草 2
瓶耳小草科 2
皱叶忍冬 117
皱叶狗尾草 187
皱叶椒草 31
皱果珍珠茅 178
益母草 144
益智 152
盐地鼠尾粟 188
盐肤木 99
破布叶 56
破布叶 128
离根草 127
离瓣寄生 90
秤星树 88
秤钩风 30

积雪草　102
窄叶台湾榕　85
窄叶柃　51
窄叶翅子树　58
笔罗子　98
笔筒树　5
笔管榕　86
粉叶轮环藤　30
粉叶菝葜　157
粉叶蕨　7
粉扑花　70
粉鸟赫蕉　151
粉鸟蝎尾蕉　151
粉防己　30
粉花杜鹃　103
粉单竹　188
粉背菝葜　157
粉萼鼠尾草　146
绢毛杜英　57
绢毛飘拂草　176
绣球花科　66
缺萼枫香　80
翅子藤科　90
翅果菊　124
翅茎白粉藤　93
翅荚决明　72
胭脂树　84
能高鳞毛蕨　11
臭矢菜　33
臭牡丹　141
臭茉莉　141
臭根子草　179
艳山姜　152
艳红赫蕉　151
艳黄赫蕉　151
荷花　28
荷花玉兰　21

荷威棕　165
荷莲豆　37
荻芦　184
莎草科　173
莠竹　184
莠狗尾草　187
莪术　152
莲子草　41
莲座紫金牛　104
莴苣　123
莹蔺　177
莺哥凤梨　150
莼菜　28
蚊母树　80
蚌壳蕨科　4
蚕茧草　39
袖珍椰子　165
被子植物　21
豹耳秋海棠　48
豺皮樟　26
透明荸荠　175
通奶草　63
通城虎　31
通泉草　133
通脱木　101
酒饼簕　94
酒瓶兰　163
酒瓶椰子　165
钳喙兰　171
铁丁兔儿风　118
铁刀木　72
铁力木　56
铁十字秋海棠　49
铁冬青　88
铁包金　91
铁芒萁　3
铁苋菜　60

铁角蕨科　10
铁线蕨　7
铁线蕨科　7
铁青树科　90
铁榄　104
高山榕　85
高良姜　152
高秆珍珠茅　178
高秆莎草　174
高野黍　183
高斑叶兰　171
鸭公树　27
鸭舌草　156
鸭跖草　148
鸭跖草科　147
鸭嘴花　138
鸳鸯茉莉　128

十一画

假丁香蓼　43
假九节　116
假大羽铁角蕨　10
假马齿苋　132
假马鞭　142
假叶树　157
假叶树科　157
假地豆　75
假肉桂　27
假芒萁　3
假杜鹃　137
假连翘　142
假泽兰　123
假苹婆　58
假茉莉　141
假俭草　183
假柿木姜子　26
假烟叶　129

十一画

假臭草 121	常春藤 101	猪屎豆 74
假淡竹叶 180	常绿臭椿 95	猪笼草 31
假菠菜 40	康乃馨 37	猪笼草科 31
假蛇尾草 183	彩云阁 63	猪菜藤 131
假紫苏 144	彩叶芋 158	猫爪藤 136
假蒟 32	彩虹肖竹芋 153	猫尾木 136
假槟榔 164	悬铃花 59	猫尾豆 80
假蹄盖蕨 8	排钱草 79	球子蕨科 10
假鹰爪 23	接骨木 117	球兰 111
剪刀股 122	斜叶榕 86	球序韭 161
匙叶茅膏菜 37	斜基粗叶木 115	球花毛麝香 132
匙叶黄杨 81	断线蕨 14	球果猪屎豆 74
匙叶鼠曲草 122	旋花科 130	球果蔊菜 34
匙羹藤 111	旋药阔蕊兰 172	球柱草 173
商陆 40	旌节花科 80	球根秋海棠 49
商陆科 40	曼陀罗 129	球菊 121
堇菜 35	望冬草 185	球穗扁莎 177
堇菜科 35	望江南 72	甜叶算盘子 64
宿根画眉草 182	桫椤 5	甜瓜 47
寄生藤 91	桫椤科 5	甜根子草 187
密子豆 79	梅 67	甜麻 56
密毛乌口树 117	梵天花 60	甜槠栲 82
密毛蕨 6	淡竹叶 184	甜橙 94
密节坡油甘 79	淡绿短肠蕨 8	畦畔莎草 174
密花石豆兰 169	深山含笑 22	畦畔飘拂草 176
密花豆 79	深波叶补血草 126	盒果藤 132
密花树 105	深绿卷柏 1	盘苞牵牛 131
密花胡颓子 92	深裂号角树 84	眼子菜科 147
密苞山姜 152	清风藤 98	眼树莲 111
崇澍蕨 11	清风藤科 98	粗毛玉叶金花 115
崖角藤 159	渐尖毛蕨 8	粗毛鸭嘴草 184
崖爬藤 93	牻牛儿苗科 41	粗毛野桐 64
崖姜 15	犁头尖 160	粗叶木 115
巢蕨 10	猕猴桃 52	粗叶耳草 115
常山 66	猕猴桃科 52	粗叶悬钩子 68
常春卫矛 89	猪仔笠 76	粗叶榕 85
常春油麻藤 78	猪肚木 113	粗齿紫萁 2

十一画

粗脉樟　25
粗喙秋海棠　48
粗糠柴　65
粘木　60
粘木科　60
续断菊　124
绵毛葡萄　93
绵枣儿　154
绶草　172
绿竹　189
绿杆铁角蕨　10
绿花安兰　168
绿花羊耳蒜　171
绿花崖豆藤　77
绿豆　80
绿宝石喜林芋　159
绿帚　39
绿萝　159
绿蓟　120
绿锭带唇兰　172
绿穗莎草　174
舶梨榕　86
菅　188
菊芋　122
菊花　121
菊科　118
菜心　34
菜王椰　166
菜豆树　136
菜蕨　8
菝葜　156
菝葜科　156
菟丝子　130
菠菜　40
菩提榕　86
菱　44
菱科　44

菲岛福木　56
萍蓬草　28
萝卜　34
萝芙木　110
萝藦科　110
蛇王藤　46
蛇瓜　48
蛇目菊　120
蛇舌兰　170
蛇尾草　185
蛇足石杉　1
蛇莓　67
蛇婆子　58
蛇菰科　91
蛇葡萄　92
蛇藤　69
蛇鞭菊　123
蛋黄果　104
象草　186
象腿丝兰　164
象腿树　33
象腿蕉　150
趾叶栝楼　48
距花黍　183
野木瓜　29
野甘草　133
野生紫苏　145
野百合　74
野芋　158
野含笑　23
野牡丹　54
野牡丹科　54
野苋　41
野鸡尾　7
野茄　129
野青树　77
野茼蒿　120

野莴苣　123
野鸭椿　98
野菰　134
野黄桂　25
野黄菊　121
野葵　59
野慈姑　147
野漆树　99
野蕉　150
野魔芋　158
铜锤玉带草　127
银毛野牡丹　55
银叶树　58
银叶菊　119
银边吊兰　154
银边草　178
银边翠　63
银合欢　70
银羽枬花竹芋　153
银杉　17
银杏　17
银杏科　17
银花苋　41
银线虾脊兰　169
银背藤　130
银钟花　105
银柴　61
银桦　45
银穗湖瓜草　177
隐穗苔草　173
隐囊蕨　7
雀舌草　38
雀梅藤　92
雀稗　186
雪花木　61
雪铁芋　160
鹿角草　122

鹿角锥　82
鹿角蕨　15
鹿角蕨科　15
鹿藿　79
麻竹　189
麻栎　83
麻疯树　64
麻楝　96
黄毛五月茶　61
黄毛白鹤藤　130
黄毛野扁豆　76
黄毛榉木　100
黄毛榕　85
黄牛木　56
黄牛奶树　106
黄风铃花　136
黄兰　22
黄叶耳草　115
黄瓜　47
黄瓜菜　123
黄皮　94
黄龙船花　115
黄丽鸟赫蕉　151
黄杞　99
黄杨　81
黄杨科　81
黄芩蒿　118
黄花小二仙草　44
黄花马樱丹　142
黄花马蹄莲　160
黄花夹竹桃　110
黄花远志　35
黄花美人蕉　153
黄花倒水莲　35
黄花狸藻　134
黄花稔　59
黄花蔺　146

黄花翼萼　134
黄连木　99
黄果厚壳桂　25
黄茅　183
黄金间碧竹　189
黄金榕　86
黄青冈栎　82
黄独　162
黄绒润楠　26
黄脉刺桐　76
黄荆　143
黄药　92
黄钟花　136
黄桐　62
黄珠子草　65
黄常山　36
黄眼草　149
黄眼草科　149
黄麻　56
黄棉木　112
黄琼草　137
黄葵　58
黄槐决明　72
黄瑞木　50
黄睡莲　28
黄腺羽蕨　12
黄鹌菜　126
黄蝉　109
黄槿　59
黄樟　25
黄醉蝶花　33
黄藤　425

十二画

割鸡芒　176
博落回　33
喜树　100

寒兰　170
戟叶堇菜　35
掌叶花烛　158
掌叶线蕨　14
掌叶海金沙　3
散尾棕　164
散尾葵　165
散沫花　43
散穗弓果黍　180
斑叶朱砂根　104
斑叶垂椒草　32
斑叶秋海棠　49
斑叶露兜树　167
斑鸠菊　125
斑茅　186
斑种草　128
普洱茶　50
普通针毛蕨　9
景天科　35
棉叶麻疯树　64
棒花落地生根　36
棒柄花　62
棕叶芦　188
棕叶狗尾草　187
棕竹　166
棕榈　167
棕榈科　164
棱果蒲桃　53
棱枝槲寄生　91
椰子　165
椰榆　84
款冬　125
湖北蓟　120
湿地松　17
犀角　111
猩猩草　63
猴欢喜　57

十二画

猴耳环 70	短萼仪花 73	紫花短筒苣苔 135
琴叶珊瑚 64	短葶仙茅 167	紫花藿香蓟 118
琴叶喜林芋 159	短颖马唐 181	紫苏 145
琴叶紫菀 118	短穗刺蕊草 146	紫苏草 133
琴叶榕 86	短穗画眉草 182	紫刺卫矛 89
琼棕 165	短穗鱼尾葵 164	紫果蔺 174
番木瓜科 49	硫磺菊 120	紫罗兰 34
番石榴 53	硬斗石栎 83	紫茉莉 45
番杏 38	硬毛草胡椒 31	紫茉莉科 45
番杏科 38	硬叶吊兰 170	紫金牛 104
番茄 129	硬枝老鸦嘴 139	紫金牛科 104
番荔枝 23	硬骨凌霄 136	紫背天葵 48
番荔枝科 23	硬穗飘拂草 175	紫背金盘 143
番薯 131	筋骨草 143	紫荆 72
疏毛白绒草 144	粟米草 38	紫荆木 104
疏花卫矛 89	粟米草科 38	紫草科 128
疏花马蓝 139	粤里白 3	紫鸭跖草 148
疏裂凤尾蕨 6	粤港耳草 114	紫弹树 83
疏穗画眉草 182	粤港谷精草 149	紫萁 2
短小蛇根草 116	粤紫萁 2	紫萁科 2
短叶水蜈蚣 176	粪箕笃 30	紫萍 160
短叶决明 72	紫万年青 149	紫雪茄花 42
短叶罗汉松 19	紫心牵牛 131	紫麻 87
短叶茳芏 174	紫叶水竹草 149	紫黄蝉 108
短叶雀舌兰 150	紫叶李 67	紫葳科 135
短叶裂稃草 187	紫叶槿 59	紫薇 43
短叶黍 185	紫玉兰 21	紫檀 79
短舌紫菀 119	紫玉盘 24	紫藤 80
短序润楠 26	紫玉盘柯 83	缘毛卷柏 1
短齿假糙苏 145	紫竹 190	缘毛珍珠茅 177
短冠东风菜 121	紫纹兜兰 172	腊肠树 72
短柄半边莲 127	紫花地丁 35	腋花蓼 40
短柄禾叶蕨 16	紫花羊耳蒜 171	萱草 155
短柄翅子藤 90	紫花芭蕉 151	萼距花 43
短柄紫珠 139	紫花前胡 101	落地生根 36
短柱茶 50	紫花香薷 144	落羽杉 18
短距苞叶兰 168	紫花铁兰 150	落花生 73

落葵 41
落葵科 41
落瓣油茶 51
葛 79
葡萄 94
葡萄科 92
葡蟠 84
董棕 165
葫芦 47
葫芦科 47
葫芦茶 80
葱 160
葱兰 161
葱草 149
萎叶 32
蛤蟆叶秋海棠 49
蛤蟆草 88
裂叶羽衣甘蓝 34
裂叶牵牛 132
裂叶秋海棠 49
裂叶蒲葵 166
越南山矾 106
越南叶下珠 65
越南安息香 105
越南野古草 179
越南篦齿苏铁 17
越橘科 103
酢浆草 42
酢浆草科 42
量天尺 49
铺地柏 19
铺地草 63
铺地黍 185
铺地蜈蚣 1
铺地蝙蝠草 74
链荚豆 73
链珠藤 109

锈毛木莓 69
锈穗飘拂草 175
锐尖山香圆 98
阔叶十大功劳 29
阔叶山麦冬 155
阔叶丰花草 113
阔叶乌蕨 5
阔叶补血草 126
阔叶建兰 170
阔叶沼兰 172
阔叶彩叶凤梨 150
阔叶猕猴桃 52
阔叶短肠蕨 8
阔苞菊 124
阔荚合欢 70
阔裂叶羊蹄甲 71
阔鳞鳞毛蕨 11
雅榕 85
韩信草 146
鹅毛玉凤花 171
鹅掌柴 101
鹅掌楸 21
鹅掌藤 101
黑叶观音莲 157
黑叶谷木 55
黑老虎 23
黑足鳞毛蕨 12
黑枔 51
黑籽水蜈蚣 176
黑籽荸荠 175
黑面神 61
黑莎草 176
黑桫椤 5
黑紫藜芦 156
黑藻 147

十三画

鼎湖血桐 64

叠穗莎草 174
幌伞枫 101
感应草 42
慈姑 147
新几内亚凤仙花 42
新月蕨 9
暗色菝葜 157
暗罗 24
椴树科 56
椿叶花椒 95
楝叶吴茱萸 94
楝科 96
楠藤 115
榀木 100
榆树 69
榄仁树 55
榆科 83
榉树 84
槐叶苹科 16
槐叶萍 16
溪边凤尾蕨 6
溪边假毛蕨 9
满天星 37
满江红 16
满江红科 16
滨虹豆 80
滨海月见草 44
滨海槭 97
滨盐肤木 99
滨菊 123
瑞香科 44
睡莲 29
睡莲科 28
睡菜科 126
睫毛蕨 10
睫毛蕨科 10
矮扁莎 177

十三画 — 十四画

矮莎草　174
矮琼棕　165
碎米荠　34
碎米莎草　174
碗蕨科　5
福建观音座莲　2
福建柏　18
福建茶　128
福建假卫矛　89
福禄桐　101
稗　181
稗荩　187
简轴草　186
腰果　98
腺叶山矾　106
腺叶桂樱　67
腺叶藤　132
腺茉莉　141
腺柄山矾　106
蒜　160
蒜香藤　136
蒲包花　132
蒲瓜　47
蒲草　176
蒲桃　54
蒲葵　166
蒺藜草　179
蓖齿苏铁　17
蓖麻　65
蓝耳草　148
蓝花丹　127
蓝花西番莲　46
蓝花参　127
蓝花楹　136
蓝果树科　100
蓝香草　141
蓝桉　53

蓝翅西番莲　46
蓬莱葛　107
蜂腰洒金榕　62
蜈蚣草　7
蜈蚣草　183
蜈蚣藤　159
裸子植物　17
裸子蕨科　7
裸花水竹叶　148
裸花华紫珠　140
裸柱菊　124
赪桐　141
辐叶鹅掌柴　101
辐射砖子苗　177
锡兰肉桂　25
锡兰橄榄　57
锡叶藤　45
锥叶榕　86
锦叶龙舌兰　163
锦绣杜鹃　103
锦葵　59
锦葵科　58
锦熟黄杨　81
雏菊　119
雷神　163
雾水葛　88
鼓槌石斛　170
鼠妇草　182
鼠曲草　122
鼠尾粟　187
鼠尾囊颖草　187
鼠李科　91
鼠刺　66
鼠刺科　66

十四画

嘉兰　154

嘉陵花　24
嘉赐树　46
弊草　183
截叶胡枝子　77
截裂毛蕨　9
截穗苔草　173
榕叶冬青　88
榛叶黄花稔　60
槟榔　164
槭叶瓶干树　57
槭树科　97
漆树科　98
瑶山谷精草　149
瘦风轮　144
碧冬茄　129
楔颖草　178
端红彩叶凤梨　150
算盘子　64
管苞瓶蕨　4
蓟柊　46
箬叶竹　189
箬竹　189
棕巴箬竹　189
罂粟科　33
翠云草　2
翠柏　18
翠菊　119
翡翠景天　36
聚花草　148
膜蕨科　4
舞草　74
蓸奥　93
蓼科　39
蕇莱　35
蕉草　177
蔓九节　116
蔓千斤拔　77

十四画 — 十六画

蔓马缨丹 142
蔓生莠竹 184
蔓花生 73
蔓赤车 87
蔓茎堇菜 35
蔓胡颓子 92
蔓荆 143
蔓草虫豆 74
蔷薇风铃花 136
蔷薇科 67
蜀葵 59
蜘蛛兰 161
蜘蛛抱蛋 154
蜜甘草 65
蜡烛果 104
蜻蜓凤梨 149
蝉翼藤 35
褐叶星蕨 15
褐叶柄果木 97
褐苞薯蓣 162
褐穗狗尾草 187
福氏隔距兰 169
豨莶 124
赛葵 59
辣木科 33
辣椒 128
酸叶胶藤 110
酸豆 73
酸浆 129
酸橙 94
酸藤子 105
锲叶豆梨 68

十五画

墨兰 170
撑篙竹 189
槲蕨 15

槲蕨科 15
樟叶朴 83
樟叶泡花树 98
樟树 24
樟科 24
横经席 56
橄榄 95
橄榄科 95
橡胶榕 85
潺槁木姜子 25
澳洲坚果 45
瘤足蕨 3
瘤足蕨科 3
瘤果槲寄生 91
瘤蕨 15
磕藤子 70
篌竹 190
薹树 80
蕉芋 153
蕊木 109
蕨 6
蕨叶苔草 173
蕨科 6
蕨类植物 1
蝙蝠西番莲 46
蝙蝠草 74
蝴蝶兰 172
蝴蝶花 142
蝴蝶花 162
蝴蝶花豆 74
蝴蝶果 62
蝴蝶树 58
蝶形花科 73
蝶花荚蒾 117
趣蝶莲 36
醉鱼草 107
醉香含笑 22

醉蝶花 33
靠脉肋毛蕨 12
鲫鱼胆 105
鲫鱼草 181
鲫鱼草 182
鹤顶兰 172
鹤望兰 151
黎檬 94
橘 94
橙红五色梅 142
橙黄卡特兰 169
橙黄玉凤花 171
燕子掌 36
燕尾山槟榔 166

十六画

磨盘草 59
糖胶树 109
糖棕 164
糙叶丰花草 112
糙叶树 83
糙叶斑鸠菊 125
蕹菜 131
薄叶卷柏 1
薄叶润楠 26
薄叶假耳草 116
薄叶猴耳环 70
薄叶碎米蕨 7
薄叶瘤足蕨 3
薄果猴欢喜 57
薄荷 145
薇甘菊 123
薏苡 180
薛荔 86
薯莨 162
薯蓣 162
薯蓣科 162

蹄盖蕨科　8
镜面草　88
鞘花寄生　90
颠茄　128
鹧鸪草　183

十七画

戴星草　124
擘蓝　33
檀香科　91
爵床　138
爵床科　137
穗序木蓝　77
穗序鹅掌柴　101
穗花杉　20
簪竹　190
簇花清风藤　98
簇花粟米草　38
簕仔树　70
簕古子　167
簕党花椒　95
繁缕　38
翼核果　92
臀果木　68
臀形草　179
鳄梨　27
黛粉叶　158

十八画

檫木　27

檵木　81
瞿麦　37
翻白叶树　58
藜　40
藜芦　156
藜科　40
藤竹草　185
藤金合欢　69
藤麻　88
藤黄科　56
藤黄檀　75
藤槐　74
藤榕　85
蟛蜞菊　125
镰叶铁角蕨　10
镰羽贯众　11
镰形假毛蕨　9
镰形觿茅　181
镰翅羊耳蒜　171
鞭叶铁线蕨　7
鞭花卷瓣兰　169
鹰爪花　23

十九画

攀援星蕨　15
藿香蓟　118
蟹爪兰　50
麒麟角　63

二十画

糯米团　87
糯米条　117
蘘荷　152
魔芋　158
鳝藤　109
鳞子莎　176
鳞毛蕨科　11
鳞花草　138
鳞始蕨　5
鳞始蕨科　5
鳞秕泽米苏铁　17
鼗萌　82

二十一画

露子草　185
露水草　148
露兜树　167
露兜树科　167
露兜草　167
霸王棕　164
鳢肠　121

二十二画以上

蘘颖草　187
蠹刺　187

拉丁学名

A

Abelia chinensis R. Br. 117
Abelia uniflora R. Br. 117
Abelmoschus moschatus (L.) Medic. 58
Abrus cantoniensis Hance 73
Abrus mollis Hance 73
Abrus precatorius L. 73
Abutilon indicum (L.) Sweet 59
Abutilon striatum Dickson 59
Abutilon theophrastii Medic. 59
Acacia auriculiformis A. Cunn. ex Benth. 69
Acacia concinna (Willd.) DC. 69
Acacia confusa Merr. 69
Acacia farnesiana (L.) Willd. 69
Acacia mangium Willd. 69
Acacia pennata (L.) Willd. 69
Acalypha australis L. 60
Acalypha hispida Burm. f. 60
Acalypha wilkesiana Muell. Arg. 60
Acampe rigida (Buch. -Ham. ex J. E. Smith) P. F. Hunt 168
Acanthaceae 137
Acanthopanax gracilistylus W. W. Smith 100
Acanthopanax trifoliatus (L.) Merr. 100
Acanthus ilicifolius L. 137
Acer buergerianum Miq. 97
Acer decandrum Merr. 97
Acer fabri Hance 97
Acer lucidum Metc. 97

Acer maluanshanensis X. M. Wang, R. H. Miau et W. B. Liao 97
Acer palmatum Thunb. 97
Acer sino-oblongum Metc. 97
Acer tutcheri Duthie 97
Aceraceae 97
Achyranthes aspera L. 40
Achyranthes bidentata Bl. 40
Acmena acuminatissima (Bl.) Merr. et Perry 52
Acorus gramineus Soland. 157
Acorus tatarinowii Schott 157
Acroceras munroanum (Balansa) Henrard 178
Acronychia pedunculata (L.) Miq. 94
Acrostichum 7
Acrostichum aureum L. 7
Actinidia cinerascens C. F. Liang 52
Actinidia deliciosa C. F. Liang 52
Actinidia eriantha Benth. 52
Actinidia latifolia (Gardn. et Champ.) Merr. 52
Actinidiaceae 52
Adenanthera pavonina L. var. microsperma (Teijsm. et Binnend.) Nielsen 69
Adenium obesum (Forssk.) Roem. et Schult. 108
Adenophora hunanensis Nannf. 127
Adenosma glutinosum (L.) Druce 132
Adenosma indianum (Lour.) Merr. 132
Adenostemma lavenia (L.) O. Kuntze

118
Adiantaceae   7
Adiantum capillus-veneris L.   7
Adiantum caudatus L.   7
Adiantum flabellulatum L.   7
Adiantum philippense L.   7
Adina pilulifera (Lam.) Franch. ex Drake   112
Adina polycephala Benth.   112
Adinandra glischroloma Hand.-Mazz.   50
Adinandra hainanensis Hayata   50
Adinandra millettii (Hook. et Arn.) Benth. et Hook. f. ex Hance   50
Aechmea fasciata (Lindl.) Baker   149
Aechmea lueddemanniana (K. Koch) Brongn. ex Mez   149
Aegiceras corniculatum (L.) Blanco   104
Aeginetia indica L.   134
Aeonium tabuliforme Webb et Berth.   35
Aeschynanthus acuminatus Wall. ex A. DC.   134
Aeschynanthus lobbianus Hook.   134
Aeschynanthus speciosus Hook.   135
Agapanthus africanus (L.) Hoffmanns   160
Agathis dammara (Lamb.) Rich.   19
Agavaceae   162
Agave americana L.   162
Agave attenuata Salm-Dyck   162
Agave potatorum Zucc. var. *verschaffeltii* (Lem.) Berger   163
Agave schidigera Lemm.   163
Agave sisalana Perr. ex Engelm   163
Agave victoriae-reginae T. Moore   163

Ageratum conyzoides L.   118
Ageratum houstonianum Mill.   118
Aglaia duperreana Pierre   96
Aglaia odorata Lour.   96
Aglaonema commutatum Schott   157
Aglaonema crispum (Pitcher et Manda) Nicolson   157
Aglaonema modestum Schott ex Engl.   157
Aidia canthioides (Champ. ex Benth.) Masamune   112
Aidia cochinchinensis Lour.   112
Aidia pycnantha (Drake) Tirvenz   112
Ailanthus fordii Noot.   95
Ailanthus triphysa (Dennst.) Alston   95
Ainsliaea fragrans Champ.   118
Ainsliaea macroclinidioides Hayata   118
Ainsliaea rubrifolia Franch.   118
Aizoaceae   38
Ajuga decumbens Thunb.   143
Ajuga nipponensis Makino   143
Akebia trifoliata (Thunb.) Koidz. ssp. *australis* (Diels) T. Shimizu   29
Alangiaceae   100
Alangium chinense (Lour.) Harms   100
Alangium kurzii Craib   100
Albizia chinensis (Osbeck) Merr.   69
Albizia corniculata (Lour.) Druce   70
Albizia falcataria (L.) Fosberg   70
Albizia julibrissin Durazz.   70
Albizia kalkora (Roxb.) Prain   70
Albizia lebbeck (L.) Benth.   70
Alchornea rugosa (Lour.) Muell.-Arg.   60

Alchornea trewioides (Benth.) Muell. - Arg. 61
Aleurites moluccana (L.) Willd. 61
Alisma orientale (Samuel.) Juz. 147
Alismataceae 147
Allamanda blanchetii A. DC. 108
Allamanda cathartica L. 108
Allamanda cathartica L. var. *hendersonii* (Bull ex Dombr.) Bailey et Raffill 108
Allamanda schottii Pohl 109
Allantodia dilatata (Bl.) Ching 8
Allantodia matthewii (Cop.) Ching 8
Allantodia virescens (Kunze) Ching 8
Alleizettella leucocarpa (Champ. ex Benth.) Tirvenz 112
Allium chinense G. Don 160
Allium fistulosum L. 160
Allium sativum L. 160
Allium thunbergii G. Don 161
Allium tuberosum Rottl. ex Spreng. 161
Alloteropsis semialata (R. Br.) Hitchc. 178
Alniphyllum fortunei (Hemsl.) Makino 105
Alocasia amazonica (Lour.) Schott 157
Alocasia cucullata (Lour.) Schott 157
Alocasia macrorrhiza (L.) Schott 157
Aloe mitriformis Mill. 153
Aloe vera L. var. *chinensis* (Haw.) Berger 153
Alopecurus aequalis Sobol. 178
Alpinia galanga (L.) Willd. 151
Alpinia hainanensis K. Schum. 152
Alpinia japonica (Thunb.) Miq. 152
Alpinia oblongifolia Hayata 151

Alpinia officinalis Hance 152
Alpinia oxyphylla Miq. 152
Alpinia pumila Hook. f. 152
Alpinia stachyoides Hance 152
Alpinia zerumbet (Pers.) Burtt et Smith 152
Alsophila spinulosa (Wall. ex Hook.) Tryon 5
Alstonia scholaris (L.) R. Br. 109
Alternanthera bettzickiana (Regel) Yoss 40
Alternanthera philoxeroides (Mart.) Griseb. 41
Alternanthera sessilis (L.) DC. 41
Alternanthera tenella Colla 41
Althaea rosea (L.) Cavan. 59
Altingia chinensis (Champ.) Oliv. ex Hance 80
Alysicarpus vaginalis (L.) DC. 73
Alyxia sinensis Champ. ex Benth. 109
Alyxia vulgaris Tsiang 109
Amaranthaceae 40
Amaranthus spinosus L. 41
Amaranthus tricolor L. 41
Amaranthus viridis L. 41
Amaryllidaceae 160
Amentotaxus argotaenia (Hance) Pilger 20
Amischotolype hispida (Less et A. Rich.) Hong 147
Ammannia arenaria H. B. K. 42
Ammannia baccifera L. 42
Amorphophallus dunnii Tutcher 157
Amorphophallus oncophyllus Prain ex Book. f. 158
Amorphophallus rivieri Durieu ex Riviere 158

Amorphophallus variabilis Bl.　158
Ampelopsis cantoniensis (Hook. et Arn.) Planch.　92
Ampelopsis chaffanjoni (Levl.) Rehd.　92
Ampelopsis sinica (Miq.) W. T. Wang　92
Amygdalus persica L.　67
Anacardiaceae　98
Anacardium occidentale L.　98
Ananas comosus (L.) Merr.　149
Andrographis paniculata (Burm. f.) Nees　137
Androsace umbellata (Lour.) Merr.　126
Angelica decursiva (Miq.) Franch.　101
Angiopteridaceae　2
Angiopteris fokiensis Hieron　2
Angiospermae　21
Aniseia biflora (L.) Choisy　130
Anisomeles indica (L.) O. Kuntze　143
Anisopappus chinensis (L.) Hook. et Arn.　118
Annona muricata L.　23
Annona squamosa L.　23
Annonaceae　23
Anodendron affine (Hook. et Arn.) Druce　109
Anodendron howii Tsiang　109
Anoectochilus roxburghii (Wall.) Lindl.　168
Antenoron filiforme (Thunb.) Rob. et Vant.　39
Anthurium andraeanum Linden　158
Anthurium clarinervium Matuda　158
Anthurium crystallinum Linden ex Andre　158
Anthurium pedato-radiatum Schott　158
Anthurium scherzerianum Schott　158
Antidesma bunius (L.) Spreng.　61
Antidesma fordii Hemsl.　61
Antidesma ghaesembilla Gaertn.　61
Antidesma japonicum Sieb. et Zucc.　61
Antidesma pseudomicrophyllum Croiz.　61
Antidesma venosum E. Mey. ex Tul.　61
Antigonon leptopus Hook. et Arn.　39
Antirhea chinensis (Champ. ex Benth.) Forbes et Hemsl.　112
Antirrhinum majus L.　132
Aphanamixis grandifolia Bl.　96
Aphanamixis polystachya (Wall.) R. N. Parker　96
Aphananthe aspera (Thunb.) Planch.　83
Aphananthe cuspidata (Bl.) Planch.　83
Apium graveolens L.　102
Apluda mutica L.　178
Apocopis wrightii Munro　178
Apocynaceae　108
Aponogeton lakhonensis A. Camus　147
Aponogetonaceae　147
Aporosa dioica (Roxb.) Muell.-Arg.　61
Appendicula cornuta Bl.　168
Aquifoliaceae　88
Aquilaria sinesis (Lour.) Spreng.　44
Araceae　157
Arachis duranensis Krapov. et W. C. Gregory　73
Arachis hypogaea L.　73
Arachniodes chinensis (Ros.) Ching　11

A

Arachniodes exilis (Hance) Ching 11
Aralia armata (Wall.) Seem. 100
Aralia chinensis L. 100
Aralia decaisneana Hance 100
Araliaceae 100
Araucaria bidwillii Hook. 19
Araucaria cunninghamii Sweet 19
Araucaria heterophylla (Salisb.) Franco 19
Araucariaceae 19
Archiboehmeria atrata (Gagnep.) C. J. Chen 87
Archidendron clypearia (Jack.) Nielsen 70
Archidendron lucidum (Benth.) Nielsen 70
Archidendron utile (Chun et How) Nielsen 70
Archontophoenix alexandrae (F. Muell.) Wendl. et Drude 164
Arctium lappa L. 118
Ardisia crenata Sims 104
Ardisia elegans Andr. 104
Ardisia hanceana Mez 104
Ardisia japonica (Thunb.) Bl. 104
Ardisia lindleyana D. Dietr. 104
Ardisia mamillata Hance 104
Ardisia primulifolia Gardn. et Champ. 104
Ardisia quinquegona Bl. 104
Areca catechu L. 164
Areca triandra Roxb. ex Buch.-Ham. 164
Arecaceae 164
Arenaria seropyllifolia L. 37
Arenga engleri Becc. 164
Arenga pinnata (Wurmb.) Merr. 164

Arenga westerhoutii Griffith 164
Argyranthemum frutescens (L.) Sch.-Bip 118
Argyreia acuta Lour. 130
Argyreia capitata (Vahl) Choisy 130
Argyreia mollis (Burm. f.) Choisy 130
Argyreia nervosa (Burm. f.) Boj. 130
Argyreia obtusifolia Lour. 130
Arisaema cordatum N. E. Br. 158
Arisaema erubeacens (Wall.) Schott 158
Aristida chinensis Munro 178
Aristolochia fangchi Y. C. Wu ex L. D. Chow et S. M. Hwang 31
Aristolochia fordiana Hemsl. 31
Aristolochia kaempferi Willd. 31
Aristolochia westlandii Hemsl. 31
Aristolochiaceae 31
Armeniaca mume Sieb. 67
Arrhenatherum elatius (L.) Presl var. bulbosum (Willd.) Hyland f. variegatum Hitchc. 178
Artabotrys hexapetalus (L. f.) Bhandari 23
Artabotrys hongkongensis Hance 23
Artemisia annua L. 118
Artemisia argyi Levl. et Vant. 118
Artemisia chingii Pamp. 118
Artemisia lactiflora Wall. ex DC. 118
Arthraxon hispidus (Thunb.) Makino 178
Arthraxon lanceolatus (Roxb.) Hochst. 178
Artocarpus altilis (Park.) Fosberg 84
Artocarpus heterophyllus Lam. 84
Artocarpus hypargyreus Hance 84
Artocarpus nitidus ssp. lingnanensis

(Merr.) Jarr. 84
Artocarpus styracifolius Pierre 84
Artocarpus tonkinensis A. Chev. ex Gagnep. 84
Arundina graminifolia (D. Don) Hochr. 168
Arundinella barbinodis Keng ex B. S. Sun et Z. H. Hu 178
Arundinella cochinchinensis Keng 179
Arundinella decenmppedalis (O. Kuntze) Janow. 179
Arundinella hirta (Thunb.) Tanaka 179
Arundinella nepalensis Trin. 179
Arundinella setosa Trin. 179
Arundo donax L. 179
Asarum caudigerum Hance 31
Asclepiadaceae 110
Asclepias curassavica L. 110
Asparagus cochinchinensis (Lour.) Merr. 153
Asparagus densiflorus (Kunth) Jessop 154
Asparagus macowanii Baker 154
Asparagus myriocladus Baker 154
Asparagus officinalis L. 154
Asparagus plumosus Baker 154
Aspidiaceae 12
Aspidistra elatior Bl. 154
Aspidistra minutiflora Stapf 154
Aspleniaceae 10
Asplenium austro-chinense Ching 10
Asplenium crinicaule Hance 10
Asplenium excisum Presl 10
Asplenium falcatum Lam. 10
Asplenium neolaserpitiifolium Tard. -Blot et Ching 10

Asplenium normale Don 10
Asplenium obscurum Bl. 10
Asplenium prolongatum Hook. 10
Asplenium pseudolaserpitiifolium Ching 10
Aster ageratoides Turcz. 118
Aster baccharoides (Benth.) Steetz. 118
Aster panduratus Nees ex Walp. 118
Aster sampsonii (Hance) Hemsl. 119
Asystasiella chinensis (S. Moore) E. Hoss. 137
Atalantia buxifolia (Poir.) Oliv. ex Benth. 94
Ataxipteris sinii (Ching) Holttum 12
Athyriaceae 8
Athyriopsis japonica (Thunb.) Ching 8
Athyriopsis petersenii (Kunze) Ching 8
Atropa belladonna L. 128
Aucuba chinensis Benth. 99
Aucuba chinensis Benth. var. *angusta* Wang 100
Averrhoa carambola L. 42
Avicennia marina (Forsk.) Vierh. 139
Axonopus affinis Chase 179
Axonopus compressus (Swartz) Beauv. 179
Azolla imbricata (Roxb.) Nakai 16
Azollaceae 16

B

Bacopa monnieri (L.) Wettst. 132
Baeckea frutescens L. 52
Balanophora harlandii Hook. f. 91
Balanophora hongkongensis K. M. Lau 91
Balanophoraceae 91

Balsaminaceae 42
Bambusa chungii McClure 188
Bambusa eutuldoides McClure 189
Bambusa flexuosa Munro 188
Bambusa gibba McClure 188
Bambusa lapidea McClure 188
Bambusa multiplex (Lour.) Raeusch. ex J. A. et J. H. Schult. 'Fernleaf' 189
Bambusa multiplex (Lour.) Raeusch. ex J. A. et J. H. Schult. var. *riviereorum* R. Maire 189
Bambusa oldhami Munro 189
Bambusa pervariablilis McClure 189
Bambusa sinospinosa McClure 189
Bambusa textilis McClure 189
Bambusa tuldoides Munro 189
Bambusa ventricosa McClure 189
Bambusa vulgaris Schrad. ex Wendl. 'Vittata' 189
Barleria cristata L. 137
Barthea barthei (Hance ex Benth.) Krass. 54
Basella alba L. 41
Basellaceae 41
Bauhinia apertilobata Merr. et Metc. 71
Bauhinia blakeana Dunn 71
Bauhinia championii (Benth.) Benth. 71
Bauhinia corymbosa Roxb. 71
Bauhinia glauca (Wall. ex Benth.) Benth. 71
Bauhinia kwangtungensis Merr. 71
Bauhinia purpurea L. 71
Bauhinia variegata L. 71
Bauhinia variegata L. var. *candida* (Roxb.) Voigt 71
Begonia bowerae Ziesenh. 48
Begonia crassirostris Irmsch. 48
Begonia elatior Hort. ex Steud. 48
Begonia fimbristipula Hance 48
Begonia fischeri Schrank 48
Begonia heracleifolia Cham. et Schldl. 49
Begonia imperialis Lem. 49
Begonia maculata Raddi 49
Begonia masoniana Iremsch. 49
Begonia palmata D. Don 49
Begonia rex Putz. 49
Begonia rex-cultorum Bailey 49
Begonia semperflorens Link et Otto 49
Begonia tuberhybrida Voss. 49
Begoniaceae 48
Beilschmiedia delicata S. Lee et Y. T. Wei 24
Beilschmiedia tsangii Merr. 24
Belamcanda chinensis (L.) DC. 161
Bellis perennis L. 119
Benincasa hispida (Thunb.) Cogn. 47
Benincasa hispida (Thunb.) Cogn. var. *chieh-qua* How 47
Berberidaceae 29
Berchemia floribunda (Wall.) Brongn. 91
Berchemia lineata (L.) DC. 91
Berchemia polyphylla Wall. ex Lawson var. *leioclada* Hand.-Mazz. 91
Bergia ammanioides Roxb. ex Roth 37
Bergia capensis L. 37
Beta vulgaris L. var. *cicla* L. 40
Bidens alba (L.) DC. 119
Bidens bipinnata L. 119

Bidens biternata (Lour.) Merr. et Sherff. 119
Bidens pilosa L. 119
Bignoniaceae 135
Billbergia pyramidalis (Sims) Lindl. 149
Biophytum sensitivum (L.) DC. 42
Bischofia javanica Bl. 61
Bischofia polycarpa (Levl.) Airy-Shaw 61
Bismarckia nobilis Hildebr. et H. Wendl. 164
Bixa orellana L. 46
Bixaceae 46
Blainivillea acmella (L.) Phillipson 119
Blastus cochinchinensis Lour. 54
Blastus conginiauxii Stapf 54
Blechnaceaeu 10
Blechnum orientale L. 10
Blumea balsamifera (L.) DC. 119
Blumea clarkei Hook. f. 119
Blumea fistulosa (Roxb.) Kurz 119
Blumea megacephala (Randeria) Chang et Tseng 119
Blumea oblongifolia Kitam. 119
Blumea sessiliflora Decne. 119
Blyxa japonica (Miq.) Maxim. 146
Boehmeria macrophylla Hornem. 87
Boehmeria nivea (L.) Gaud. 87
Boehmeria nivea (L.) Gaud. var. tenacissima (Gaudich) Miq. 87
Boeica guileana B. L. Burtt 135
Bolbitidaceae 13
Bolbitis heteroclita (Presl) Ching 13
Bolbitis subcordata (Cop.) Ching 13
Bombacaceae 58

Bombax ceiba L. 58
Boraginaceae 128
Borassus flabellifer L. 164
Borreria articularia (L. f.) F. N. Will. 112
Borreria latifolia (Aubl.) K. Schum. 113
Borreria stricta (L. f.) G. E. W. Mey. 113
Bothriochloa bladhii (Retz.) S. T. Blake 179
Bothriospermum tenellum (Hornem.) Fisch. et Mey. 128
Bougainvillea glabra Choisy 45
Bougainvillea spectabilis Willd. 45
Bowringia callicarpa Champ. ex Benth. 74
Brachiaria eruciformis (J. E. Smith) Griseb. 179
Brachiaria ramosa (L.) Stapf 179
Brachiaria subquadripara(Trin.) Hitchc. 179
Brachiaria villosa (L.) A. Camus 179
Brachychiton acerifolius (Cunn.) F. Muell. 57
Brachychiton rupestris (Lindl.) Schum. 57
Brachycorythis galeandra (Rchb. f.) Summ. 168
Brainea insignis (Hook.) J. Sm. 11
Brasenia schreberi J. F. Gmel. 28
Brassica alboglabra L. H. Bailey 33
Brassica caulorapa DC. ex Laveille 33
Brassica chinensis L. 33
Brassica integrifolia (H. West) O. E. Schulz 33
Brassica juncea (L.) Coss. var.

megarrhiza Tsen et Lee 34
Brassica juncea (L.) Czern. et Coss. 33
Brassica napobrassica Mill. 34
Brassica oleracea L. var. *acephala* f. *partita* Hort. 34
Brassica oleracea L. var. *acephala* f. *tricolor* Hort. 34
Brassica oleracea L. var. *capitata* L. 34
Brassica parachinensis Bailey 34
Brassica pekinensis (Lour.) Skeels 34
Breynia fruticosa (L.) Hook. f. 61
Breynia nivosa (W. C. Sm.) Small 61
Breynia vitis-idaea (Burm. f.) C. E. C. Fisch. 61
Bridelia insulana Hance 61
Bridelia stipularis (L.) Bl. 61
Bridelia tomentosa Bl. 62
Bromeliaceae 149
Broussonetia kazinoki Sieb. et Zucc. 84
Broussonetia papyrifera (L.) L'Herit. ex Vent. 84
Brucea javanica (L.) Merr. 95
Bruguiera gymnorrhiza (L.) Lam. 55
Brunfelsia latifolia Benth. 128
Bryophyllum pinnatum (L. f.) Oken 36
Buchnera cruciata Hamilt 132
Buddleja asiatica Lour. 107
Buddleja lindleyana Fort. 107
Bulbophyllum affine Lindl. 168
Bulbophyllum ambrosia (Hance) Schltr. 168
Bulbophyllum kwangtungense Schltr. 169

Bulbophyllum odoratissimum (J. E. Smith) Lindl. 169
Bulbostylis barbata (Rottb.) Kunth 173
Bulbostylis densa (Wall.) Hand. -Mazz. 173
Burmannia itoana Makino 168
Burmanniaceae 168
Burseraceae 95
Butia capitata (Mart.) Becc. 164
Butomaceae 146
Buxaceae 81
Buxus harlandii Hance 81
Buxus henryi Mayr. 81
Buxus sempervirens L. 81
Buxus sinica (Rehd. et Wils.) Cheng 81
Buxus stenophylla Hance 81
Byttneria aspera Col. 57

C

Cactaceae 49
Caesalpinia bonduc (L.) Roxb. 71
Caesalpinia crista L. 71
Caesalpinia decapetala (Roth.) Alston 72
Caesalpinia magnifoliolata Metc. 72
Caesalpinia millettii Hook. et Arn. 72
Caesalpinia minax Hance 72
Caesalpinia pulcherrima (L.) Sw. 72
Caesalpinia sappan L. 72
Caesalpinia sinensis (Hemsl.) J. E. Vidal 72
Caesalpinia vernalis (L.) Champ. ex Benth. 72
Caesalpiniaceae 71
Cajanus cajan (L.) Millsp. 74

Cajanus scarabaeoides (L.) Thou. 74
Caladium bicolor Vent. 158
Caladium hortulanum Birdsey 158
Calamagrostis epigejos (L.) Roth 179
Calamus rhabdocladus Burret 164
Calamus tetradactylus Hance 164
Calamus thysanolepis Hance 164
Calanthe argenteo-striata C. Z. Tang et S. J. Cheng 169
Calanthe discolor Lindl. 169
Calanthe formosana Rolfe 169
Calanthe reflexa (O. Kuntze) Maxim. 169
Calanthe triplicata (Willem.) Ames 169
Calathea makoyana E. Morr. et Boom 153
Calathea roseopicta (Linden) Regel 153
Calceolaria crenatiflora Cav. 132
Calendula officinalis L. 119
Calliandra haematocephala Hassk. 70
Calliandra riparia Pittier 70
Calliaspidia guttata (T. S. Brand.) Bremek. 137
Callicarpa brevipes (Benth.) Hance 139
Callicarpa candicans (Burm. f.) Hochr. 139
Callicarpa cathayana H. T. Chang 140
Callicarpa dichotoma (Lour.) K. Koch 140
Callicarpa formosana Rolfe 140
Callicarpa giraldii Hesse ex Rehd. 140
Callicarpa integerrima Champ. 140

Callicarpa kochiana Makino 140
Callicarpa kwangtuangensis Chun 140
Callicarpa macrophylla Vahl 140
Callicarpa nudiflora Hook. et Arn. 140
Callicarpa rubella Lindl. 140
Callicarpa rubella Lindl. f. *angustata* Péi 140
Callicarpa rubella Lindl. f. *crenata* Péi 140
Callipteris esculenta (Retz.) J. Sm. ex Moore et Houlst. 8
Callistemon pinifolius Sweet 52
Callistemon rigidus R. Br. 52
Callistemon viminalis G. Don ex Loud. 53
Callistephus chinensis (L.) Nees 119
Callitrichaceae 44
Callitriche oryzetorum Petr. 44
Callitriche stagnalis Scop. 44
Calocedrus macroleps Kurz var. *macrolepis* Kurz 18
Calogyne pilosa R. Br. ssp. *chinensis* (Benth.) H. S. Kiu 127
Calophanoides chinensis (Champ.) C. Y. Wu et H. S. Lo (*Adhatoda chinensis* Champ. , *Justicia championi* T. Anders.) 137
Calophanoides hainanensis C. Y. Wu 137
Calophanoides quadrifaria Ridl. 137
Calophyllum membranaceum Gardn. et Champ. 56
Camellia assamica (Mast.) Chang 50
Camellia assimilis Champ. et Benth. 50
Camellia brevistyla (Hayata) Coh. St.

50
Camellia caudata Wall. 50
Camellia chekiang-oleosa Hu 50
Camellia crapnelliana Tutch. 50
Camellia euryoides Lindl. 50
Camellia granthamiana Sealy 50
Camellia japonica L. 51
Camellia kissi Wall. 51
Camellia nitidissima Chi 51
Camellia oleifera Abel 51
Camellia salicifolia Champ. ex Benth. 51
Camellia sasanqua Thunb. 51
Camellia semiserrata Chi 51
Camellia sinensis (L.) O. Kuntze 51
Campanulanceae 127
Campanumoea javanica Bl. 127
Campsis grandiflora (Thunb.) Schum. 135
Camptotheca acuminata Decne. 100
Canarium album (Lour.) Raeusch. 95
Canarium tramdenum Dai et Yakovl. 96
Canavalia cathartica Thou. 74
Canavalia maritima (Aubl.) Thou. 74
Canna edulis Ker 153
Canna generalis Bailey 153
Canna indica L. 153
Canna indica L. var. *flava* Roxb. 153
Canna orchioides Bailey 153
Cannaceae 153
Cansjera rheedii J. F. Gmelin 90
Canthium dicoccum (Gaertn.) Teysmenn et Binnedijk. 113
Canthium horridum Bl. 113

Capillipedium assimile (Steud.) A. Camus 179
Capillipedium parviflorum (R. Br.) Stapf 179
Capparia cantoniensis Lour. 33
Capparidaceae 33
Caprifoliaceae 117
Capsella bursa-pastoris (L.) Medic. 34
Capsicum annuum L. 128
Capsicum annuum L. var. *cerasiformis* Irish 128
Capsicum frutescens L. 128
Carallia brachiata (Lour.) Merr. 55
Cardamine flexuosa With. 34
Cardamine hirsuta L. 34
Cardamine scutata Thunb. 34
Cardiospermum halicacabum L. 96
Carex baccans Nees 173
Carex chinensis Retz. 173
Carex cruciata Wahlenb. 173
Carex cryptostachys Brongn. 173
Carex filicina Nees 173
Carex manca Boott 173
Carex nemostachys Steud. 173
Carex scaposa C. B. Clarke 173
Carex teinogyna Boott 173
Carex truncatigluma C. B. Clarke 173
Carica papaya L. 49
Caricaceae 49
Carmona microphylla (Lam.) G. Don 128
Caryophyllaceae 37
Caryopteris incana (Thunb.) Miq. 141
Caryota maxima Bl. 165
Caryota mitis Lour. 164
Caryota monostachya Becc. 165

Caryota urens L.   165
Casearia glomerata Roxb.   46
Casearia villilimba Merr.   46
Cassia alata L.   72
Cassia bicapsularis L.   72
Cassia fistula L.   72
Cassia leschenaultiana DC.   72
Cassia mimosoides L.   72
Cassia occidentalis L.   72
Cassia siamea Lam.   72
Cassia surattensis Burm. f.   72
Cassia tora L.   72
Cassytha filiformis L.   24
Castanea mollissima Bl.   81
Castanopsis carlesii (Hemsl.) Hayata   82
Castanopsis chinensis Hance   82
Castanopsis eyrei (Champ.) Tutch.   82
Castanopsis fabri Hance   82
Castanopsis fargesii Franch.   82
Castanopsis fissa (Champ. ex Benth.) Rehd. et Wils.   82
Castanopsis hystrix A. DC.   82
Castanopsis kawakamii Hayata   82
Castanopsis lamontii Hance   82
Castanopsis tibetana Hance   82
Castanospermum australe A. Cunn. ex Mudie   74
Casuarina cunninghamiana Miq.   83
Casuarina equisetifolia Forst.   83
Casuarinaceae   83
Catharanthus roseus (L.) G. Don   109
Cathaya argyrophylla Chun et Kuang   17
Cattleya bicolor Lindl.   169
Cattleya citrina Lindl.   169
Cattleya labiata Lindl.   169

Cattleya labiata Lindl. var. *trianae* (Linden et Reichb. f.) Regel   169
Catunaregam spinosa (Thunb.) Tirv.   113
Cayratia corniculata (Benth.) Gagnep.   93
Cayratia japonica (Thunb.) Gagnep.   93
Cecropia adenopus Mast. ex Miq.   84
Ceiba insignis (Kunth) Gibbs et Semir   58
Ceiba pentandra (L.) Gaertn.   58
Celastraceae   89
Celastrus aculeatus Merr.   89
Celastrus hindsii Benth.   89
Celastrus kusanoi Hayata   89
Celastrus monospermus Roxb.   89
Celosia argentea L.   41
Celosia cristata L.   41
Celosia cristata L. var. *plumosa* Hort.   41
Celtis austro-sinensis Chun   83
Celtis biondii Pamp.   83
Celtis sinensis Pers.   83
Celtis timorensis Span   83
Cenchrus echinatus L.   179
Centaurea cineraria L.   119
Centella asiatica (L.) Urban   102
Centipeda minima (L.) A. Br. et Aschers   119
Centotheca lappacea (L.) Desv.   180
Centranthera cochinchinensis (Lour.) Merr.   132
Cephalotaxaceae   19
Cephalotaxus fortunei Hook. f.   19
Cerasus cerasoides (D. Don) S. Ya. Sokolov   67

Ceratophyllum 28
Ceratophyllum demersum L. 28
Ceratopteris thalictroides (L.) Brongn.
  7
Cerbera manghas L. 109
Cercis chinensis Bunge 72
Cereus fernambucensis Lem. 49
Ceropegia woodii Schldl. 110
Cestrum nocturnum L. 129
Chamaecyparis obtusa (Sieb. et Zucc.) Endl. 18
Chamaedorea elegans Mart. 165
Chamaedorea metallica O. F. Cook ex H. E. Moore 165
Championella tetrasperma (Champ. ex Benth.) Brem. 137
Cheilosoria chusana (Hook.) Ching et K. H. Shing 7
Cheilosoria tenuifolia (Burm.) Trev. 7
Chenopodiaceae 40
Chenopodium acuminatum Willd. ssp. *virgatum* (Thunb.) Kitam. 40
Chenopodium ambrosioides L. 40
Chenopodium ficifolium Smith 40
Chenopodium glaucum L. 40
Chieniopteris harlandii (Hook.) Ching 11
Chirita eburnea Hance 135
Chirita fimbrisepala Hand.-Mazz. 135
Chirita sinensis Lindl. 135
Chirita sinensis Lindl. var. *angustifolia* Dunn 135
Chloranthaceae 32
Chloranthus erectus (Buch.-Ham.) Verdc. 32
Chloranthus holostegius Hand.-Mazz. 32
Chloranthus monostachys R. Br. 32
Chloranthus multistachys Pei 32
Chloranthus spicatus (Thunb.) Makino 32
Chloris barbata Sw. 180
Chloris formosana (Honda) Keng 180
Chlorophytum bichetii (Karrer) Backer 154
Chlorophytum capense (L.) O. Kuntze var. *variegatum* Hort. 154
Chlorophytum comosum Baker 154
Chlorophytum laxum R. Br. 154
Choerospondias axillaris (Roxb.) Burtt. et Hill 98
Christia obcordata (Poir.) Bakh. f. 74
Christia vespertilionis (L. f.) Bahn. f. 74
Chrysalidocarpus lutescens Wendl. 165
Chrysanthemum paludosum Poir. 119
Chrysanthemum segetum L. 119
Chrysopogon aciculatus (Retz.) Trin. 180
Chrysopyllum lanceolatum A. DC. Var. *stellatocarpon* Van Royen 103
Chukrasia tabularis A. Juss. 96
Chukrasia tabularis A. juss. var. *velutina* (Wall.) King 96
Chuniophoenix hainanensis Burret 165
Chuniophoenix nana Burret 165
Cibotium barometz (L.) J. Sm. 4
Cinnamomum appelianum Schewe 24
Cinnamomum aromaticum Nees 24
Cinnamomum burmannii (C. G. et Th. Nees) Bl. 24
Cinnamomum camphora (L.) Presl

24
Cinnamomum japonicum Sieb. 24
Cinnamomum jensenianum Hand. -Mazz. 25
Cinnamomum micranthum (Hayata) Hayata 25
Cinnamomum porrectum (Roxb.) Kosterm. 25
Cinnamomum validinerve Hance 25
Cinnamomum zeylanicum Nees 25
Cirrhopetalum delitescens (Hance) Rolfe 169
Cirsium chinense Gardn. et Champ. 120
Cirsium hupehense Pamp. 120
Cirsium japonicum Fisch. ex DC. 120
Cirsium lineare (Thunb.) Sch. -Bip. 120
Cissus hexangularis Thorel ex Planch. 93
Cissus repens Lam. 93
Citrullus lanatus (Thunb.) Mats. et Nakai 47
Citrus aurantium L. 94
Citrus grandis (L.) Osbeck 94
Citrus limon (L.) Burm. f. 94
Citrus limonia Osbeck 94
Citrus medica L. 94
Citrus medica L. var. *sarcodactylis* (Noot.) Swingle 94
Citrus reticulata Blanco 94
Citrus sinensis (L.) Osbeck 94
Cladium mariscus Pohl ssp. *jamaicense* (Crantz.) Kukenthal 173
Claoxylon indicum (Reinw. ex Bl.) Hassk. 62
Clausena lansium (Lour.) Skeels 94

Cleidiocarpon cavaleriei (Levl.) Airy Shaw 62
Cleidion brevipetiolatum Pax et Hoffm. 62
Cleisostoma fordii Hance 169
Cleisostoma rostratum (Lodd.) Seidenf. ex Averyanov 169
Cleisostoma simondii (Gagnep.) Seidenf. 169
Cleisostoma simondii (Gagnep.) Seidenf. var. *guangdongense* Z. H. Tsi 169
Cleistocalyx operculatus (Roxb.) Merr. et Perry 53
Clematis chinensis Osbeck 27
Clematis crassifolia Benth. 27
Clematis filamentosa Dunn 27
Clematis finetiana Levl. et Vant. 27
Clematis henryi Oliv. 27
Clematis meyeniana Walp. 27
Clematis uncinata Champ. 27
Cleome lutea Hook. 33
Cleome spinosa Jacq. 33
Cleome viscosa L. 33
Clerodendranthus spicatus (Thunb.) C. Y. Wu ex H. W. Li 143
Clerodendrum bungei Steud. 141
Clerodendrum canescens Wall. 141
Clerodendrum colebrookianum Walp. 141
Clerodendrum cyrtophyllum Turcz. 141
Clerodendrum fortunatum L. 141
Clerodendrum inerme (L.) Gaertn. 141
Clerodendrum japonicum (Thunb.) Sweet 141
Clerodendrum kwangtungense Hand. - Mazz. 141

Clerodendrum philippinum Schauer 141
Clerodendrum philippinum Schauer var. *simplex* Moldenke 141
Clerodendrum speciosissimum Van Geert 141
Clerodendrum splendens G. Don 141
Clerodendrum thomsonae Balf. 142
Clerodendrum ugandense Prain 142
Cleyera japonica Thunb. 51
Clinopodium chinensis (Benth.) O. Kuntze 143
Clinopodium gracile (Benth.) Matsum. 144
Clitoria ternatea L. 74
Clivia miniata Regel 161
Coccothrinax crinita Becc. 165
Cocculus orbiculatus (L.) DC. 30
Cocos nucifera L. 165
Codariocalyx gyroides (Roxb. ex Link.) Hassk. 75
Codariocalyx motorius (Houtt.) Ohashi 74
Codiaeum variegatum (L.) A. Juss. var. *pictum* (Lodd.) Muell.-Arg. 62
Codiaeum variegatum (L.) A. Juss. var. *pictum* (Lodd.) Muell.-Arg. f. *ambigum* Pax 62
Codiaeum variegatum (L.) A. Juss. var. *pictum* (Lodd.) Muell.-Arg. f. *appendiculatum* Pax 62
Codiaeum variegatum (L.) A. Juss. var. *pictum* (Lodd.) Muell.-Arg. f. *platylla* Pax 62
Codonacanthus pauciflorus Nees 137
Codonopsos lanceolata (Sieb. et Zucc.) Trautv. 127

Coelachne simpliciuscula (Wight et Arn.) Munro ex Benth. 180
Coelogyne fimbrata Lindl. 170
Coffea arabica L. 113
Coix lacryma-jobi L. 180
Coleus scutellarioides (L.) Benth. 144
Coleus scutellarioides (L.) Benth. Var *crispiplus* (Merr.) H Keng 144
Colocasia antiquorum Schott 158
Colocasia eaculenta (L.) Schott 158
Colysis digitata (Baker) Ching 14
Colysis elliptica (Thunb.) Ching 14
Colysis hemionitidea (Wall. ex Mett.) C. Presl 14
Colysis pothifolia (D. Don) C. Presl 14
Combretaceae 55
Combretum alfredii Hance 55
Commelina auriculata Bl. 148
Commelina bengalensis L. 148
Commelina communis L. 148
Commelina diffusa Burm. f. 148
Commelina paludosa Bl. 148
Commelinaceae 147
Compositae 118
Connaraceae 99
Consolida ajacis (L.) Schur 28
Convolvulaceae 130
Conyza bonariensis (L.) Cronq. 120
Conyza canadensis (L.) Cronq. 120
Conyza leucantha (D. Don) Ludlow er Raven 120
Coptosapelta diffusa (Champ. ex Benth.) Van Steenis 113
Corchorus aestuans L. 56
Corchorus capularis L. 56
Cordia dichotoma Forst. f. 128
Cordyline fruticosa (L.) A. Cheval.

163

Coreopsis lanceolata L.   120
Coreopsis tinctoria Nutt.   120
Coriandrum sativum L.   102
Cornaceae   99
Corypha umbraculifera L.   165
Corypha utan Lam.   165
Cosmos bipinnatus Cav.   120
Cosmos sulphureus Cav.   120
Costus speciosus (Koen.) Smith   152
Craibiodendron scleranthum (Dop) Judd. var. kwangtungense (S. Y. Hu) Judd.   102
Crassocephalum crepidioides (Benth.) S. Moore   120
Crassula arborescens Willd.   36
Crassula lycopodioides Lam.   36
Crassula ovata (Mill.) Druce   36
Crassulaceae   35
Crateva religiosa G. Forst.   33
Crateva trifoliata (Roxb.) Sun   33
Cratoxylum cochinchinense (Lour.) Bl.   56
Crepidomanes bipunctatum (Poir.) Cop.   4
Crepidomanes insigne (v. d. Bosch) Fu   4
Crepidomanes racemulosum (v. d. Bosch) Ching   4
Crescentia alata H. B. K.   136
Crinum amabile Donn.   161
Crinum asiaticum L. var. sinicum (Roxb. ex Herb.) Baker   161
Crossandra infundibuliformis Nees   137
Crossostephium chinense (L.) Makino   120
Crotalaria albida Heyne ex Roth   74
Crotalaria assamica Benth.   74
Crotalaria calycina Schrank   74
Crotalaria pallida Ait.   74
Crotalaria retusa L.   74
Crotalaria sessiliflora L.   74
Crotalaria uncinella Lam.   74
Crotalaria zanzibarica Benth.   75
Croton crassifolius Geisel.   62
Croton lachnocarpus Benth.   62
Croton tiglium L.   62
Cruciferae   33
Cryptanthus acaulis (Lindl.) Beer   150
Cryptanthus zonatus Beer   150
Cryptocarya chinensis (Hance) Hemsl.   25
Cryptocarya concinna Hance   25
Cryptolepis sinensis (Lour.) Merr.   111
Ctenanthe oppenheimiana (E. Morr.) K. Schum.   153
Ctenitis costulisora Ching   12
Ctenitis rhodolepis (Clarke) Ching   12
Cucumis melo L.   47
Cucumis melo L. var. conomon (Thunb.) Makino   47
Cucumis sativus L.   47
Cucurbita moschata (Duch. ex Lam.) Duch. ex Poir.   47
Cucurbita pepo L. var. ovifera (L.) Harz   47
Cucurbitaceae   47
Cudrania cochinchinensis (Lour.) Kudo et Masam.   84
Cudrania tricuspidata (Carr.) Bureau ex Lavallee   85
Cunninghamia lanceolata (Lamb.) Hook.   18
Cuphea balsamona Cham. et Schlecht.

42
Cuphea hookeriana Walp.   43
Cuphea hyssopifolia H. B. K.   43
Cuphea platycentra Lemarie   42
Cupressaceae   18
Cupressus funebris Endl.   18
Curculigo breviscapa S. C. Chen   167
Curculigo capitulata (Lour.) O. Kuntze   167
Curculigo glabrescens (Ridl.) Merr.   167
Curculigo orchioides Gaertn.   167
Curcuma aromatica Salisb.   152
Curcuma kwangsiensis Lee et Liang   152
Curcuma longa L.   152
Curcuma phaeocaulis Val.   152
Cuscuta chinensis Lam.   130
Cuscuta japonica Choisy   130
Cyanotis arachnoidea C. B. Clarke   148
Cyanotis vaga (Lour.) Roem. et Schult.   148
Cyatheaceae   5
Cyathula prostrate (L.) Bl.   41
Cycadaceae   17
Cycas elongata (Leandri) D. Y. Yang et T. Chen   17
Cycas fairylakea D. Y. Wang   17
Cycas furfuracea W. V. Fitzg.   17
Cycas hainanensis C. J. Chen   17
Cycas parvulus S. L. Yang   17
Cycas pectinata Griff.   17
Cycas revolut a Thunb.   17
Cycas sexseminifera F. N. Wei   17
Cycas taiwaniana Carruth.   17
Cyclamen persicum Mill.   126
Cyclea barbata Miers   30

Cyclea hypoglauca (Schauer) Diels   30
Cyclea racemosa Oliv.   30
Cyclobalanopsis championii (Benth.) Oerst.   82
Cyclobalanopsis fleuryi (Hick. et A. Camus) Chun   82
Cyclobalanopsis myrsinifolia (Bl.) Oerst.   82
Cyclobalanopsis neglecta Schott.   82
Cyclobalanopsis pachyloma (Seem.) Schott.   82
Cyclosorus acuminatus (Houtt.) Nakai   8
Cyclosorus dentatus (Forsk.) Ching   8
Cyclosorus heterocarpus (Bl.) Ching   9
Cyclosorus interruptus (Willd.) H. Ito   9
Cyclosorus latipinnus (Benth.) Tard. - Blot   9
Cyclosorus parasiticus (L.) Farw.   9
Cyclosorus truncatus (Poir.) Farw.   9
Cymbidium dayanum Rchb. f.   170
Cymbidium ensifolium (L.) Sw.   170
Cymbidium ensifolium (L.) Sw. var. munronianum (King et Prantl) Tang et Wang   170
Cymbidium faberi Rolfe   170
Cymbidium goeringii (Rchb. f.) Rchb. f.   170
Cymbidium kanran Makino   170
Cymbidium lancifolium Hook.   170
Cymbidium simulans Rolfe   170
Cymbidium sinense (Jackson ex Andr.) Willd.   170
Cymbopogon caesius (Nees) Stapf   180

Cymbopogon distans (Nees) A. Camus 180
Cymbopogon hamatulus (Nees ex Hook. et Arn.) A. Camus 180
Cymbopogon nardus (L.) Rendle 180
Cynanchum auriculatum Royle ex Wight 111
Cynanchun corymbosum Wight 111
Cynodon dactylon (L.) Pers. 180
Cyperaceae 173
Cyperus compressus L. 174
Cyperus difformis L. 174
Cyperus diffusus Vahl 174
Cyperus exaltatus Retz. 174
Cyperus flabelliformis Rottb. 174
Cyperus haspan L. 174
Cyperus imbricatus Retz. 174
Cyperus iria L. 174
Cyperus malaccensis Lam. 174
Cyperus malaccensis Lam. var. *brevifolius* Bocklr. 174
Cyperus microiria Steud. 174
Cyperus nipponicus Franch. et Sav. 174
Cyperus papyrus L. 174
Cyperus pilosus Vahl 174
Cyperus polystachyus (Rottb.) P. Beauv. 174
Cyperus pygmaeus Rottb. 174
Cyperus rotundus L. 174
Cyrtococcum accrascens (Trin.) Stapf 180
Cyrtococcum patens (L.) A. Camus 180
Cyrtomium balansae (Christ) C. Chr. 11
Cyrtomium caryotideum (Wall. ex Hook. et Grev.) Presl 11
Cyrtomium falcatum (L. f.) Presl 11
Cyrtomium fortunei J. Sm. 11
Cyrtostachys renda Bl. 165

D

Dactyloctenium aegyptium (L.) Beauv. 180
Daemonorops margaritae (Hance) Becc. 165
Dahlia coccinea Cav. 120
Dahlia jaurezii Hort. ex Sasaki 120
Dahlia merckii Lehm. 120
Dahlia pinnata Cav. 120
Dalbergia balansae Prain 75
Dalbergia benthamii Prain 75
Dalbergia candenatensis (Dennst.) Prain 75
Dalbergia hancei Benth. 75
Dalbergia millettii Benth. 75
Dalbergia mimosoides Franch. 75
Dalbergia odorifera T. Chen 75
Damnacanthus indicus (L.) Gaertn. f. 113
Daphne papyracea Wall. ex Steud. 44
Daphniphyllaceae 66
Daphniphyllum calycinum Benth. 66
Daphniphyllum macropodum Miq. 66
Daphniphyllum oldhamii (Hemsl.) Rosenth. 66
Datura metel L. 129
Datura stramonium L. 129
Daucus carota L. var. *sativa* Hoffm. 102
Davallia austro-sinica Ching 13
Davallia formosana Hayata 13
Davalliaceae 13

Delonix regia (Hook.) Raf. 72
Dendranthema indicum (L.) Des Moul. 121
Dendranthema morifolium (Ramat.) Tzvel. 121
Dendrobenthamia hongkongensis (Hemsl.) Hutch. 100
Dendrobium acinaciforme Roxb. 170
Dendrobium chrysotoxum Lindl. 170
Dendrobium loddigesii Rolfe 170
Dendrobium nobilis Lindl. 170
Dendrocalamopsis beecheyana (Munro) Keng f. var. *pubescens* (P. F. Li) Keng f. 189
Dendrocalamus latiflorus Munro 189
Dendropanax dentigerus (Harms) Merr. 100
Dendropanax proteus (Champ. ex Benth.) Benth. 100
Dendrophthoe pentandra (L.) Miq. 90
Dendrotrophe varians Miq. 91
Dennstaedtiaceae 5
Derris alborubra Hemsl. 75
Derris elliptica (Roxb.) Benth. 75
Derris fordii Oliv. 75
Derris fordii Oliv. var. *lucida* How 75
Derris trifoliata Lour. 75
Desmodium caudatum (Thunb.) DC. 75
Desmodium gangeticum (L.) DC. 75
Desmodium heterocarpon (L.) DC. 75
Desmodium heterophyllum (Willd.) DC. 76
Desmodium laxiflorum DC. 76
Desmodium microphyllum (Thunb.) DC. 76
Desmodium multiflorum DC. 76

Desmodium reticulatum Champ. et Benth. 76
Desmodium styracifolium (Osb.) Merr. 76
Desmodium triflorum (L.) DC. 76
Desmodium velutinum (Willd.) DC. 76
Desmos chinensis Lour. 23
Dianella ensifolia (L.) DC. 154
Dianthus caryophyllus L. 37
Dianthus chinensis L. 37
Dianthus japonicus Thunb. 37
Dianthus superbus L. 37
Dichondra micrantha Urban 130
Dichroa febrifuga Lour. 36
Dichroa febrifuga Lour. 66
Dichrocephala integrifolia (L. f.) O. Kuntze 121
Dicksoniaceae 4
Dicliptera chinensis (L.) Juss. 137
Dicotyledoneae 21
Dicranopteris glaucum (Thunb. ex Houtt.) Naikai 3
Dicranopteris linearis (Burm. f.) Underw. 3
Dicranopteris pedata (Houtt.) Nakai 3
Dicranopteris splendida (Hand.-Mazz.) Tagawa 3
Dictyocline wilfordii (Hook.) J. Sm. 9
Didymostigma obtusum (Clarke) W. T. Wang 135
Dieffenbachia amoena Nichols. 158
Dieffenbachia maculata (Lodd.) G. Don 158
Dieffenbachia seguine (Jacq.) Schott 158

Digitaria adscendens (H. B. K.) Henrard 180
Digitaria chrysoblephara Fig. 181
Digitaria ciliaris (Retz.) Koel. 181
Digitaria heterantha (Hook. f.) Merr. 181
Digitaria longiflora (Retz.) Pers. 181
Digitaria microbachne (Presl.) Hitchc. 181
Digitaria radicosa (Presl) Miq. 181
Digitaria sanquinalis (L.) Scop. 181
Digitaria violascens Link 181
Dillenia indica L. 45
Dillenia turbinata Finet et Gagnep. 45
Dilleniaceae 45
Dimeria flacata Hack. 181
Dimocarpus longan Lour. 97
Dionaea muscipula J. Ellis 37
Dioon spinulosum Dyer 17
Dioscorea alata L. 162
Dioscorea benthamii Prain et Burkill 162
Dioscorea bulbifera L. 162
Dioscorea cirrhosa Lour. 162
Dioscorea fordii Prain et Burkill 162
Dioscorea hispida Dennst. 162
Dioscorea pentaphylla L. 162
Dioscorea persimilis Prain et Burkill 162
Dioscorea polystachya Turcz. 162
Dioscoreaceae 162
Diospyros eriantha Champ. ex Benth. 103
Diospyros kaki Thunb. 103
Diospyros morrisiana Hance 103
Diospyros strigosa Hemsl. 103

Diospyros tsangii Merr. 103
Diospyros tutcheri Dunn 103
Diospyros vaccinioides Lindl. 103
Diplazium donianum (Mett.) Tard. - Blot 8
Diplazium neilingdingensis Miau et W. B. Liao 8
Diplazium subsinuatum (Wall. ex Hook. et Grev.) Tagawa 8
Diploclisia affinis (Oliv.) Diels 30
Diploclisia glaucescens (Bl.) Diels 30
Diploclisia renincarpa Miau et W. B. Liao 30
Diplopanax stachyanthus Hand.-Mazz. 100
Diploprora championii (Lindl.) Hook. f. 170
Diplopterygium cantonensis (Ching) Nakai 3
Diplopterygium chinense (Ros.) De Vol 3
Diplopterygium laevissimium (Christ) Nakai 3
Diplospora dubia (Lindl.) Masamune 113
Dipteridaceae 14
Dipteris chinensis Christ 14
Dipterocarpaceae 52
Dischidia chinensis Champ. ex Benth. 111
Disporum cantoniense (Lour.) Merr. 154
Disporum nantouense S. S. Ying 154
Distylium racemosum Sieb. et Zucc. 80
Dodonaea viscosa (L.) Jacq. 97
Doelligeria scaber (Thunb.) Nees

121
Doellingeria marchandii (H. Lev.) Ling 121
Dolichandrone cauda-felina (Hance) Benth. et Hook. f. 136
Dopatrium junceum (Roxb.) Buch.- Ham. ex Benth. 133
Dracaena cambodiana Pierre ex Gagnep. 163
Dracaena fragrans (L.) Ker-Grawl. 163
Dracaena godseffiana Hort. ex Baker 163
Dracaena reflexa Lam. 163
Dracaena sanderiana Sander ex M. T. Mast. 163
Dracontomelon duperreanum Pierre 99
Drosera burmannii Vahl 37
Drosera spathulata Labill. 37
Drosera spathulata Labill. var. *loureirii* (Hook. et Arn.) Y. Z. Ruan 37
Droseraceae 37
Drymaria cordata (L.) Willd. 37
Drymoglossum piloselloides (L.) C. Presl 14
Drynaria fortunei (Kunze) J. Sm. 15
Drynariaceae 15
Dryoathyrium boryanum (Willd.) Ching 8
Dryopteridaceae 11
Dryopteris championii (Benth.) C. Chr. ex Ching 11
Dryopteris chinensis (Baker) Koidz. 11
Dryopteris costalisora Tagawa 11
Dryopteris decipiens (Hook.) O. Kuntze 12

Dryopteris fuscipes C. Chr. 12
Dryopteris podophylla (Hook.) O. Kuntze 12
Dryopteris scottii (Bedd.) Ching ex C. Chr. 12
Dryopteris tenuicula Matthew et Christ. 12
Dryopteris varia (L.) O. Kuntze 12
Drypetes obtusa Merr. et Chun 62
Duchesnea indica (Andr.) Focke 67
Dunbaria fusca (Wall.) Kurz. 76
Dunbaria podocarpa Kurz. 76
Dunbaria punctata (Wight et Arn.) Benth. 76
Duranta erecta L. 142
Dyckia brevifolia Hort. ex Baker 150
Dypsis decaryi (H. Jumelle) Beentje et J. Dransf. 165
Dysosma pleiantha (Hance) Woodson 29
Dysosma versipellis (Hance) M. Cheng ex Ying 29
Dysoxylum hongkongense (Tutch.) Merr. 96

E

Ebenaceae 103
Echeveria glauca Baker 36
Echinochloa colona (L.) Link 181
Echinochloa crusgalli (L.) Beauv. 181
Echinochloa crusgalli (L.) Beauv. var. *cruspavonis* (H. B. K.) Hitchc. 181
Echinochloa crusgalli (L.) Beauv. var. *mitis* (Pursh) Peterm. 181
Echinochloa glabrescens Munro ex Hook.

f. *hispidula* (Retz.) Nees　181
Echinodorus paleafolius (Nees et Mart.) Macbr.　147
Echinopsis multiplex Pfeiff. et Otto　49
Eclipta prostrata (L.) L.　121
Egenolfia appendiculata (Willd.) J. Sm.　13
Egenolfia sinensis (Baker) Maxon　13
Ehretia longiflora Champ. ex Benth.　128
Ehretia thyrsiflora (Sieb. et Zucc.) Nakai　128
Eichhornia crassipes (Mart.) Solms　156
Elaeagnaceae　92
Elaeagnus conferta Roxb.　92
Elaeagnus glabra Thunb.　92
Elaeagnus gonyanthes Benth.　92
Elaeagnus loureiri Champ.　92
Elaeagnus pungens Thunb.　92
Elaeagnus tucherii Dunn　92
Elaeis guineensis N. J. Jacq.　165
Elaeocarpaceae　57
Elaeocarpus apiculatus Mast.　57
Elaeocarpus chinensis (Gardn. et Champ.) Hook. f. ex Benth.　57
Elaeocarpus decipiens Hemsl.　57
Elaeocarpus dubius A. DC.　57
Elaeocarpus hainanensis Oliv.　57
Elaeocarpus japonicus Sieb. et Chun　57
Elaeocarpus nitentifolius Merr. et Chun　57
Elaeocarpus petiolatus (Jack) Wall. ex Kurz　57
Elaeocarpus serratus L.　57
Elaeocarpus sylvestris (Lour.) Poir.　57

Elaphoglossaceae　13
Elaphoglossum yoshinagae (Yatabe) Makino　13
Elatinaceae　37
Elatine triandra Schkuhr　37
Elatostema lineolatum Wight var. *majus* Wedd　87
Elatostema macintyrei Dunn　87
Eleocharis acicularis (L.) Roem. et Schult.　175
Eleocharis atropurpurea (Retz.) Presl　174
Eleocharis congesta D. Don　175
Eleocharis dulcis (Burm. f.) Trin. ex Hensch.　175
Eleocharis geniculata (L.) Roem. et Schult.　175
Elephantopus scaber L.　121
Elephantopus tomentosus L.　121
Eleusine indica (L.) Gaertn.　181
Elsholtzia argyi L vl.　144
Elsholtzia ciliata (Thunb.) Hyl.　144
Elsholtzia splendens N. akai ex F. Maekawa　144
Embelia laeta (L.) Mez　105
Embelia parviflora Wall. ex A. DC.　105
Embelia ribes Burm. f.　105
Embelia vestita Roxb.　105
Emilia sonchifolia (L.) DC.　121
Encephalartos ferox Bertol. f.　17
Endospermum chinense Benth.　62
Engelhardtia fenzelii Merr.　99
Engelhardtia roxburghiana Lindl.　99
Enkianthus quingqueflorus Lour.　102
Enkianthus serrulatus (Wils.) Schneid.　102

Ensete glaucum (Roxb.) Cheesm. 150
Entada phaceoloides (L.) Merr. 70
Epaltes australis Less. 121
Epiphyllum oxypetalum (DC.) Haw. 49
Epipremnum pinnatum (L.) Engl. 159
Equisetaceae 2
Equisetum arvense L. 2
Equisetum debile Roxb. 2
Equisetum ramosissimum Desf. ex Vauch. 2
Eragrostis amabilis (L.) Wight et Arn. 181
Eragrostis atrovirens (Desf.) Trin. ex Steud. 182
Eragrostis chariis (Schult.) Hitchc. 182
Eragrostis cylindrica (Roxb.) Nees ex Hook. et Arn. 182
Eragrostis ferruginea (Thunb.) P. Beauv. 182
Eragrostis japonica (Thunb.) Trin. 182
Eragrostis longispicula S. C. Sun et H. Q. Wang 182
Eragrostis minor Host. 182
Eragrostis nevinii Hance 182
Eragrostis perennans Keng 182
Eragrostis perlaxa Keng 182
Eragrostis pilosa (L.) Beauv. 182
Eragrostis pilosissima Link 182
Eragrostis reflexa Hack. 182
Eragrostis tenella (L.) P. Beauv. ex Roem. et Schult. 182
Eragrostis tephrosanthus Schult. 182
Eragrostis unioloides (Retz.) Nees ex Steud. 182
Eragrostis zeylanica Nees et Mey. 182
Eranthemum pulchellum Andrews. 137
Erechtites hieracifolia (L.) Raf. ex DC. 121
Eremochloa ciliaris (L.) Merr. 183
Eremochloa ophiuroides (Munro) Hack. 183
Eremochloa procera (Retz) Hubb. 183
Eria corneri Rchb. f. 170
Eria coronaria (Lindl.) Rchb. f. 171
Eria lasiopetala (Willd.) Ormerod 171
Eria pusilla (Griff.) Lindl. 171
Eriachne pallescens R. Br. 183
Erianthus formosanus Stapf 183
Ericaceae 102
Eriobotrya cavaleriei (Levl.) Rehd. 67
Eriobotrya fragrans Champ. ex Benth. 67
Eriobotrya japonica (Thunb.) Lindl. 67
Eriocaulaceae 149
Eriocaulon buergerianum Koern. 149
Eriocaulon cinereum R. Br. 149
Eriocaulon luzulifolium Mart. 149
Eriocaulon sexangulare L. 149
Eriocaulon yaoshanense Ruhl. 149
Eriocaulon yuegangensis F. W. Xing et Y. X. Zhang 149
Eriosema chinensis Vog. 76
Erycibe obtusifolia Benth. 131
Eryngium foetidum L. 102
Erythrina corallodendron L. 76
Erythrina crista-galli L. 76
Erythrina variegata L. 76

Erythrina variegata L. var. *picta* Graf. 76
Erythrodes chinensis (Rolfe) Schltr. 171
Erythrophleum fordii Oliv. 73
Escalloniaceae 66
Eucalyptus camaldulensis Dehnh. 53
Eucalyptus citriodora Hook. f. 53
Eucalyptus globulus Labill. 53
Eucalyptus robusta Smith 53
Eucalyptus tereticornis J. E. Smith 53
Eugenia uniflora L. 53
Eulalia quadrinervis (Hack.) O. Kuntze 183
Eulalia speciosa (Debeaux) O. Kuntze 183
Eulophia graminea Lindl. 171
Eulophia zollingerii (Rchb. f.) J. J. Smith 171
Euonymus angustatus Sprague 89
Euonymus hederaceus Champ. ex Benth. 89
Euonymus laxiflorus L. 89
Euonymus nitidus Benth. 89
Euonymus oblongifolius Loes. et Rehd. 89
Eupatorium catarium Veldk. 121
Eupatorium chinensis L. 121
Eupatorium japonicum Thunb. 121
Eupatorium lindleyanum DC. 121
Eupatorium odoratum L. 121
Euphorbia aggregata A. Berger 62
Euphorbia cotinifolia L. 62
Euphorbia cyathophora Murray 63
Euphorbia heterophylla L. 63
Euphorbia hirta L. 63
Euphorbia humifusa Willd. 63
Euphorbia hypericifolia L. 63
Euphorbia marginata Pursh 63
Euphorbia milli Ch. des Moul. 63
Euphorbia neriifolia L. ' Cristata ' 63
Euphorbia prostrata Ait. 63
Euphorbia pulcherrima Willd. ex Klotzsch 63
Euphorbia thymifolia L. 63
Euphorbia tirucalli L. 63
Euphorbia trigona Haw. 63
Euphorbiaceae 60
Eurya auriformis Chang 51
Eurya chinensis R. Br. 51
Eurya ciliata Merr. 51
Eurya distichophylla Hemsl. 51
Eurya groffii Merr. 51
Eurya loquaiana Dunn 51
Eurya macartneyi Champ. 51
Eurya muricata Dunn 51
Eurya nitida Korth. 51
Eurya stenophylla Merr. 51
Eurya trichocarpa Korth. 51
Eurycorymbus cavaleriei (Levl.) Rehd. et Hand. -Mazz. 97
Euscaphis japonica (Thunb.) Kanitz 98
Eustigma oblongifolium Gardn. et Champ. 80
Evodia glabrifolia (Champ. ex Benth.) Huang 94
Evodia lepta (Spreng.) Merr. 94
Evolvulus alsinoides (L.) L. 131
Excoecaria agallocha L. 63
Excoecaria cochinchinensis Lour. 63

F

Fagaceae 81

Fagerlindia depauperata (Drake) Tirv. 113
Fagraea ceilanica Thunb. 107
Farfugium japonicum (L. f.) Kitam. 122
Fatoua villosa (Thunb.) Nakai 85
Fatsia japonica (Thunb.) Decne. et Planch. 100
Ficus altissima Bl. 85
Ficus auriculata Lour. 85
Ficus benjamina L. 85
Ficus benjamina L. Golden Princess' 85
Ficus carica L. 85
Ficus concinna (Miq.) Miq. 85
Ficus elasticar Roxb. ex Hornem. 85
Ficus erecta Thunb. var. *beecheyana* (Hook. et Arn.) King 85
Ficus esquiroliana Levl. 85
Ficus fistulosa Reinw ex Bl. 85
Ficus formosana Maxim. 85
Ficus formosana Maxim. var. *shimadai* (Hayata) W. C. Chen 85
Ficus hederacea Roxb. 85
Ficus heteromorpha Hemsl. 85
Ficus hirta Vahl 85
Ficus hispida L. f. 85
Ficus lyrata Warb. 86
Ficus microcarpa L. f. 86
Ficus microcarpa L. f. 'Golden Leaves' 86
Ficus nervosa Heyne ex Roth. 86
Ficus pandulata Hance var. *holophylla* Migo 86
Ficus pandurata Hance 86
Ficus pumila L. 86
Ficus pyriformis Hook. et Arn. 86

Ficus religiosa L. 86
Ficus sagittata Vahl 86
Ficus samentosa Buch. -Ham. ex J. E. Sm. 86
Ficus stenophylla Hemsl. 86
Ficus subulata Bl. 86
Ficus superba Miq. var. *japonica* Miq. 86
Ficus tikoua Bur. 86
Ficus tinctoria ssp. *gibbosa* (Bl.) Corner 86
Ficus variegata Bl. var. *chlorocarpa* (Benth.) King 86
Ficus variolosa Lindl. ex Benth. 86
Ficus vasculosa Wall. ex Miq. 87
Ficus virens Ait. var. *sublanceolata* (Miq.) Corner 87
Fimbristylis aestivalis (Retz.) Vahl 175
Fimbristylis bisumbellata (Forst.) Bubani 175
Fimbristylis complanata (Retz.) Link 175
Fimbristylis consanguinea Kunth 175
Fimbristylis cymosa (Lam.) R. Br. 176
Fimbristylis dichotoma (L.) Vahl 175
Fimbristylis dichotoma (L.) Vahl f. *tomentosa* (Vahl) Ohwi 175
Fimbristylis ferruginea (L.) Vahl 175
Fimbristylis insignis Thw. 175
Fimbristylis leptoclada Benth. 175
Fimbristylis longistipitata Tang et Wang 175
Fimbristylis miliacea (L.) Vahl 175
Fimbristylis schoenoides (Retz.) Vahl 176

Fimbristylis sericea (Poir.) R. Br. 176
Fimbristylis subbispicata Nees et Meyen 176
Fimbristylis tetragona R. Br. 176
Fimbristylis thomsonii Bocklr. 176
Fimbristylis velata R. Br. 176
Firmiana hainanensis Kosterm. 58
Firmiana simplex (L.) W. F. Wight 58
Fissistigma glaucescens (Hance) Merr. 23
Fissistigma oldhamii (Hemsl.) Merr. 24
Fissistigma polyanthum (Hook. f. et Thoms.) Merr. 24
Fissistigma uonicum (Dunn.) Merr. 24
Fittonia verschaffeltii (Lemaire) Van Houtte 138
Fittonia verschaffeltii (Lemaire) Van Houtte var. *argyroneura* Nichols. 138
Flacourtiaceae 46
Flemingia macrophylla (Willd.) Prain 77
Flemingia prostrata Roxb. f. ex Roxb. 77
Floscopa scandens Lour. 148
Foeniculum vulgare Mill. 102
Fokienia hodginsii (Dunn) Henry et Thomas 18
Fortunella hindsii (Champ. ex Benth.) Swingle 94
Fortunella japonica (Thunb.) Swingle 94
Fortunella margarita (Lour.) Swingle 94
Fragaria ananassa Duch. 67
Fraxinus chinensis Roxb. 107
Fraxinus insularis Hemsl. 107
Freesia refracta Klatt 161
Fuchsia hybrida Hort. ex Sieb. et Voss. 43
Fuirena ciliaris (L.) Roxb. 176
Fuirena umbellata Rottb. 176
Furcraea foetida (L.) Haw. 163

G

Gahnia tristis Nees 176
Garcinia multiflora Champ. ex Benth. 56
Garcinia oblongifoli a Champ. et Benth. 56
Garcinia subelliptica Merr. 56
Gardenia jasminoides Ellis 113
Gardenia jasminoides Ellis var. *angustifolia* Makino 113
Gardenia jasminoides Ellis var. *fortuniana* (Lindl.) Hara 113
Gardenia jasminoides Ellis var. *grandiflora* Makino 113
Gardenia jasminoides Ellis var. *radicans* (Thunb.) Makino 113
Gardneria multiflora Makino 107
Gelsemium elegans (Gardn. et Champ.) Benth. 107
Gentiana loureirii (D. Don) Griseb. 126
Gentianaceae 126
Geophila herbacea (Jacq.) K. Schum. 114
Geraniaceae 41
Gerbera jamesonii Bolus 122

Gerberia piloselloides (L.) Cass. 122
Gesneriaceae 134
Ginkgo biloba L. 17
Ginkgoaceae 17
Gironniera subaequalis Planch. 83
Gladiolus gandavensis van Houtte. 162
Glechoma longituba (Nakai) Kupr. 144
Gleditsia australis Hemsl. 73
Gleditsia fera (Lour.) Merr. 73
Gleditsia sinensis Lam. 73
Glehnia littoralis F. Schmidt ex Miq. 102
Gleicheniaceae 3
Glinus oppositifolius (L.) A. DC. 38
Glinus stricta L. 38
Glochidion eriocarpum Champ. ex Benth. 63
Glochidion hirsutum (Roxb.) Voigt 63
Glochidion lanceolarium (Roxb.) Voigt 63
Glochidion philippicum (Cav.) C. B. Rob. 64
Glochidion puberum (L.) Hutch. 64
Glochidion wrightii Benth. 64
Glochidion zeylanicum (Gaertn.) A. Juss. 64
Gloriosa superba L. 154
Glossocardia bidens (Retz.) Veldk. 122
Glottiphyllum linguiforme N. E. Br. 38
Glycine max (L.) Merr. 77
Glycosmis parviflora (Sims) Kurz 95
Glyptostrobus pensilis (Staunt.) K. Koch 18
Gmelina hainanensis Oliv. 142

Gnaphalium affine D. Don 122
Gnaphalium pensylvanicum Willd. 122
Gnaphalium polycaulon Pers. 122
Gnetaceae 20
Gnetum lofuense C. Y. Cheng 20
Gnetum parvifolium (Warb.) C. Y. Cheng ex Chun 20
Gomphocarpus fruticosus (L.) W. T. Aiton 111
Gomphostemma chinense Oliv. 144
Gomphrena celosioides Mart. 41
Gomphrena globosa L. 41
Gomphrena globosa L. var. *rubra* Hort. 41
Gonocormus minutus (Bl.) v. d. Bosch 4
Gonostegia hirta (Bl.) Miq. 87
Goodeniaceae 127
Goodyera procera (Ker-Gewl.) Hook. 171
Gordonia axillaris (Roxb.) Dietr. 51
Gramineae 178
Grammitidaceae 16
Grammitis dorsipila (Christ) C. Chr. et Tard. 16
Grammitis hirtella (Bl.) Tuyama 16
Grammitis lasiosora (Bl.) Ching 16
Graphistemma pictum (Benth.) B. D. Jacks 111
Grevillea robusta Cunn. ex R. Br. 45
Grewia biloba D. Don 56
Grewia biloba G. Don var. *parviflora* (Bunge) Hand.-Mazz. 56
Guttiferae 56
Gymnanthera oblonga (Burm. f.) P. S. Green 112
Gymnema sylvestre (Retz.) Schult.

111
Gymnogrammitidaceae 14
Gymnogrammitis dareiformis (Hook.) Ching ex Tard. -Blot et C. Chr. 14
Gymnopetalum chinensis (L.) Merr. 47
Gymnospermae 17
Gymnosphaera gigantea (Wall. ex Hook.) J. Sm. 5
Gymnosphaera hancockii (Cop.) Ching 5
Gymnosphaera podophylla (Hook.) Cop. 5
Gynostemma pentaphyllum (Thunb.) Makino 47
Gynura bicolor (Roxb. ex Willd.) DC. 122
Gynura divaricata (L.) DC. 122
Gynura japonica (L. f.) Juel 122
Gypsophila paniculata L. 37

H

Habenaria dentata (Sw.) Schltr. 171
Habenaria leptoloba Benth. 171
Habenaria linguella Lindl. 171
Habenaria rhodocheila Hance 171
Haemanthus multiflorus Martyn 161
Halesia macgregorii Chun 105
Haloragaceae 44
Haloragis chinensis (Lour.) Merr. 44
Haloragis micrantha (Thunb.) R. Br. ex Sieb. et Zucc. 44
Hamamelidaceae 80
Hamelia patens Jacq. 114
Haworthia cymbiformis (Haw.) Duval var. *translucens* Triebner et Poelln. 155

Haworthia fasciata (Willd.) Haw. 155
Haworthia margaritifera Haw. 155
Hedera helix L. 100
Hedera nepalensis K. Koch var. *sinensis* (Tobl.) Rehd. 101
Hedychium coronarium Koen. 152
Hedyotis acutangula Champ. ex Benth. 114
Hedyotis ampliflora Hance 114
Hedyotis auricularis L. 114
Hedyotis biflora (L.) Lam. 114
Hedyotis caudatifolia Merr. et Metcalf 114
Hedyotis chrysotricha (Palib.) Merr. 114
Hedyotis consanguinea Hance 114
Hedyotis corymbosa (L.) Lam. 114
Hedyotis diffusa Willd. 114
Hedyotis hedyotidea (DC.) Merr. 114
Hedyotis loganioides Benth. 114
Hedyotis ovata Thunb. ex Maxim. 114
Hedyotis pinifolia Wall. ex G. Don 114
Hedyotis tenelliflora Bl. 114
Hedyotis tetrangularia (Korth.) Walp. 114
Hedyotis uncinella Hook. et Arn. 114
Hedyotis vachellii Hook. 115
Hedyotis verticillata (L.) Lam. 115
Hedyotis xanthochroa Hance 115
Helianthus annuus L. 122
Helianthus tuberosus L. 122
*Helicia cochinchinensi*s Lour. 45
Helicia kwangtungensis W. T. Wang 45
Helicia reticulata W. T. Wang 45

Heliconia bihai L. 151
Heliconia collinsiana Griggs 151
Heliconia humilis Jacq. 151
Heliconia platystachys Baker 151
Heliconia rostrata Ruiz et Pav. 151
Heliconia subulata Ruiz et Pav. 151
Heliconia wagneriana Petersen 151
Helicteres angustifolia L. 58
Heliotropium indicum L. 128
Helixanthera parasitica Lour. 90
Helixanthera sampsoni (Hance) Danser 90
Hemerocallis fulva (L.) L. 155
Hemiboea subcapitata Clarke var. *guangdongensis* (Z. Y. Li) Z. Y. Li 135
Hemigramma decurrens (Hook.) Cop. 12
Hemionitidaceae 7
Hemistepta lyrata (Bunge) Bunge 122
Heritiera angustata Pierre 58
Heritiera littoralis Dryand. 58
Heritiera parvifolia Merr. 58
Heteropanax fragrans (Roxb.) Seem. 101
Heteropholis cochinchinensis (Lour.) C. E. Hubb. 183
Heteropogon contortus (L.) Beauv. ex Roem. et Schult. 183
Heterosmilax chinensis Wang 156
Heterosmilax japonica var. *gaudichaudiana* (Kunth) Maxim. 156
Hevea brasiliensis (Willd. ex A. Juss.) Muell. Arg. 64
Hewittia malabarica (L.) Suresh 131
Hibiscus acetosella Welw. ex Hiern 59
Hibiscus cannabinus L. 59

Hibiscus coccineus (Medic.) Walt. 59
Hibiscus hamabo Sieb. et Zucc. 59
Hibiscus indicus (Burm. f.) Hochr. 59
Hibiscus mutabilis L. 59
Hibiscus schizopetalus (Mast.) Hook. f. 59
Hibiscus syriacus L. 59
Hibiscus tiliaceus L. 59
Hippeastrum vittatum Herb. 161
Hippocarateaceae 90
Hiptage benghalensis (L.) Kurz 60
Histiopteris incisa (Thunb.) J. Sm. 6
Hodgsonia macrocarpa (Bl.) Cogn. 47
Holarrhena antidysenterica Wall. ex A. DC. 109
Holmskioldia sanguinea Retz. 142
Homalium austro-chinense G. S. Fan 46
*Homalium cochinchinensi*s (Lour.) Druce 46
Homalium hainanense Gagnep. 46
Homalocladium platycladum (F. Muell.) Bailey 39
Homalomena occulta (Lour.) Schott 159
Hordeum vulgare L. 183
Hosta plantaginea (Lam.) Aschers. 155
Houttuynia cordata Thunb. 32
Hovenia dulcis Thunb. 91
Howea belmoreana Becc. 165
Hoya carnosa (L. f.) R. Br. 111
Humata repens (L. f.) Diels 13
Humata tyermannii T. Moore 14
Hunteria zelanica (Retz.) Garden. ex

Thwaites 109
Huperzia serrata (Thunb.) Trev. 1
Huperziaceae 1
Hyacinthus orientalis L. 155
Hydnocarpus hainanensis (Merr.) Sleum. 46
Hydrangea macrophylla (Thunb.) Ser. 66
Hydrangeaceae 66
Hydrilla verticillata (L. f.) Royle 147
Hydrocharitaceae 146
Hydrocotyle nepalensis Hook. 102
Hydrocotyle sibthorpioides Lam. 102
Hydrocotyle wilfordii Maxim. 102
Hydrolea zeylanica (L.) Vahl 128
Hydrophyllaceae 128
Hygrophila salicifolia (Vahl) Nees 138
Hylocereus undatus (Haw.) Britt. et Rose 49
Hymenachne assamica (Hook. f.) Hitchc. 183
Hymenachne aurita (Presl) Balansa 183
Hymenaea courbaril L. 73
Hymenocallis americana M. Roem. 161
Hymenophyllaceae 4
Hymenophyllum barbatum (v. d. Bosch) Baker 4
Hyophorbe lagenicaulis (L. H. Bailey) H. E. Moore 165
Hypericaceae 56
Hypericum japonicum Thunb. ex Murray 56
Hypericum monogynum L. 56
Hypoestes purpurea (L.) R. Br. 138

Hypolepidaceae 6
Hypolepis punctata (Thunb.) Mett. 6
Hypolytrum hainanense (Merr.) Tang et Wang 176
Hypolytrum nemorum (Vahl) Spreng. 176
Hypoxidaceae 167
Hypoxis aurea Lour. 168
Hypserpa nitida Miers 30
Hyptis suaveolens (L.) Poiteau 144

I

Icacinaceae 90
Ichnanthus vicinus (F. M. Bailey) Merr. 183
Ilex asprella (Hook. et Arn.) Champ. ex Benth. 88
Ilex cornuta Lindl. et Paxt. 88
Ilex ficoidea Hemsl. 88
Ilex graciliflora Champ. ex Benth. 88
Ilex memecylifolia Champ. ex Benth. 88
Ilex pubescens Hook. et Arn. 88
Ilex rotunda Thunb. 88
Ilex rotunda Thunb. var. *microcarpa* (Lindl. et Paxt.) S. Y. Hu 89
Ilex triflora Bl. 89
Ilex viridis Champ. ex Benth. 89
Illiciaceae 23
Illicium ternstroemioides A. C. Smith 23
Illigera celebica Miq. 27
Illigeraceae 27
Impatiens balsamina L. 42
Impatiens chinensis L. 42
Impatiens hawkeri W. Bull 42
Impatiens hongkongensis Grey-Wilson

42
Impatiens walleriana Hook. f.   42
Imperata cylindrica (L.) Beauv. var. major (Nees) C. E. Hubb.   183
Indigofera decora Lindl.   77
Indigofera hirsuta L.   77
Indigofera spicata Forsk.   77
Indigofera suffruticosa Mill.   77
Indigofera trifoliata L.   77
Indocalamus herklotsii McClure   189
Indocalamus longiauritus Hand. -Mazz.   189
Indocalamus tesselatus (Munro) Keng f.   189
Inula cappa (Buch. -Ham.) DC.   122
Ipomoea aquatica Forsk.   131
Ipomoea batatas (L.) Lam.   131
Ipomoea cairica (L.) Sweet   131
Ipomoea digitata L.   131
Ipomoea fistulosa Mart. ex Choisy   131
Ipomoea obscura (L.) Ker-Gawl.   131
Ipomoea pes-caprae (L.) Sweet   131
Ipomoea pileata Roxb.   131
Ipomoea triloba L.   131
Iresine herbstill Hook. f. ex Lindl.   41
Iridaceae   161
Iris japonica Thunb.   162
Iris tectorum Maxim.   162
Isachne albens Trin.   183
Isachne dispar Trin.   183
Isachne globosa (Thunb.) O. Kuntze   184
Ischaemum aristatum L.   184
Ischaemum barbatum Retz.   184
Ischaemum indicum (Houtt.) Merr.   184
Ischaemum rugosum Salisb.   184

Isodon amethystoides (Benth.) Hara   144
Itea chinensis Hook. et Arn.   66
Itea oblonga Hand. -Mazz.   66
Ixeridium gracile (DC.) Shih   122
Ixeris gracilis Stebb.   122
Ixeris japonica (Burm. f.) Nakai   122
Ixeris repens (L.) A. Gray   122
Ixonanthaceae   60
Ixonanthes chinensis Champ.   60
Ixora chinensis Lam.   115
Ixora coccinea L.   115
Ixora henryi Levl.   115
Ixora lutea (Veitch) Hutch.   115
Ixora parviflora Vahl   115

J

Jacaranda mimosifolia D. Don   136
Jacquemontia paniculata (Burm. f.) Hall. f.   131
Jasminum elongatum (Bergius) Willd.   107
Jasminum lanceolarium Roxb.   107
Jasminum mesnyi Hance   107
Jasminum sambac (L.) Ait.   107
Jasminum sinensis Hemsl.   108
Jatropha curcas L.   64
Jatropha gossypifolia L.   64
Jatropha integerrima Jacq.   64
Jatropha podagrica Hook.   64
Juglandaceae   99
Juncaceae   173
Juncus articulatus L.   173
Juncus effusus L.   173
Juncus leschenaultii Gay   173
Justicia adhodata L.   138
Justicia gendarussa Burm. f.   138

Justicia procumbens L.　138
Justicia ventricosa Wall.　138

K

Kadsura coccinea (Lem.) A. C. Smith　23
Kadsura heteroclita (Roxb.) Craib　23
Kadsura longipedunculata Finet et Gagnep.　23
Kalanchoe ceratophylla Haw.　36
Kalanchoe daigremontiana Hamet et Perr.　36
Kalanchoe lanciniata (L.) DC.　36
Kalanchoe synsepala Baker　36
Kalanchoe thyrsiflora Harv.　36
Kalanchoe tubiflora (Harvey) Raym.-Hamet　36
Kalanchoe verticillata Elliot　36
Kalimeris indica (L.) Sch.-Bip.　123
Kandelia candel (L.) Druce　55
Keteleeria fortunei (Murr.) Carr.　17
Khaya senegalensis (Desr.) A. Juss.　96
Kigelia africana (Lam.) Benth.　136
Kochia scoparia Schrad. var. *sieversiana* (Pall.) Ulbr. f. *trichophylla* (Hort.) Schinz et Thell.　39
Koelreuteria bipinnata Franch.　97
Koelreuteria elegans (Seem.) A. C. Smith ssp. *formosana* (Hayata) Meyer　97
Kopsia lancibracteolata Merr.　109
Korthalsella japonica (Thunb.) Engl.　90
Kummerowia striata (Thunb.) Schindl.　77
Kyllinga brevifolia Rottb.　176
Kyllinga melanosperma Nees　176
Kyllinga monocephala Rottb.　176

L

Labiatae　143
Lablab purpureus (L.) Sweet　77
Lactuca indica L.　123
Lactuca sativa L.　123
Lactuca sativa L. var. *romana* L. H. Bailey　123
Lactuca seriola Torner (*L. serriola* Torner)　123
Lagenaria siceraria (Molina) Standl.　47
Lagenaria siceraris (Molina) Standl. var. *hispida* (Thunb.) Hara　47
Lagerstroemia indica L.　43
Lagerstroemia speciosa (L.) Pers.　43
Lagerstroemia subcostata Koehne　43
Laggera alata (D. Don) Sch.-Bip. ex Hochst.　123
Lantana camara L.　142
Lantana camara L. var. *flava* Mold.　142
Lantana camara L. var. *hybrida* Mold.　142
Lantana camara L. var. *mista* L. H. Bailey　142
Lantana montevidensis (Spreng.) Briq.　142
Lardizabalaceae　29
Lasianthus chinensis (Champ.) Benth.　115
Lasianthus henryi Hutch.　115
Lasianthus wallichii (Wight et Arn.) Wight　115
Latania lontaroides (J. Gaertn.) H. E.

Moore 165	*Lespedeza chinensis* G. Don 77
Launaea sarmentosa (Willd.) Merr. et Chun 123	*Lespedeza cuneata* (Dum. -Cours.) G. Don 77
Lauraceae 24	*Lespedeza formosa* (Vog.) Koehne 77
Laurocerasus marginata (Dunn) Yu et Lu 67	*Leucaena leucocephala* (Lam.) de Wit 70
Laurocerasus phaeosticta (Hance) S. K. Schneid. 67	*Leucanthemum vulgare* H. S. La 123
Laurocerasus zippeliana (Miq.) T. T. Yu et L. T. Lu 67	*Leucas mollissima* Wall. var. *chinensis* Benth. 144
Lawsonia inermis L. 43	*Leucas zeylanica* (L.) R. Br. 144
Leea glabra C. L. Li 93	*Liatris spicata* Willd. 123
Leea guineensis G. Don 93	*Licuala fordiana* Becc. 166
Leersia hexandra Swartz 184	*Licuala grandis* (Bull.) H. A. Wendl. 166
Lemmaphyllum microphyllum C. Presl 14	*Licuala spinosa* Thunb. 166
Lemna minor L. 160	*Ligularia japonica* (Thunb.) Less. 123
Lemnaceae 160	
Lentibulariaceae 134	*Ligustrum amamianum* Koidz. 108
Leonurus japonicus Houtt. 144	*Ligustrum lianum* Hsu 108
Lepidagathis incurva Buch. -Ham. ex D. Don 138	*Ligustrum lucidum* Ait. 108
	Ligustrum sinense Lour. 108
Lepidogrammitis diversa (Ros.) Ching 14	*Ligustrum vicaryi* Rehd. 108
	Liliaceae 153
Lepidogrammitis drymoglossoides (Baker) Ching 14	*Lilium brownii* F. E. Brown ex Miellez 155
Lepidogrammitis rostrata (Bedd.) Ching 14	*Lilium tigrinum* Ker-Gawl. 155
	Limnocharis flava (L.) Buch. 146
Lepidosperma chinense Nees 176	*Limnophila aromatica* (Lam.) Merr. 133
Lepironia articulata (Retz.) Dom. 176	*Limonium latifolium* O. Kuntze 126
Lepisorus thunbergianus (Kaulf.) Ching 14	*Limonium sinense* (Girard) O. Kuntze 126
Leptochloa chinensis (L.) Nees 184	*Limonium sinuatum* (L.) Mill. 126
Leptochloa panicea (Retz.) Ohwi 184	Linaceae 41
Leptosiphonium venustum (Hance) E. Hossain 138	*Lindera aggregata* (Sims) Kosterm. 25
	Lindera aggregata (Sims) Kosterm.

var. *playfairii* (Hemsl.) H. P. Tsui   25
Lindera communis Hemsl.   25
Lindera glauca (Sieb. et Zucc.) Bl.   25
Lindera nacusua (D. Don) Merr.   25
Lindernia anagallis (Burm. f.) Pennell   133
Lindernia angustifolia (Benth.) Wettst.   133
Lindernia antipoda (L.) Alston   133
Lindernia ciliata (Colsm.) Pennell   133
Lindernia crustacea (L.) F. Muell.   133
Lindernia procumbens (Krock.) Philcox   133
Lindernia ruellioides (Colsm.) Pennell   133
Lindernia urticifolia (Hance) Bonati   133
Lindsaea ensifolia Sw.   5
Lindsaea heterophylla Dry.   5
Lindsaea odorata Roxb.   5
Lindsaea orbiculata (Lam.) Mett. ex Kuhn   5
Lindsaeaceae   5
Liparis bootanensis Griff.   171
Liparis nervosa (Thunb. ex A. Murray) Lindl.   171
Liparis nigra Seidenf.   171
Liparis viridiflora (Bl.) Lindl.   171
Lipocarpha chinensis (Osb.) Kern   176
Lipocarpha sengalensis (Lam.) Dandy   177
Liquidambar acalycina H. T. Chang   80
Liquidambar formosana Hance   80
Liriodendron chinense (Hemsl.) Sarg.   21
Liriodendron tulipifera L.   21
Liriope graminifolia (L.) Baker   155
Liriope muscari (Decne.) Bailey   155
Liriope spicata (Thunb.) Lour.   155
Litchi chinensis Sonn.   97
Lithocarpus corneus (Lour.) Rehd.   82
Lithocarpus elizabethae (Tutch.) Rehd.   82
Lithocarpus glaber (Thunb.) Nakai   82
Lithocarpus hancei (Benth.) Rehd.   83
Lithocarpus litseifolius (Hance) Chun   83
Lithocarpus paniculatus Hand. -Mazz.   83
Lithocarpus quercifolius Huang et Y. T. Chang   83
Lithocarpus uvariifolius (Hance) Rehd.   83
Litsea acutivena Hayata   25
Litsea cubeba (Lour.) Pers.   25
Litsea glutinosa (Lour.) C. B. Rob.   25
Litsea kwangsiensis H. T. Chang   26
Litsea lancifolia (Roxb. ex Nees) Benth. ex Hook. f.   26
Litsea monopetala (Roxb.) Pers.   26
Litsea rotundifolia (Nees) Hemsl.   26
Litsea rotundifolia (Nees) Hemsl. var. *oblongifolia* (Nees) Allen   26
Litsea subcoriacea Yang et P. H. Huang   26

Litsea verticillata Hance 26
Livistona chinensis (Jacq.) R. Br. 166
Livistona decipiens Becc. 166
Lobelia alsinoides Lam. 127
Lobelia chinensis Lour. 127
Lobelia zeylanica L. 127
Lobeliaceae 127
Lobularia maritima (L.) Desv. 34
Loeseneriella concina A. C Smith 90
Loganiaceae 107
Lonicera confusa (Sweet) DC. 117
Lonicera japonica Thunb. 117
Lonicera japonica Thunb. var. *chinensis* (Wats.) Baker 117
Lonicera longiflora (Lindl.) DC. 117
Lonicera macrantha (D. Don) Spreng. 117
Lonicera rhytidophylla Hand. -Mazz. 117
Lophatherum gracile Brongn. 184
Lophostemon confertus (R. Br.) P. G. Wilson et J. T. Waterhouse 54
Loranthaceae 90
Loropetalum chinense (R. Br.) Oliv. 81
Loropetalum chinense (R. Br.) Oliv. f. *rubrum* H. T. Chang 81
Loxogrammaceae 16
Loxogramme salicifolia (Makino) Makino 16
Ludisia discolor (Ker-Gawl.) A. Rich. 171
Ludwigia adscendens (L.) Hara 43
Ludwigia hyssopifolia (G. Don) Exell 43
Ludwigia octovalvis (Jacq.) Raven 43
Ludwigia peploides (Kunth) Raven 44

Ludwigia perennis L. 44
Ludwigia prostrata Roxb. 44
Luffa acutangula (L.) Roxb. 47
Luffa aegyptica Mill. 47
Lycianthes biflora (Lour.) Bitter 129
Lycium chinensis Mill. 129
Lycopersicon esculentum Mill. 129
Lycopodiastrum casuarinoides (Spring) Holub 1
Lycopodium japonicum Thunb. ex Murray 1
Lycoris aurea (L'Her.) Herb. 161
Lycoris radiata (L'Her.) Herb. 161
Lygisma inflexum (Cost.) Kerr 111
Lygodiaceae 3
Lygodium conforme C. Chr. 3
Lygodium flexuosum (L.) Sw. 4
Lygodium japonicum (Thunb.) Sw. 4
Lygodium microstachyum Desv. 4
Lygodium scandens (L.) Sw. 4
Lysidice brevicalyx Wei 73
Lysidice rhodostegia Hance 73
Lysimachia candida Lindl. 126
Lysimachia christinae Hance 126
Lysimachia clethroides Duby 126
Lysimachia fortunei Maxim. 126
Lysionotus pauciflorus Maxim. 135
Lythraceae 42
Lythrum salicaria L. 43

M

Macadamia integrifolia Maiden et Betche 45
Macadamia terrifolia F. Muell. 45
Macaranga auriculata (Merr.) Airy-Shaw 64
Macaranga sampsonnii Hance 64

Macaranga tanarius (L.) Muell. -Arg. 64
Macfadyena unguis-cati (L.) A. Gentry 136
Machilus breviflora (Benth.) Hemsl. 26
Machilus chekiangensis S. K. Lee 26
Machilus chinensis (Champ. ex Benth.) Hemsl. 26
Machilus grijsii Hance 26
Machilus leptophylla Hand. -Mazz. 26
Machilus pauhoi Kanehira 26
Machilus salicina Hance 26
Machilus thunbergii Sieb. et Zucc. 26
Machilus velutina Champ. ex Benth. 26
Macleaya cordata (Willd.) R. Br. 33
Macropanax decandrus Hoo 101
Macrosolen cochinchinensis (Lour.) Van Tregh 90
Macrothelypteris torresiana (Gaud.) Ching 9
Madhuca pasquieri (Dubard) Lam. 104
Maesa japonica (Thunb.) Moritzi ex Zoll. 105
Maesa perlaria (Lour.) Merr. 105
Maesa salcifolia Walker 105
Magnolia championii Benth. 21
Magnolia coco (Lour.) DC. 21
Magnolia delavayi Franch. 21
Magnolia denudata Desr. 21
Magnolia grandiflora L. 21
Magnolia liliflora Desr. 21
Magnolia officinalis Rehd. et Wils. ssp. *biloba* (Rehd. et Wils.) Cheng et Law 21
Magnolia paenetalauma Dandy 21
Magnolia soulangeana Soul. -Bod. 21
Magnoliaceae 21
Mahonia bealeibealii (Fort.) Carr. 29
Mahonia bodinieri Gagnep. 29
Mahonia japonica (Thunb.) DC. 29
Mahonia oiwakensis Hayata 29
Malachium aquaticum (L.) Fries 37
Malaisia scandens (Lour.) Planch. 87
Malaxis finetii (Gagnap.) T. Tang et F. T. Wang 172
Malaxis latifolia J. E. Smith 172
Mallotus apelta (Lour.) Muell. -Arg. 64
Mallotus hookerianus (Seem.) Muell. -Arg. 64
Mallotus paniculatus (Lam.) Muell. Arg. 64
Mallotus peltatus (Geisel.) Muell. -Arg. 64
Mallotus philippinensis (Lam.) Muell. -Arg. 65
Mallotus repandus (Willd.) Muell. -Arg. 65
Malpighiaceae 60
Malva sinensis Cav. 59
Malva verticillata L. 59
Malvaceae 58
Malvastrum coromandelianum (L.) Garcke 59
Malvaviscus arboreus Cav. var. *penduliflorus* (DC.) Schery 59
Mandevilla sanderi (Hemsl.) Woodson 109
Mangifera indica L. 99
Mangifera persiciformis C. Y. Wu et T.

L. Ming 99
Manglietia chingii Dandy 21
Manglietia fordiana Oliv. 21
Manglietia glauca Bl. 21
Manglietia grandis Hu et Cheng 22
Manglietia hainanensis Dandy 22
Manglietia insignis (Wall.) Bl. 22
Manglietia kwangtungensis (Merr.) Dandy 22
Manglietia maguanica Chang et B. L. Chen 22
Manglietia megaphylla Hu et Cheng 22
Manglietia yuyuanensis Y. W. Law 22
Manihot esculentus Crantz 65
Manilkara zapota (L.) Van Royen 104
Mappianthus iodoides Hand.-Mazz. 90
Maranta arundinacea L. 153
Maranta bicolor Ker-Gawl. 153
Marantaceae 153
Mariscus javanicus (Houtt.) Merr. et Metc. 177
Mariscus radians (Nees et Meyen) Tang et Wang 177
Mariscus sumatrensis (Retz.) Koyama 177
Marsilea quadrifolia L. 16
Marsileaceae 16
Matteuccia orientalis (Hook.) Trev. 10
Matthiola incana (L.) R. Br. 34
Mayodendron ignea (Kurz) Kurz 136
Mazus pumilus (Burm. f.) Steenis 133
Mecodium microsorum (v. d. Bosch) Ching 4
Medinilla magnifica Lindl. 54

Melaleuca parviflora Lindl. 53
Melaleuca quinquenervia (Cav.) S. T. Blake 53
Melastoma affine D. Don 54
Melastoma candidum D. Don 54
Melastoma dodecandrum Lour. 54
Melastoma intermedium Dunn 54
Melastoma normale D. Don 54
Melastoma sanguineum Sims 54
Melastomataceae 54
Melia azedarach L. 96
Meliaceae 96
Meliosma fordii Hemsl. ex Forb. et Hemsl. 98
Meliosma rigida Sieb. et Zucc. 98
Meliosma squamulata Hance 98
Meliosma thorellii Lec. 98
Melochia corchorifolia L. 58
Melodinus fusiformis Champ. ex Benth. 109
Melodinus suaveolens Champ. ex Benth. 109
Memecylon ligustrifolium Champ. ex Benth. 55
Memecylon nigrescens Hook. et Arn. 55
Menispermaceae 30
Mentha haplocalyx Briq. 145
Menyanthaceae 126
Merremia hederacea (Burm. f.) Hall. f. 131
Merremia hirta (L.) Merr. 131
Merremia tridentata (L.) Hall. f. ssp. *hastata* (Desr.) Oststr. 131
Merremia umbellata (L.) Hall. f. 132
Merremia umbellata Hall. f. ssp. *orientalis* (Hall. f.) Oststr. 132

Mesona chinensis Benth. 145
Mesua ferrea L. 56
Metasequoia glyptostroboides Hu et Cheng 18
Michelia alba DC. 22
Michelia balansae (A. DC.) Dandy 22
Michelia champaca L. 22
Michelia chapensis Dandy 22
Michelia figo (Lour.) Spreng. 22
Michelia foveolata Merr. ex Dandy 22
Michelia hedyosperma Y. W. Law 22
Michelia macclurei Dandy 22
Michelia maudiae Dunn 22
Michelia shiluensis Chun et Y. F. Wu 22
Michelia skinneriana Dunn 23
Michelia yunnanensis Franch. 23
Microcarpaea minima (Koenig) Merr. 133
Microcos paniculata L. 56
Microdesmis caseraiifolia Planch. ex Hook. 66
Microglossa pyrifolia (Lam.) O. Kuntze 123
Microlepia hancei Prantl 5
Microlepia hookeriana (Wall. ex Hook.) Presl 5
Microlepia marginata (Houtt.) C. Chr. 5
Microsorium buergerianum (Miq.) Ching 15
Microsorium fortunei (T. Moore) Ching 15
Microsorium pteropus (Bl.) Cop. 15
Microsorium punctatum (L.) Cop. 15
Microsorium superficiale (Bl.) Ching 15
Microsorium zippelii (Bl.) Ching 15
Microstegium ciliatum (Trin.) A. Camus 184
Microstegium nodosum (Kom.) Trvel. 184
Microstegium vagans (Nees ex Steud.) A. Camus 184
Microstegium vimineum (Trin.) A. Camus 184
Microtropis biflora Merr. et Freem. 89
Microtropis fokienensis Dunn 89
Microtropis obscurinervia Merr. et Freem. 89
Microtropis reticulata Dunn 89
Mikania cordata (Burm. f.) B. L. Robinson 123
Mikania scandens (L.) Willd. 123
Millettia championi Benth. 77
Millettia dielsiana Harms ex Diels 77
Millettia nitida Benth. 77
Millettia nitida Benth. var. *hirsutissima* Z. Wei 77
Millettia oraria Dunn 78
Millettia pachycarpa Benth. 78
Millettia pulchra (Benth.) Kurz 78
Millettia reticulata Benth. 78
Millettia speciosa Champ. ex Benth. (*M. speciosa* Champ.) 78
Mimosa bimucronata (DC.) O. Kuntze 70
Mimosa diplotricha C. Wrigth ex Sauvalle 71
Mimosa pudica L. 71
Mimosaceae 69
Mimusops elengii L. 104
Mirabilis jalapa L. 45

Miscanthus floridulus (Labill.) Warb. ex Suhum. et Lauterb. 184
Miscanthus sacchariflorus (Maxim.) Benth. 184
Miscanthus sinensis Anderss. 184
Mischocarpus pentapetalus (Roxb.) Radlk. 97
Mitrasaceme pygmaea R. Br. 107
Molluginaceae 38
Momordica charantia L. 48
Momordica cochinchinenses (Lour.) Spreng. 47
Momordica cochinchinensis (Lour.) Spreng. 48
Monochoria korsakowii Regel et Maack 156
Monochoria vaginalis (Burm. f.) Kunth 156
Monocotyledoneae 146
Monstera deliciosa Liebm. 159
Moraceae 84
Morinda parvifolia Bartl. ex DC. 115
Morinda umbellata L. 115
Moringa drouhardii Jumelle 33
Moringaceae 33
Morus alba L. 87
Mosla dianthera (Buch.-Ham. ex Roxb.) Maxim. 145
Mosla scabra (Thunb.) C. Y. Wu et II. W. Li 145
Mucuna birdwoodiana Tutch. 78
Mucuna championi Benth. 78
Mucuna hainanensis Hayata 78
Mucuna macrocarpa Wall. 78
Mucuna pruriens (L.) DC. 78
Mucuna sempervirens Hemsl. ex Forb. et Hemsl. 78

Muntingia colabura L. 56
Murdannia bracteata (Clarke) J. K. Morton ex Hong 148
Murdannia loriformis (Hassk.) R. Rao et Kamm. 148
Murdannia macrocarpa Hong 148
Murdannia nudiflora (L.) Brenan. 148
Murdannia simplex (Vahl) Brenan. 148
Murdannia vaginata (L.) Bruckn 148
Murraya paniculata (L.) Jack. 95
Musa acuminata Colla 150
Musa balbisiana Colla 150
Musa basjoo Sieb. et Zucc. 150
Musa coccinea Andr. 150
Musa nana Lour. 151
Musa ornata Roxb. 151
Musaceae 150
Musella lasiocarpa (Franch.) C. Y. Wu ex W. W. Li 151
Mussaenda erosa Champ. 115
Mussaenda erythrophylla Schum. et Thonn. 115
Mussaenda esquirolii Levl. 115
Mussaenda hirsutula Miq. 115
Mussaenda kwangtungensis Li 115
Mussaenda pubescens Ait. f. 116
Myrica rubra (Lour.) Sieb. et Zucc. 81
Myricaceae 81
Myrsinaceae 104
Myrtaceae 52
Mytilaria laosensis Lec. 81

N

Nageia fleuryi (Hickel) Laubenf. 19

Nageia nagi (Thunb.) O. Kuntze 19
Nandina domestica Thunb. 29
Narcissus tazetta L. var. *chinensis* Roem. 161
Nasturtium officinale R. Br. 34
Nauclea officinalis (Pierre ex Pitard) Merr. et Chun 116
Neanotis hirsuta (L. f.) Lewis 116
Nechamandra alternifolia (Roxb. ex Wight) Thw. 147
Nelumbo lutea Pers. 28
Nelumbo nucifera Gaertn. 28
Neohusnotia tonkinensis A. Camus 184
Neolitsea cambodiana Lec. var. *glabra* Allen 27
Neolitsea chuii Merr. 27
Neolitsea kwangsiensis Liou 27
Neolitsea levinei Merr. 27
Neolitsea phanerophlebia Merr. 27
Neolitsea zeylanica (Ness) Merr. 27
Neomarica gracilis Spragne 162
Neoregelia carolinae (Beer) L. B. Sm. 150
Neoregelia concentrica (Vell.) L. B. Sm. 150
Neoregelia spectabilis (T. Moore) L. B. Sm. 150
Neottopteris antrophyoides (Christ) Ching 10
Neottopteris nidus (L.) J. Smith 10
Neottopteris phyllitidis (Don) J. Smith 10
Nepenthaceae 31
Nepenthes mirabilis (Lour.) Druce 31
Nephelaphyllum tenuiflorum Bl. 172
Nephelium lappaceum L. 97
Nephrolepidacea 13
Nephrolepis auriculata (L.) Trimen 13
Nephrolepis biserrata (Sw.) Schott 13
Nephrolepis hirsutula (Forst.) Presl 13
Nerium oleander L. 109
Neyraudia arundinacea (L.) Henrard 185
Neyraudia montana Keng 185
Neyraudia reynaudiana (Kunth) Keng ex A. S. Hitchc. 185
Nicotiana sanderae W. Watson 129
Nicotiana tabacum L. 129
Nolina recurvata (Lem.) Hemsl. 163
Nopalxochia ackermennii (Haw.) F. M. Knuth 49
Notholaena hirsuta (Poir.) Desv. 7
Nouelia insignis Franch. 123
Nuphar pumilum (Timm.) DC. 28
Nyctaginaceae 45
Nymphacaceae 28
Nymphaea alba L. 28
Nymphaea alba L. var. *rubra* Lonnr. 28
Nymphaea lotus L. 28
Nymphaea lotus L. var. *pubescens* (Willd) Hook. f. et Thoms. 28
Nymphaea mexicana Zucc. 28
Nymphaea nouchali Burm. f. 29
Nymphaea odorata Ait. 28
Nymphaea rubra Roxb. ex Salisb. 28
Nymphaea tetragona Georgi 29
Nymphoides peltatum (Gmel.) O. Kuntze 126
Nyssaceae 100

O

Ochrosia borbonica Gmel. 109

Ocimum basilicum L. 145
Ocimum gratissimum L. var. *suave* (Willd.) Hook. f. 145
Oenanthe javanica (Bl.) DC. 102
Oenothera drummondii Hook. 44
Olacaceae 90
Olea cuspidate Wall. et G. Don 108
Olea dioica Roxb. 108
Oleaceae 107
Oleandra cumingii J. Sm. 13
Oleandraceae 13
Onagraceae 43
Onocleaceae 10
Onychium japonicum (Thunb.) Kunze 7
Operculina turpethum (L.) Manso 132
Ophioglossaceae 2
Ophioglossum pedunculosum Desv. 2
Ophioglossum pendula Presl 2
Ophioglossum vulgatum L. 2
Ophiopogon intermedius D. Don 155
Ophiopogon japonicus (L. f.) Ker-Gawl. 155
Ophiopogon reversus Hwang 155
Ophiorrhiza cantoniensis Hance 116
Ophiorrhiza japonica Bl. 116
Ophiorrhiza pumila Champ. ex Benth. 116
Ophiuros exaltatus (L.) O. Kuntze 185
Opiliaceae 90
Oplismenus compositus (L.) Beauv. 185
Oplismenus compositus (L.) Beauv. var. *intermedius* (Honda) Ohwi 185

Oplismenus undulatifolius (Ard.) Roem. et Schult. 185
Opuntia stricta (Haw.) Haw. var. *dillenii* (Ker-Gawl.) L. D. Benson 49
Orchidaceae 168
Oreocharis benthamii Clarke 135
Oreocharis benthamii Clarke var. *reticulata* Dunn 135
Oreocnidea frutescens (Thunb.) Miq. 87
Ormosia emarginata (Hook. et Arn.) Benth. 78
Ormosia indurata L. Chen 78
Ormosia pinnata (Lour.) Merr. 78
Ormosia semicastrata Hance 78
Orobanchaceae 134
Oroxylum indicum (L.) Benth. ex Kurz 136
Oryza sativa L. 185
Osbeckia chinensis L. 55
Osmanthus fragrans (Thunb.) Lour. 108
Osmanthus fragrans (Thunb.) Lour. var. *aurantiacus* Makino 108
Osmanthus fragrans Lour. (Thunb.) var. *thunbergii* Makino 108
Osmanthus matsumuranus Hayata 108
Osmunda angustifolia Ching 2
Osmunda banksiifolia (Presl) Kuhn 2
Osmunda japonica Thunb. 2
Osmunda mildei C. Chr. 2
Osmunda vachellii Hook. 3
Osmundaceae 2
Ottelia alismoides (L.) Pers. 147
Ottochloa malabarica (L.) Dandy 185
Ottochloa nodosa (Kunth) Dandy 185

Oxalidaceae 42
Oxalis bowiei Lindl. 42
Oxalis corniculata L. 42
Oxalis corymbosa DC. 42
Oxyceros sinensis Lour. 116

P

Pachira macrocarpa (Cham. et Schlcht.) Walp. 58
Pachyrhizus erosus (L.) Urb. 78
Pachystachys lutea Nees 138
Paederia scandens (Lour.) Merr. 116
Paederia scandens var. *tomentosa* (Bl.) Hand. -Mazz. 116
Paederia stenobotrya Merr. 116
Palhinhaea cernua (L.) A. Franco et Vasc. 1
Paliurus ramosissimus (Lour.) Poir. 91
Pandaceae 66
Pandanaceae 167
Pandanus austrosinensis T. L. Wu 167
Pandanus forceps Martelli 167
Pandanus furcatus Roxb. 167
Pandanus tectorius Sol. 167
Pandanus utilis Bory 167
Pandanus veitchi (Dall.) Hort. 167
Panicum brevifolium L. 185
Panicum dichotomiforum Michaux 185
Panicum incomtum Trin. 185
Panicum maximum Jacq. 185
Panicum notatum Retz. 185
Panicum repens L. 185
Papaveraceae 33
Paphiopedilum purpuratum (Lindl.) Stein 172
Papilionaceae 73

Paraixeris denticulata (Houtt.) Nakai 123
Parakmeria lotungensis (Chun et Tsoong) Y. W. Law 23
Parakmeria yunnanensis Hu 23
Paramichelia baillonii (Pierre) Hu 23
Paraphlomis albida Hand. -Mazz var. *brevidens* Hand. -Mazz. 145
Paraphlomis javanica (Bl.) Prain var. *angustifolia* (C. Y. Wu) C. Y. Wu et H. W. Li 145
Parathelypteris angulariloba (Ching) Ching 9
Parathelypteris glanduligera (Kunze) Ching 9
Paris polyphylla Smith 156
Parkeriaceae 7
Parnassia wightiana Wall. ex Wight et Arn. 37
Parthenocissus dalzielii Gagnep. 93
Parthenocissus semicordatus (Wall.) Planch. 93
Parthenocissus tricuspidata (Sieb. et Zucc.) Planch. 93
Paspalum conjugatum Berg. 185
Paspalum dilatatum Poir. 186
Paspalum distichum L. 186
Paspalum hirsutum Retz. 186
Paspalum scrobiculatum L. var. *orbiculare* (G. Forst.) Hack. 186
Paspalum thunbergii Kunth ex Steud. 186
Paspalum vaginatum Sw. 186
Passiflora alatocaerulea Lindl. 46
Passiflora capsularis L. 46
Passiflora cochichinensis Spreng. 46
Passiflora coerulea L. 46

Passiflora edulis Sims 46
Passiflora foetida L. 46
Passifloraceae 46
Pavetta hongkongensis Bremek. 116
Pedilanthus tithymaloides (L.) Poit. 65
Pelargonium hortorum Bailey 41
Pelargonium graveolens L'Her. 41
Peliosanthes macrostegia Hance 155
Pellionia grijsii Hance 87
Pellionia radicans (Sieb. et Zucc.) Wedd. 87
Pellionia repens (Lour.) Merr. 87
Pellionia scabra Benth. 87
Peltophorum pterocarpum (DC.) Baker ex K. Heyn. 73
Pennisetum alopecuroides (L.) Spreng. 186
Pennisetum polystachyon (L.) Schult. 186
Pennisetum purpureum Schum. 186
Pentaphylacaceae 52
Pentaphylax euryoides Gardn. et Champ. 52
Pentas lanceolata (Forsk.) Schum. 116
Pentasacme caudatum Wall. ex Wight 111
Peperomia argyreia E. Morren 31
Peperomia blanda (Jacq.) Kunth 31
Peperomia caperata Yunck. 31
Peperomia cavaleriei C. DC. 31
Peperomia incaua A. Dietr. 31
Peperomia obtusifolia (L.) A. Dietr. 31
Peperomia pellucida (L.) Kunth 31
Peperomia serpens C. DC. Variegata' 32
Peperomia tetraphylla (G. Forst.) Hook. et Arn. 32
Pereskia aculeata Mill. 50
Pericallis hybrida B. Nord. 124
Pericampylus glaucus (Lam.) Merr. 30
Perilla frutescens (L.) Britt. 145
Perilla frutescens (L.) Britt. var. *crispa* (Thunb.) Hand. -Mazz. 145
Perilla frutescens (L.) Britt. var. *purpurascens* (Hayata) H. W. Li 145
Periplocaceae 112
Peristylus lacertiferus (Lindl.) J. J. Sm. 172
Perotis indica (L.) O. Kuntze 186
Persea americana Mill. 27
Pertusadina hainanensis (How) Ridsd. 116
Petunia hybrida (Hook.) Vilm. 129
Peucedanum praeruptorum Dunn 102
Phaius tankervilleae (Banks ex L'Her.) Bl. 172
Phalaenopsis aphrodite Rchb. f. 172
Pharbitis indica (Burm.) R. C. Fang 132
Pharbitis nil (L.) Choisy 132
Pharbitis purpurea (L.) Voigt 132
Philodendron emerald Duke 159
Philodendron erubescens C. Koch et Aug. 159
Philodendron grandifolium Schott 159
Philodendron imbe Schott ex Engl. 159
Philodendron oxycardium Schott 159
Philodendron panduraeforme Kunth 159
Philodendron pittieri Engl. 159
Philodendron selloum C. Koch 159

Philydraceae 168
Philydrum lanuginosum Banks et Sol. ex Gaertn. 168
Phlegmariurus fordii (Baker) Ching 1
Phoebe faberi (Hemsl.) Chun 27
Phoenix canariensis Hort. ex Chabaud 166
Phoenix dactylifera L. 166
Phoenix hanceana Naudin 166
Phoenix paludosa Roxb. 166
Phoenix roebelenii O. Brien 166
Pholidota cantonensis Rolfe 172
Pholidota chinensis Lindl. 172
Photinia benthamiana Hance 67
Photinia raupingensis Kuan 67
Phragmites australis Trin. ex Steud. 186
Phragmites karka (Retz.) Trin. ex Steud. 186
Phrynium rheedei Suresh et Nichols. 153
Phyla nodiflora (L.) Greene 142
Phyllanthus cochinchinensis (Lour.) Spreng. 65
Phyllanthus emblica L. 65
Phyllanthus niruri L. 65
Phyllanthus reticulatus Poir. 65
Phyllanthus reticulatus Poir. var. *glaber* Muell.-Arg. 65
Phyllanthus urinaria L. 65
Phyllanthus ussuriensis Rupr. ex Maxim. 65
Phyllanthus virgatus Forst. f. 65
Phyllodium elegans (Lour.) Desv. 79
Phyllodium pulchellum (L.) Desv. 79
Phyllostachys aurea Carr. ex Cam. et Riv. 189
Phyllostachys nidularia Munro 190
Phyllostachys nigra (Lodd. ex Lindl.) Munro 190
Phymatodes longissima (Bl.) J. Smith 15
Phymatodes scolopendria (Burm. f.) Ching 15
Physalis alkekengi L. 129
Physalis alkekengi L. var. *francheti* (Mast.) Makino 129
Physalis minima L. 129
Phytolacca acinosa Roxb. 40
Phytolacca americana L. 40
Phytolaccaceae 40
Picrasma quassioides (D. Don) Benn. 95
Picria fel-terrae Lour. 133
Pilea angulata (Bl.) Bl. 88
Pilea cadierei Gagnep. et Guill. 88
Pilea microphylla (L.) Liebm. 88
Pilea mollis Hemsl. 88
Pilea peltata Hance 88
Pilea peperomioides Diels 88
Pilea repens Liebm. 88
Pileostegia tomentella Hand.-Mazz. 66
Pileostegia viburnoides Hook. f. et Thoms. 66
Pinaceae 17
Pinanga sinii Burret 166
Pinus elliottii Engelm. 17
Pinus kwangtungensis Chun ex Tsiang 18
Pinus latteri Mason 18
Pinus massoniana Lamb. 18
Pinus tabulaeformis Carr. 18
Piper arboricola C. DC. 32
Piper austrosinense Y. C. Tseng 32

Piper betle L. 32
Piper hancei Maxim. 32
Piper hongkongense Hatusima 32
Piper longum L. 32
Piper nigrum L. 32
Piper sarmentosum Roxb. 32
Piperaceae 31
Pistacia chinensis Bunge 99
Pistia stratiotes L. 159
Pitcairnia muscosa Mart. ex Schult. f. 150
Pithecellobium dulce (Roxb.) Benth. 71
Pittosporaceae 45
Pittosporum glabratum Lindl. 45
Pittosporum glabratum var. *neriifolium* Rehd. et Wils. 45
Pittosporum tobira (Thunb.) Ait. 45
Pityrogramma calomelanos (L.) Link 7
Plagiogyria adnata (Bl.) Bedd. 3
Plagiogyria euphlebia Mett. 3
Plagiogyria stenoptera (Hance) Diels 3
Plagiogyria tenuifolia Cop. 3
Plagiogyriaceae 3
Plantaginaceae 127
Plantago asiatica L. 127
Platyceriaceae 15
Platycerium bifurcatum (Cav.) C. Chr. 15
Platycerium wallichii Hook. 15
Platycerium willinckii T. Moore 16
Platycladus orientalis (L.) Franco 18
Platycodon grandiflorus (Jacq.) A. DC. 127
Pleioblastus amarus (Keng) Keng f.

190
Pleocnemia winitii Holtt. 12
Pleurosoriopsidaceae 10
Pleurosoriopsis makinoi (Maxim.) Fomin 10
Pluchea indica (L.) Less. 124
Plumbaginaceae 126
Plumbago auriculata Lam. 127
Plumbago zeylanica L. 127
Plumeria rubra L. 109
Plumeria rubra L. ' Acutifolia ' 109
Podocarpaceae 19
Podocarpus macrophyllus (Thunb.) D. Don 19
Podocarpus macrophyllus (Thunb.) D. Don var. *maki* Endl. 19
Podocarpus neriifolius D. Don 19
Pogonatherum crinitum (Thunb.) Kunth 186
Pogostemon auricularius (L.) Hassk. 145
Pogostemon cablin (Blanco) Benth. 146
Pogostemon championii Prain 146
Polyalthia suberosa (Roxb.) Thw. 24
Polycarpaea corymbosa (L.) Lam. 38
Polycarpon prostratum (Forssk.) Aschers. et Schweinw. 38
Polygala arillata Buch. -Ham. ex D. Don 35
Polygala fallax Hemsl. 35
Polygala glomerata Lour. 35
Polygala hongkongensis Hemsl. 35
Polygalaceae 35
Polygonaceae 39
Polygonatum cyrtonema Hua 155
Polygonum aviculare L. 39

Polygonum barbatum L.   39
Polygonum barbatum L. var. *gracile* (Danser) Steward   39
Polygonum caespitosum Bl.   39
Polygonum chinense L.   39
Polygonum cuspidatum Sieb. et Zucc.   39
Polygonum glabrum Willd.   39
Polygonum hastato-sagittatum Makino   39
Polygonum hydropiper L.   39
Polygonum japonicum Meisn.   39
Polygonum juncundum Meisn.   39
Polygonum lapathifolium L.   39
Polygonum lapathifolium L. var. *salicifolium* Sibth.   39
Polygonum multiflorum Thunb.   40
Polygonum muricatum Meissn.   40
Polygonum orientale L.   40
Polygonum perfoliatum L.   40
Polygonum plebeium R. Br.   40
Polygonum viscosum Ham. Buch. -Ham. ex D. Don   40
Polypodiaceae   14
Polyscias balfouriana (Hort. ex Sander) Bailey   101
Polyscias fruticosa (L.) Harms   101
Polyscias guilfoylei (Cogn. et March.) Bailey   101
Polystichum eximium (Mett.) C. Chr.   12
Pongamia pinnata (L.) Merr.   79
Pontederiaceae   156
Popowia pisocarpa (Bl.) Endl.   24
Portulaca grandiflora Hook.   38
Portulaca oleracea L.   38
Portulaca pilosa L.   38

Portulacaceae   38
Portulacaria afra (L.) Jacq.   38
Potamogeton malaianus Miq.   147
Potamogetonaceae   147
Pothos chinensis (Raf.) Merr.   159
Pothos repens (Lour.) Druce   159
Pottsia laxiflora (Bl.) O. Kuntze   110
Pouteria campechiana (Kunth) Baehni   104
Pouzolzia sanguinea (Bl.) Merr.   88
Pouzolzia zeylanica (L.) Benn.   88
Pratia nummularia (Lam.) A. Br. et Aschers.   127
Premna microphylla Turcz.   142
Premna obtusifolia R. Br.   142
Primula polyantha Mill.   126
Primulaceae   126
Procris wightiana Wall. ex Wedd.   88
Pronephrium aspera (Presl) W. C. Shieh et J. L. Tsai   9
Pronephrium lakhimpurense (Ros.) Holtt.   9
Pronephrium simplex (Hook.) Holtt.   9
Pronephrium triphyllum (Sw.) Holtt.   9
Proteaceae   45
Prunella vulgaris (L.)   146
Prunus cerasifera Ehrh. var. *atropurea* Jacq.   67
Prunus salicina Lindl.   68
Pseudocyclosorus ciliatus (Wall. ex Benth.) Ching   9
Pseudocyclosorus falcilobus (Hook.) Ching   9
Pseudodrynaria coronans (Wall. ex Mett.) Ching   15

Pseudosasa amabilis (McClure) Keng f. 190
Pseudosasa cantorii (Munro) Keng f. 190
Pseudosasa hindsii (McClure) C. D. Chu et C. S. Chao 190
Psidium guajava L. 53
Psilotaceae 1
Psilotum nudum (L.) Beauv. 1
Psophocarpus tetragonolobus (L.) DC. 79
Psychotria asiatica L. 116
Psychotria serpens L. 116
Psychotria tutcheri Dunn 116
Pteridaceae 6
Pteridiaceae 6
Pteridium aquilinum (L.) Kuhn var. *latiusculum* (Desv.) Underw. ex Heller 6
Pteridium esculentum (Forst.) Cokayne 6
Pteridium revolutum (Bl.) Nakai 6
Pteridophyta 1
Pteridrys australis Ching 12
Pteris cadieri Christ 6
Pteris cretica L. var. *nervosa* (Thunb.) Ching et S. H. Wu 7
Pteris dispar Kunze 6
Pteris ensiformis Burm. f. 6
Pteris excelsa Gaud. 6
Pteris fauriei Hieron. 6
Pteris finotii Christ 6
Pteris linearis Poir. 6
Pteris multifida Poir. 6
Pteris semipinnata L. 7
Pteris vittata L. 7
Pterocarpus indicus Willd. 79

Pteroceltis tatarinowii Maxim. 84
Pterocypsela indica (L.) Shih 124
Pterospermum heterophyllum Hance 58
Pterospermum lanceaefolium Roxb. 58
Ptychosperma macarthurii Nichols. 166
Pueraria lobata (Willd.) Ohwi 79
Pueraria lobata (Willd.) Ohwi var. *montana* (Lour.) Van der Maesen 79
Pueraria lobata (Willd.) Ohwi. var. *thomsonii* (Benth.) Van der Maesen 79
Pueraria phaseoloides (Roxb.) Benth. 79
Punica granatum L. 43
Punica granatum L. Multiplex' 43
Punicaceae 43
Pycnospora lutescens (Poir.) Schindl. 79
Pycreus flavidus (Retz.) Koyama 177
Pycreus polystachyos (Rottb.) P. Beauv. 177
Pycreus pumilus (L.) Domin 177
Pycreus sanguinolentus (Vahl) Nees 177
Pygeum topengii Merr. 68
Pyracantha fortuneana (Maxim.) Li 68
Pyrethrum cinerariifolium Trev. 124
Pyrostegia venusta (Ker-Gawl.) Miers 136
Pyrrosia adnascens (Sw.) Ching 15
Pyrrosia lingua (Thunb.) Farw. 15
Pyrus calleryana Decne. 68
Pyrus calleryana Decne. var. *koehnei* (Schneid.) Yu 68

Pyrus pyrifolia (Burm. f.) Nakai 68

Q

Quamoclit pennata (Desr.) Bojer 132
Quercus acutissima Carruth. 83
Quisqualis indica L. 55

R

Radermachera hainanensis Merr. 136
Radermachera sinica (Hance) Hemsl. 136
Ranunculaceae 27
Ranunculus cantoniensis DC. 28
Ranunculus chinensis Bunge 28
Ranunculus japonicus Thunb. 28
Ranunculus sceleratus L. 28
Rapanea linearis (Lour.) S. Moore 105
Rapanea neriifolia (Sieb. et Zucc.) Mez 105
Raphanus sativus L. 34
Raphanus sativus L. var. *longipinnatus* Bailey 34
Raphiolepis indica (L.) Lindl. 68
Raphiolepis lanceolata Hu 68
Raphiolepis salicifolia Lindl. 68
Rauvolfia tetraphylla L. 110
Rauvolfia verticillata (Lour.) Baill 110
Ravenala madagascariensis Sonn. 151
Ravenea rivularis Jum. et H. Perrier 166
Reevesia thyrsoideas Lindl. 58
Rehderodendron kwangtungensise Chun 105
Reineckia carnea (Andr.) Kunth 156
Reinwardtia indica Dumort 41

Rhamnaceae 91
Rhamnus brachypoda C. Y. Wu ex Y. L. Chen 91
Rhamnus crenata Sieb. et Zucc. 92
Rhamnus napalensis (Wall.) Lawson 92
Rhaphidophora hongkongensis Schott 159
Rhapis excelsa (Thunb.) Henry ex Rehd. 166
Rhapis gracilis Burret 166
Rhapis multifida Burret 166
Rhizophoraceae 55
Rhododendron championae Hook. 102
Rhododendron farrerae Tate ex Sweet 102
Rhododendron hongkongense Hutch. 103
Rhododendron indicum (L.) Sweet 103
Rhododendron moulmainense Hook. 103
Rhododendron mucronatum (Bl.) G. Don 103
Rhododendron mucronatum (Bl.) G. Don var. *kemono* Hort. 103
Rhododendron pulchrum Sweet 103
Rhododendron simsii Planch. 103
Rhodoleia championi Hook. f. 81
Rhodomyrtus tomentosa (Ait.) Hassk. 53
Rhopalephora scaberrima (Bl.) Faden 148
Rhus chinensis Mill. 99
Rhus chinensis Mill. var. *roxburghii* (DC.) Rehd. 99
Rhus hypoleuca Champ. ex Benth. 99

Rhynchelytrum repens (Willd.) Hubb. 186
Rhynchosia volubilis Lour. 79
Rhynchospora rubra (Lour.) Makino 177
Rhynchospora rugosa (Vahl) Gale 177
Rhynchotechum formosanum Hatusima 135
Ricinus communis L. 65
Robiquetia succisa (Lindl.) Seidenf. et Garay 172
Rohdea japonica (Thunb.) Roth 156
Rorippa cantoniensis (Lour.) Ohwi 34
Rorippa dubia (Pers.) Hara 34
Rorippa globosa (Turcz.) Vassilcz. 34
Rorippa indica (L.) Hiern 35
Rosa chinensis Jacq. 68
Rosa chinensis Jacq. var. *mimima* Voss. 68
Rosa kwangtungensis Yu et Tsai 68
Rosa laevigata Michx. 68
Rosa multiflora Thunb. 68
Rosa odorata Sweet 68
Rosa rugosa Thunb. 68
Rosa wichuraiana Crep. 68
Rosaceae 67
Rotula indica (Willd.) Koehne 43
Rotala rotundifolia (Buch.-Ham. ex Roxb.) Koehne 43
Rottboellia exaltata (L.) L. f. 186
Rourea microphylla (Hook. et Arn.) Planch. 99
Rourea minor (Gaertn.) Alston 99
Roystonea oleracea (Jacq.) O. F. Cook 166
Roystonea regia (Kunth) O. F. Cook 167
Rubiaceae 112
Rubus alceaefolius Poir. 68
Rubus gressitii Metc. 68
Rubus leucanthus Hance 68
Rubus parvifolius L. 69
Rubus pirifolius Smith 69
Rubus reflexus Ker-Gawl. 69
Rubus reflexus Ker-Gawl. var. *hui* (Diels apud Hu) Metc. 69
Rubus reflexus Ker-Gawl. var. *lancelobus* Metc. 69
Rubus rosaefolius Smith 69
Rubus sumatranus Miq. 69
Rubus swinhoei Hance 69
Rubus tsangii Merr. 69
Ruellia brittoniana Leonard 138
Rumex maritimus L. 40
Rumohra adiantiformis (G. Forst.) Tindale 12
Rungia pectinata (L.) Nees 138
Ruscaceae 157
Ruscus aculeatus L. 157
Russelia equisetiformis Schlcht. et Cham. 133
Ruta graveolens L. 95
Rutaceae 94

S

Sabia fasciculata Lec. ex L. Chen 98
Sabia japonica Maxim. 98
Sabia limoniacea Wall. ex Hook. f. var. *ardisoides* (Hook. et Arn.) L. 98
Sabia swinhoei Hemsl. 98
Sabiaceae 98

Sabina chinensis (L.) Ait. 19
Sabina chinensis (L.) Ait. 'Kaizuca' 19
Sabina procumbens (Sieb. ex Endl.) Iwata et Kusaka 19
Saccharum arundinaceum Retz. 186
Saccharum officinarum L. 187
Saccharum spontaneum L. 187
Sacciolep indica (L.) A. Chase 187
Sacciolep myosuroides (R. Br.) A. Gamus 187
Sageretia lucida Merr. 92
Sageretia thea (Osbeck) Johnst 92
Sagittaria sagittifolia L. ssp. l eucopetala (Miq.) Hartog 147
Sagittaria trifolia L. 147
Saintpaulia ionantha H. Wendl. 135
Salicaceae 81
Salix babylonica L. 81
Salomonia cantoniensis Lour. 35
Salvia chinensis Benth. 146
Salvia coccinea L. 146
Salvia farinacea Benth. 146
Salvia plebeia R. Br. 146
Salvia splendens Ker-Gawl. 146
Salvinia natans (L.) All. 16
Salviniaceae 16
Samanea saman (Jacq.) Merr. 71
Sambucus williamsii Hance 117
Samydaceae 46
Sanchezia nobilis Hook. f. 138
Sanchezia parvibracteata Sprag. et Hutch. 139
Sansevieria cylindrica Bojer 163
Sansevieria trifasciata Prain 163
Sansevieria trifasciata Prain var. *laurentii* N. E. Br. 163

Santalaceae 91
Sapindaceae 96
Sapindus saponaria L. 97
Sapium discolor (Champ. ex Benth.) Muell. -Arg. 65
Sapium sebiferum (L.) Roxb. 65
Sapotaceae 103
Saraca dives Pierre 73
Sarcandra glabra (Thunb.) Nakai 33
Sarcandra hainanensis (Pei) Swamy et Bailey 33
Sarcopsperma laurinum (Benth.) Hook. f. 104
Sarcopspermaceae 104
Sargentodoxa cuneata (Oliv.) Rehd. et Wils. 30
Sargentodoxaceae 30
Saritaea magnifica (Sprague ex van Steenis) Dugand 136
Sarracenia leucophylla Raf. 37
Sarraceniaceae 37
Sassafras tzumu (Hemsl.) Hemsl. 27
Saurauia tristyla DC. 52
Saurauiaceae 52
Sauropus bacciformis (L.) Airy-Shaw 65
Sauropus spatulifolius Beille 65
Saururaceae 32
Saururus chinensis (Lour.) Baill. 32
Saxifragaceae 36
Scaevola hainanensis Hance 128
Scaevola sericea Vahl 128
Schefflera actinophylla (Endl.) Harms 101
Schefflera arboricola Hayata 101
Schefflera delavayi (Franch.) Harms ex Diels 101

Schefflera elegantissima (Veitch. ex Mast.) Lowry et Frodin 101
Schefflera heptaphylla (L.) Frodin 101
Schima superba Gardn. et Champ. 51
Schima wallichii Choisy 52
Schisandraceae 23
Schizachyrium brevifolium (Sw.) Nees ex Buse 187
Schizachyrium sanguineum (Retz.) Alston 187
Schizostachyum pseudolima McClure 190
Schlumbergera bridgesii (Lem.) Loefgr. 50
Schlumbergera truncata (Haw.) Moran. 50
Schoepfia chinensis Gardn. et Champ. 90
Scilla scilloides (Lindl.) Druce 154
Scirpus juncoides Roxb. 177
Scirpus lacustris L. ssp. *validus* (Vahl) Koyama 177
Scirpus triqueter L. 177
Scleria biflora Roxb. 177
Scleria ciliaris Nees 177
Scleria elata Thw. var. *latior* C. B. Clarke 178
Scleria harlandii Hance 178
Scleria levis Retz. 178
Scleria lithosperma (L.) Sw. 178
Scleria parvula Steud. 178
Scleria rugosa R. Br. 178
Scleria terrestris (L.) Foss. 178
Scolopia chinensis (Lour.) Clos 46
Scolopia saeva (Hance) Hance 46
Scoparia dulcis L. 133

Scrophulariaceae 132
Scurrula parasitica L. 90
Scutellaria barbata D. Don 146
Scutellaria indica L. 146
Sebastiania chamaelea (L.) Muell.-Arg. 65
Sechium edule (Jacq.) Swartz 48
Securidaca inappendiculata Hassk. 35
Securinega virosa (Roxb. ex Willd.) Baill. 66
Sedum acre L. 36
Sedum lineare Thunb. 36
Sedum marganianum Walth. 36
Sedum mexicanum Britt. 36
Sedum sarmentosum Bunge 36
Selaginella biformis A. Br. ex Kuhn 1
Selaginella ciliaris (Retz.) Spring 1
Selaginella delicatula (Desv. ex Pior.) Alston 1
Selaginella doederleinii Hieron 1
Selaginella heterostachys Baker 1
Selaginella involvens (Sw.) Spring 1
Selaginella kraussiana (Kunze) A. Br. 2
Selaginella limbata Alston 2
Selaginella moellendorffii Hieron 2
Selaginella tamariscina (Beauv.) Spring 2
Selaginella uncinata (Desv.) Spring 2
Selaginella xipholepis Baker 2
Selaginellaceae 1
Selenodesmium siamense (Christ) Ching et C. H. Wang 4
Semecarpus gigantifolia Vidal 99
Semiliquidambar cathayensis H. T. Chang 81

Senecio scandens Buch. -Ham. ex D. Don 124
Senecio stauntonii DC. 124
Serissa japonica (Thunb.) Thunb. 117
Serissa serissoides (DC.) Druce 117
Sesbania bispinosa (Jacq.) W. F. Wight 79
Sesbania cannabina (Retz.) Poir. 79
Sesbania sesban (L.) Merr. 79
Sesuvium portulacastrum L. 38
Setaria geniculata (Lam.) P. Beauv. 187
Setaria glauca (L.) Beauv. 187
Setaria pallidifusca (Schum.) Stapf et C. E. Hubb. 187
Setaria palmifolia (Koen.) Stapf 187
Setaria plicta (Lam.) T. Cooke 187
Setaria viridis (L.) P. Beauv. 187
Setcreacea purpurea B. K. Boom 148
Sida acuta Burm. f. 59
Sida alnifolia L. 59
Sida alnifolia L. var. *microphylla* (Cavan.) S. Y. Hu 59
Sida chinensis Retz. 60
Sida cordata (Burm. f.) Boiss. 60
Sida cordifolia L. 60
Sida rhombifolia L. 60
Sida subcordata Span. 60
Siegesbeckia orientalis L. 124
Silybum marianum (L.) Gaertn. 124
Simaroubaceae 95
Sindora glabra Merr. 73
Sindora tonkinensis A. Cheval. ex K. et S. S. Larsen 73
Sinningia speciosa Benth. et Hook. 135
Sinopteridaceae 7

Sinosideroxylon wightianum (Hook. et Arn.) Aubr. 104
Sloanea leptocarpa Diels 57
Sloanea sinensis (Hance) Hemsl. 57
Smilacaceae 156
Smilax astrosperma Wang et Tang 156
Smilax china L. 156
Smilax corbularia Kunth 157
Smilax davidiana A. DC. 157
Smilax glabra Roxb. 157
Smilax hypoglauca Benth. 157
Smilax lanceifolia Roxb. 157
Smilax lanceifolia Roxb. var. *opaca* A. DC. 157
Smilax riparia A. DC. 157
Smithia conferta Smith 79
Solanaceae 128
Solandra nitida Zucc. 129
Solanum americanum Mill. 129
Solanum capsicoides Allioni 129
Solanum coagulans Forsk. 129
Solanum erianthum D. Don 129
Solanum lasiocarpum Dunal 129
Solanum lyratum Thunb. 130
Solanum macaonense Dunal 130
Solanum macranthum Sw. 130
Solanum mammosum L. 130
Solanum melongena L. 130
Solanum nigrum L. 130
Solanum suffruticosum Schousb. 130
Solanum torvum Swartz 130
Solanum tuberosum L. 130
Solene amplexicaulis (Lam.) Gandhi 48
Solidago canadensis L. 124
Solidago decurrens Lour. 124
Soliva anthemifolia (Juss.) R. Br.

124
Sonchus arvensis L. 124
Sonchus asper (L.) Hill. 124
Sonchus oleraceus L. 124
Sophora japonica L. var. *pendula* Loud. 79
Sorghum nitidum (Vahl) Pers. 187
Spathiphyllum floribundum (Linden et Andre) N. E. Br. 159
Spathiphyllum kochii Engl. et Krause 160
Spathodea campanulata Beauv. 136
Spathoglottis pubescens Lindl. 172
Spatholobus suberectus Dunn 79
Sphaeranthus africanus L. 124
Sphaerocaryum malaccense (Trin.) Pilg. 187
Sphaeropteris lepifera (J. Sm. ex Hook.) R. Tryon 5
Sphenomeris biflora (Kaulf.) Akas. 5
Sphenomeris chinensis (L.) Maxon 6
Spilanthes paniculata Wall. ex DC. 124
Spinacia oleracea L. 40
Spinifex littoreus (Burm. f.) Merr. 187
Spiraea cantoniensis Lour. 69
Spiraea chinensis Maxim. 69
Spiranthes hongkongensis S. Y. Hu et Barretto 172
Spiranthes sinensis (Pers.) Ames 172
Spirodela polyrrhiza (L.) Schleid. 160
Spondias lakonensis Pierre 99
Sporobolus fertilis (Steud.) W. D. Clayton 187
Sporobolus virginicus (L.) Kunth 188

Stachytarpheta jamaicensis (L.) Vahl 142
Stachyuraceae 80
Stachyurus chinensis Franch. 80
Stahlianthus involucratus (King ex Baker) Craib 152
Stapelia grandiflora Mass. 111
Stapelia hirsuta L. 111
Staphyleaceae 98
Stauntonia brunoniana (Wall. ex Hemsl.) Decne. ssp. *elliptica* (Hemsl.) H. N. Qin 29
Stauntonia chinensis DC. 29
Stauntonia obovata Hemsl. 29
Stellaria alsine Grinum 38
Stellaria media (L.) Cyr. 38
Stemona tuberosa Lour. 162
Stemonaceae 162
Stenotaphrum helferi Munro ex Hook. f. 188
Stephania cepharantha Hayata 30
Stephania japonica (Thunb.) Miers 30
Stephania longa Lour. 30
Stephania tetrandra S. Moore 30
Sterculia lanceolata Cav. 58
Sterculia nobilis Smith 58
Sterculiaceae 57
Sticherus laevigatus Presl 3
Stictocardia tiliifolia (Desr.) Hall. f. 132
Strelitzia reginae Aiton 151
Strelitziaceae 151
Striga asiatica (L.) O. Kuntze 134
Strobilanthes cusia (Nees) O. Kuntze 139
Strobilanthes dyerianus Mast. 139

Strobilanthes tetraspermus (Champ. ex Benth.) Druce 139
Strobilantheus divaricatus (Nees) T. Anders. 139
Strobilanthus dalzielii (W. W. Smith) R. Ben. 138
Stromanthe sanguinea (Hook.) Sonder 153
Strophanthus divaricatus (Lour.) Hook. et Arn. 110
Strychnos angustiflora Benth. 107
Strychnos cathayensis Merr. 107
Strychnos umbellata (Lour.) Merr. 107
Stylidiaceae 128
Stylidium uliginosum Swartz 128
Styracaceae 105
Styrax faberi Perk. 105
Styrax grandiflorus Griff. 105
Styrax odoratissimus Champ. et Benth. 105
Styrax serrulatus Roxb. 105
Styrax suberifolius Hook. et Arn. 105
Styrax tonkinensis (Pierre) Craib ex Hartw. 105
Suaeda australis (R. Br.) Moq. 40
Swietenia macrophylla King 96
Swietenia mahagoni (L.) Jacq. 96
Syagrus romanzoffiana (Cham.) Glassm. 167
Sycopsis dunnii Hemsl. 81
Symplocaceae 106
Symplocos adenophylla Wall. 106
Symplocos adenopus Hance 106
Symplocos chinensis (Lour.) Druce 106
Symplocos cochinchinensis (Lour.) Moore 106
Symplocos confusa Brand. 106
Symplocos crassifolia Benth. 106
Symplocos decora Hance 106
Symplocos dolichotricha Merr. 106
Symplocos dung Eberm. et Dub. 106
Symplocos fordii Hance 106
Symplocos glauca (Thunb.) Koidz. 106
Symplocos lancifolia Sieb. et Zucc. 106
Symplocos laurina (Retz.) Wall. 106
Symplocos lucida Sieb. et Zucc. 106
Symplocos paniculata (Thunb.) Miq. 106
Symplocos racemosa Roxb. 106
Symplocos stellaris Brand 107
Symplocos sumuntia Buch-Ham. ex D. Don 107
Synedrella nodiflora (L.) Gaertn. 124
Syngonium podophyllum Schott 160
Synsepalum dulcificum (A. DC.) Daniell 104
Syzygium buxifolium Hook. ex Arn. 53
Syzygium championii (Benth.) Merr. et Perry 53
Syzygium cumini (L.) Skeels 53
Syzygium euonymifolium (Metc) Merr. et Perry 53
Syzygium fluviatile (Hemsl.) Merr. et Perry 53
Syzygium grijsii (Hance) Merr. et Perry 54
Syzygium hancei Merr. et Perry 54
Syzygium jambos (L.) Alston 54
Syzygium levinei (Merr.) Merr. et Perry

54
Syzygium odoratum (Lour.) DC. 54
Syzygium rehderianum Merr. et Perry 54
Syzygium samarangense (Bl.) Merr. et Perry 54

T

Tabebuia chrysantha (Jacq.) Nichols. 136
Tabebuia rosea (Bertol.) DC. 136
Tabermaemontana divaricata (L.) R. Br. ex Roem. et Schult. 110
Tadehagi triquetrum (L.) Ohashi 80
Tagetes erecta L. 124
Tagetes patula L. 125
Tainia hongkongensis (Rolfe) Tang et Wang 168
Tainia hookeriana Ring et Prantl 168
Tainia viridifusa Hook. 172
Talinum paniculatum (Jacq.) Gaertn. 38
Tamarindus indica L. 73
Tarenna attenunata (Voigt) Hutch. 117
Tarenna mollissima (Hook. et Arn.) Rob. 117
Taxaceae 20
Taxillus chinensis (DC.) Danser 90
Taxillus sutchuenensis (Lec.) Danser 90
Taxodiaceae 18
Taxodium distichum (L.) Rich. 18
Taxodium distichum (L.) Rich. var. *imbricatum* (Bongn.) Parl. 18
Tecoma capensis (Thunb.) Lindl. 136
Tecoma stans (L.) A. L. Juss. ex Kunth 136
Tectaria decurrens (Presl) Cop. 12
Tectaria phaeocaulis (Ros.) C. Chr. 12
Tectaria subtriphylla (Hook. et Arn.) Cop. 13
Tectona grandis L. f. 142
Telosma cathayensis Merr. 111
Telosma cordata (Burm. f.) Merr. 111
Tephrosia purpurea (L.) Pers. 80
Terminalia arjuna Wight et Arn. 55
Terminalia calamansanai (Blanco) Rolfe 55
Terminalia catappa L. 55
Terminalia mantalyi H. Perrier 55
Terminalia muelleri Benth. 55
Terminalia myriocarpa Van Huerck et Muell. Arg. 55
Ternstroemia gymnanthera (Wight et Arn.) Bedd. 52
Tetracera asiatica (Lour.) Hoogl. 45
Tetragonia tetragonioides (Pall.) O. Kuntze 38
Tetrapanax papyriferus (Hook.) K. Koch 101
Tetrastigma hemsleyanum Diels et Gilg 93
Tetrastigma obtectum (Wall.) Planch. ex Franch. 93
Tetrastigma planicaule (Hook.) Gagnep. 93
Teucrium viscidum Bl. 146
Thalia dealbata J. Fraser 153
Thalictrum acutifolium (Hand. -Mazz.) Boivin 28
Thalictrum umbricola Ulbr. 28

Theaceae 50
Thelocactus lophothele Britt. et Rose 50
Thelypteridacea 8
Themeda caudata (Nees) A. Camus 188
Themeda gigantea (Cav.) Hack. 188
Themeda hookeri (Griseb.) A. Camus 188
Themeda villosa (Poir.) A. Camus 188
Thespesia populnea (L.) Soland. ex Corr. 60
Thevetia peruviana (Pers.) Schum. 110
Thladiantha cordifolia (Bl.) Cogn. 48
Thryallis gracilis O. Kuntze 60
Thunbergia erecta (Benth.) T. Anders. 139
Thunbergia fragrans Roxb. 139
Thunbergia grandiflora Roxb. 139
Thunbergia laurifolia Lindl. 139
Thymelaeaceae 44
Thysanolaena maxima (Roxb.) O. Kuntze 188
Thysanotus chinensis Benth. 156
Tibouchina aspera Aubl. var. *asperrima* Cogn. 55
Tigridiopalma magnifica C. Chen 55
Tiliaceae 56
Tillandsia cyanea Linden ex K. Koch 150
Tinospora sagittata (Oliv.) Gagnep. 30
Tinospora sinensis (Lour.) Merr. 30
Tithonia diversifolia A. Gray 125
Toddalia asiatica (L.) Lam. 95
Toona microcarpa (C. DC.) Harms 96
Toona sinensis (A. Juss.) Roem. 96
Torenia concolor Lindl. 134
Torenia flava Buch.-Ham. ex Benth. 134
Torenia fournieri Linden ex Fourn. 134
Torenia glabra Osbeck 134
Toxicodendron succedaneum (L.) O. Kuntze 99
Toxocarpus laevigatus Tsiang 111
Toxocarpus wightianus Hook. et Arn. 111
Trachelospermum jasminoides (Lindl.) Lem. 110
Trachelospermum jasminoides var. *heterophyllum* Tsiang 110
Trachycarpus fortunei (Hook.) Wendl. 167
Trachycarpus nana Becc. 167
Tradescantia fluminensis Vell. 149
Tradescantia spathacea Sw. 149
Tradescantia zebrina Hort. ex Bosse 149
Trapa bicornis Osbeck 44
Trapaceae 44
Trema angustifolia (Planch.) Bl. 84
Trema canabina Lour. 84
Trema canabina Lour. var. *dielsiana* (Hand.-Mazz.) C. J. Chen 84
Trema orientalis (L.) Bl. 84
Trichomanes birmanicum Bedd. 4
Trichosanthes anguina L. 48
Trichosanthes curcumeroides (Ser.) Maxim. 48
Trichosanthes kirilowii Maxim. 48
Trichosanthes ovigera Bl. 48
Trichosanthes pedata Merr. et Chun

48
Tricyrtis macropoda Miq. 156
Tridax procumbens L. 125
Trigonostemon chinensis Merr. 66
Trilliaceae 156
Tripterospermum nienkui (C. Marq.) C. J. Wu 126
Tristellateia australasiae A. Rich. 60
Triumfetta cana Bl. 56
Triumfetta rhomboidea Jacq. 56
Tropaeolaceae 42
Tropaeolum majus L. 42
Tropidia curculigoides Lindl. 172
Tsoongiodendron odorum Chun 23
Tulipa gesneriana L. 156
Turpinia arguta (Lindl.) Seem. 98
Turpinia montana (Bl.) Kurz var. *glaberrima* (Merr.) T. Z. Hsu 98
Turpinia montana (Bl.) Kurz 98
Tussilago farfara L. 125
Tutcheria championii Nakai 52
Tutcheria microcarpa Dunn 52
Tylophora atrofolliculata Metc. 112
Tylophora floribunda Miq. 112
Tylophora hainanensis Tsiang 112
Tylophora ovata (Lindl.) Hook. ex Steud. 112
Tylophora tenuis Bl. 112
Typha angustifolia L. 160
Typha orientalis Presl 160
Typhaceae 160
Typhonium blumei Nicols. et Sivadasan 160
Typhonium trilobatum (L.) Schott 160

U

Ulmaceae 83

Ulmus parvifolia Jacq. 84
Umbelliferae 101
Uraria crinita (L.) Desv. ex DC. 80
Uraria lagopodioides (L.) Desv. ex DC. 80
Urceola micrantha (Wall. ex G. Don) A. DC. 110
Urceola rosea (Hook. et Arn.) D. J. Middleton 110
Urena lobata L. 60
Urena lobata L. var. *chinensis* (Osbeck) S. Y. Hu 60
Urena procumbens L. 60
Urticaceae 87
Utricularia aurea Lour. 134
Utricularia bifida L. 134
Utricularia exoleta R. Br. 134
Utricularia graminifolia Vahl 134
Utricularia striatula J. Sm. 134
Utricularia uliginosa Vahl 134
Uvaria boniana Finet et Gagnep. 24
Uvaria grandiflora Roxb. ex Hornem. 24
Uvaria macrophylla Roxb. 24

V

Vacciniaceae 103
Vaccinium bracteatum Thunb. 103
Vallisneria natans (Lour.) Hara 147
Vatica mangachapoi Blanco 52
Veitchia merrillii (Becc.) H. E. Moore 167
Ventilago leiocarpa Benth. 92
Veratrum japonicum (Baker) Loes. f. 156
Veratrum nigrum L. 156
Verbena hybrida Voss. 142

Verbena officinalis L. 143
Verbena tenera Spreng. 143
Verbenaceae 139
Vernicia fordii (Hemsl.) Airy Shaw 66
Vernicia montana Lour. 66
Vernonia andersonii Clarke 125
Vernonia aspera (Roxb.) Buch.-Ham. 125
Vernonia cinerea (L.) Less. 125
Vernonia cumingiana Benth. 125
Vernonia patula (Dryand.) Merr. 125
Vernonia saligna (Wall.) DC. 125
Vernonia solanifolia Benth. 125
Veronica undulata Wall. 134
Vetiveria zizanioides (L.) Nash 188
Viburnum fordiae Hance 117
Viburnum hanceanum Maxim. 117
Viburnum odoratissimum Ker-Gawl. 117
Viburnum sempervirens K. Koch 118
Victoria amazonica Sowerby 29
Victoria cruziana Orbign 29
Vigna marina (Burm.) Merr. 80
Vigna minima (Roxb.) Ohwei et Ohashi 80
Vigna radiata (L.) Wilozek 80
Vigna unguiculata (L.) Walp. 80
Viola betonicifolia J. E. Smith 35
Viola diffusa Ging 35
Viola inconspicua Bl. 35
Viola odorata L. 35
Viola philippica Cav. ssp. *munda* W. Beck 35
Viola principis H. de Boiss. 35
Viola thomsonii Oudem. 35
Viola tricolor L. var. *hortensis* DC. 35

Viola verecunda A. Gray 35
Violaceae 35
Viscum articulatum Burm. f. 90
Viscum diospyrosicolum Hayata 91
Viscum liquidambaricolum Hayata 91
Viscum multinerve (Hayata) Hayata 91
Viscum ovalifolium DC. 91
Vitaceae 92
Vitex negundo L. 143
Vitex negundo L. var. *cannabifolia* (Sieb. et Zucc.) Hand.-Mazz. 143
Vitex negundo var. *thyrsoides* P i et S. L. Liou 143
Vitex quinata (Lour.) F. N. Will. 143
Vitex trifolia L. 143
Vitex trifolia L. var. *simplicifolia* Cham. 143
Vitis balanseana Planch. 93
Vitis bryoniifolia Bunge 93
Vitis retordii Rom. du Caill. ex Planch. 93
Vitis vinifera L. 94
Vittaria 8
Vittaria chingii B. S. Wang 8
Vriesea carinata Wawra 150
Vriesea splendens (Brongn.) Lem. 150

W

Wahlenbergia marginata (Thunb.) A. DC. 127
Waltheria americana L. 58
Washingtonia filifera (Linden) H. Wendl. 167
Wedelia biflora (L.) DC. 125
Wedelia chinensis (Osbeck) Merr. 125
Wedelia prostrata (Hook. et Arn.)

Hemsl. 125
Wedelia trilobata (L.) Hitchc. 125
Wedelia wallichii Less. 125
Wendlandia uvariifolia Hance 117
Wikstroemia indica (L.) C. A. Mey. 44
Wikstroemia monnula Hance 44
Wikstroemia nutans Champ. ex Benth. 44
Wisteria sinensis (Sims) Sweet 80
Wodyetia bifurcata A. K. Irvine 167
Wolffia arrhiza (L.) Wimm. 160
Woodwardia japonica (L. f.) J. Sm. 11
Woodwardia orientalis Sw. 11
Woodwardia prolifera Hook. et Arn. 11
Woodwardia unigemmata (Makino) Nakai 11
Wrightia pubescens R. Br. 110

X

Xanthium sibiricum Patrin ex Widder 125
Xylosma longifolium Clos 46
Xyridaceae 149
Xyris indica L. 149
Xyris pauciflora Willd. 149

Y

Youngia japonica (L.) DC. 126
Yucca aloifolia L. var. *marginata* Bommer 163
Yucca elephantipes Hort. ex Regel 164
Yucca filamentosa L. 164
Yucca gloriosa L. 164
Yucca rostrata Engelm. ex Trelease 164

Z

Zamia furfuracea L. f. 17
Zamioculcas zamifolia (Lodd.) Engl. 160
Zantedeschia aethiopica (L.) Spreng. 160
Zantedeschia elliottiana (H. Knight) Engl. 160
Zanthoxylum ailanthoides Sieb. et Zucc. 95
Zanthoxylum avicennae (Lam.) DC. 95
Zanthoxylum myriacanthum Wall. ex Hook. f. 95
Zanthoxylum nitidum (Roxb.) DC. 95
Zanthoxylum piperitum DC. 95
Zanthoxylum scandens Bl. 95
Zea mays L. 188
Zehneria indica (Lour.) Keraudren 48
Zehneria maysorensis (Wight et Arn.) Arn. 48
Zelkova serrata (Thunb.) Makino 84
Zephyranthes candida (Lindl.) Herb. 161
Zephyranthes carinata Herb. 161
Zephyranthes rosea Lindl. 161
Zeuxine affinis (Lindl.) Benth. ex Hook. f. 172
Zingiber corallinum Hance 152
Zingiber mioga (Thunb.) Rosc. 152
Zingiber officinale Rosc. 152
Zingiber striolatum Diels 152
Zingiber zerumbet (L.) Smith 152

Z

Zingiberaceae 151
Zinnia violacea Cav. 126
Ziziphus jujuba Mill. 92
Ziziphus mauritiana Lam. 92
Zornia gibbosa Span. 80

Zoysia japonica Steud. 188
Zoysia matrella (L.) Merr. 188
Zoysia sinica Hance 188
Zoysia tenuifolia Willd. 188

英文名称

A

Abacus Plant　64
Actinoleaf Schefflera　101
Aculeate Bittersweet　89
Acuminate Banana　150
Acuminate Basket Vine　134
Acuminate Cyclosorus　8
Acute Asiaglory　130
Acute Closedspurorchis　169
Acute Fieldcitron　98
Acute Sida　59
Adiantiform Rumohra　12
Adia　50
Adnate Plagiogyria　3
Africa Asparagus　154
African Evergreen　160
African Lily　160
African Mask　157
African Oilpalm　165
African Sphaeranthus　124
African Tulip Tree　136
African Violet　135
Aggregated Euphorbia　62
Aglaonema　157
Ailanthus-like Prickly Ash　95
Air Potato　162
Air-plant　36
Alderleaf Sida　59
Alfred Windmill　55
Alligator Alternanthera　41
Alluminum Plant　88
Alsine-like Evolvuls　131
Alternateleaf Nechamandra　147

Alyce Clover　73
Amazon Water Lily　29
Ambay Pumpwood　84
Ambrosia Orchid　168
Amercian Wormseed　40
American Burnweed　121
American Lily　28
American Spider Lily　161
Ample-flowered Hedyotis　114
Amur Silvergrass　184
Anderson's Veronia　125
Angled Luffa　47
Angle-stemmed Hedyotis　114
Angularflower Elaeagnus　92
Antennae Orchid　169
Antifebrile Dichroa　36
Antifebrile Dichroa　66
Antrophyum-like Bird-nest Fern　10
Apiculate Elaeocarpus　57
Apocopis　178
Appendiculate Croton　62
Appendiculed Egenolfia　13
Aquatic Malachium　37
Aquatic Panic Grass　185
Arabian Coffee　113
Arabian Jasmine　107
Arachnoid Cyanotis　148
Arenga Palm　164
Argy's Wormwood　118
Arjuna Myrobalan　55
Armgrass　179
Arnotto Dye Plant　46
Aromatic Turmeric　152
Aromaticroot　188

Arrowleaf Fig 86
Arrow-shaped Tinospora 30
Artillery Clearweed 88
Arum Lily 160
Ascendent Crabgrass 180
Ashygreyflower Lovergrass 182
Asia Belltree 136
Asian Cable Creeper 109
Asian Holly Fern 11
Asiatic Ardisia 104
Asiatic Butterfly-bush 107
Asiatic Pennywort 102
Asiatic Striga 134
Asiatic Sweet-leaf 106
Asparagus Fern 154
Assam Hymenachne 183
Assam Rattlebox 74
Assamking Begonia 49
Auriclar Hedyotis 114
Auriculata Dayflower 148
Auriculate Acacia 69
Auriculate Pogostemon 145
Auriform Plagiogyria 3
Austral Akebia 29
Australian Acalypha 60
Australian Almond 55
Australian Bluestem 179
Australian Cow-plant 111
Australian Smut-grass 187
Autumn Fern 12
Autumn Zephyr-lily 161
Avocado 27
Awned Duck-beak 184
Awned Rice Galingale 174

B

Babys-breath 37

Baillon Paramichelia 23
Balanse Michelia 22
Balfour Polyscias 101
Balloon Flower 127
Balloon Vine 96
Balsam Cuphea 42
Balsamic Blumea 119
Bamboo Orchid 168
Bamboo Palm 165
Bambooleaf Fig 86
Bambooleaf Pondweed 147
Bamboo-leaved Oak 82
Bamboo-stemmed Orchid 172
Bambusa Lapidea 188
Banana Orchid 168
Banana Shrub 22
Barbados Aloe 153
Barbate Cyclea 30
Barbate Filmy Fern 4
Barbed Skullcap 146
Bareet Grass 184
Barley 183
Barn-yard Grass 181
Basket Vine 135
Batavia Cinnamon 24
Bayberry Waxmyrtlefruit Terminalia 55
Beach Naupaka 128
Beach Primrose 44
Beach Wedelia 125
Bead Vine 109
Beaked Yucca 164
Beak-leaf Lepidogrammitis 14
Beal's Mahonia 29
Bearded Duck-beak 184
Beardless Barnyardgrass 181
Beardless Microstegium 184
Beautiful Bougainvillea 45

Beautiful Leptosiphonium 138
Beautiful Lespedeza 77
Beautiful Neoregelia 150
Beautiful Phyllodium 79
Beautiful Sweetgum 80
Beautiful Sweet-leaf 106
Beauty Berry 140
Beauty-of-the-night 45
Bedding Dahlia 120
Behtham's Rosewood 75
Bell Pepper 128
Belly-ache Bush 64
Bengal Dayflower 148
Bengal Hiptage 60
Bent-corolla Lygisma 111
Bentham Oreocharis 135
Bentham's Bitter-sweet 89
Bentham's Yam 162
Bermuda Arrowroot 153
Bermuda Grass 180
Berry-fruited Sedge 173
Berry-shaped Sauropus 65
betel Palm 164
Betel Pepper 32
Bhutan Twayblade 171
Biflor Microtropis 89
Biform Spikemoss 1
Biformed Tree-gingseng 100
Big Fiddle-leaf Fig 86
Big Ladder Orchid 172
Big Marigold 124
Big Themeda 188
Bigbract Dayflower 148
Bigcover Peliosanthes 155
Bigflower Box 81
Bigflower Fleshy Lobelia 127
Bigflower Loquat 67

Bigflower Snowbell 105
Big-flowered Blumea 119
Bigfruit Mucuna 78
Bigleaf Bergia 37
Bigleaf Cherrylaurel 67
Bigleaf Ticktrefoil 76
Big-leaved Desmodium 75
Big-leaved False Nettle 87
Big-leaved Fig 87
Big-leaved Prickly Ash 95
Bigspike Arundinella 179
Billygoat-weed 118
Bilobed Grewia 56
Biond's Hackberry 83
Bird of Paradise 151
Bird-nest Fern 10
Birdwood' Mucuna 78
Bismarck Palm 164
Bisumbellate Fluttergrass 175
Biternate Beggartick 119
Bitter Cress 34
Bitter Cucumber 48
Bitter Mustard 33
Bitterbamboo 190
Bittersweet 130
Black Bamboo 190
Black Bean 74
Black Currant Tree 61
Black Falsehellebore 156
Black Mangrove 139
Black Nightshade 130
Black Olive 96
Black Pepper 32
Black Tree-fern 5
Black-berry Lily 161
Blackleaf Memecylon 55
Black-leaf Panamiga 88

Blackseed Water-centipede 176
Black-seeded Spikesedge 175
Blanchett Allamanda 108
Bleeding Heart Glorybower 142
Blood Lily 161
Blood-flower Milkweed 110
Bloodred Melastoma 54
Blue Billygoat-weed 118
Blue Butterfly 142
Blue Eranthemum 137
Blue Grass 155
Blue Gum 53
Blue Hearts 132
Blue Indianlotus 29
Blue Jacaranda 136
Blue Sage 146
Blue Torenia 134
Bluecrown Passionflower 46
Blueflower Leadword 127
Blunt-fruit Drypetes 62
Blunt-leaf Asiaglory 130
Blushing Bromeliad 150
Blushing Philodendron 159
Bodinier Mahonia 29
Bog Starwort 38
Bona Conyza 120
Borecole 34
Boston Ivy 93
Bottle Gourd 47
Bottle Palm 163
Bottle Palm 165
Bottle Tree 33
Bottlebrush Orchid 171
Bourbon Ochrosia 109
Boxleaf Atalantia 94
Boxleaf Syzygium 53
Bracken Fern 6

Bracket plant 154
Branch Pycreus 177
Brazil Rubbertree 64
Brazilian Gloxinia 135
Brazilian Jasmine 109
Brazilian Pitcairnia 150
Breadfruit 84
Breakingfruit Chinkapin 82
Bredders Gladiolus 162
Bridal Wreath 69
Brightred Mayodendron 136
Brisbane Box 54
Bristle Beakrush 177
Bristly Canthium 113
Brittle Falsepimpernel 133
Britton Ruellia 138
Broad Colysis 14
Broad Sword-fern 13
Broadflower Dragonbamboo 189
Broadleaf Blainvillea 119
Broadleaf Liriope 155
Broadleaf Raintree 128
Broadleaf Zeuxine 172
Broad-leaved Actinidia 52
Broad-leaved Addermouth Prchid 172
Broad-leaved Bowgrass 180
Broad-leaved Microtropis 89
Broadlobed Bauhinia 71
Broad-pinna Cyclosorus 9
Broad-pinna Wedgelet Fern 5
Broadspur Cuphea 42
Bronze Banana 151
Broom Bamboo 190
Broomsedge 39
Brown Punctate Cherry 67
Brown Rock- orchid 170
Brownhair Bristlegrass 187

Brownleaf Microsorium 15
Brownleaf Mischocarp 97
Brown-stalk Halberd Fern 12
Buchananialeaf Meliosma 98
Buddha Bamboo 189
Buddhanail 36
Buddhist Bauhinia 71
Buddhist Pine 19
Buerger Maple 97
Buerger Pipewort 149
Bunch-like Reevesia 58
Bunge Swallowwart 111
Bunya-bunya 19
Bur Grass 179
Burma Conehead 139
Burmann Sundew 37
Burmareed 185
Bur-Marigold 119
Burmese Rosewood 79
Burvine 57
Bush Morning-Glory 131
Bush Tedpepper 128
Bush Thunbergia 139
Butchers Broom 157
Buttercup Orchid 172
Butterfly Pea 74
Butterflyfruit 62
Button Melon 48
Butulang Canthium 113

C

Cabbage 34
Cablin Patchouli 146
Callery Pear 68
Calys-shaped Tigernanmu 66
Calyxless Sweetgum 80
Cambodia Dragonblood 163

Camel's Foot 71
Camomileleaf Soliva 124
Camphor Tree 24
Canadian Goldenrod 124
Canary Island Date Palm 166
Candlenut Tree 61
Canton Ampelopsis 92
Canton Caper 33
Canton Diplopterygium 3
Canton Mombin 99
Canton Ophiorrhiza 116
Cantonese Fairy Bells 154
Cantor Pseudosasa 190
Cape Jasmine 113
Cape-Honeysuckle 136
Capitate Phrynium 153
Capitate-flower Asiaglory 130
Capsule-fruited Passionflower 46
Carambola 42
Cardboard Palm 17
Carib Heliconia 151
Carib Heliconia 151
Caribbean Copper Plant 62
Carier Brake 6
Carles's Chinkapin 82
Carnation 37
Carpetgrass 179
Carrot 102
Cashew 98
Cassava 65
Cassia Bark Tree 24
Castorbean 65
Cat-claw Vine 136
Cathay Poisonnut 107
Cathay Silver Fir 17
Cat's Tail Bean 80
Cat-tail Tree 136

C

Cattail-leaved Lipocarpha 176
Caudate Pentasachme 111
Caudate Wildginger 31
Cavaler Eurycorymbus 97
Cavaleri Peperomia 31
Celery 102
Celery-leaved Crowfoot 28
Celosia 41
Centipede Plant 39
Central America Mahogany 96
Centripede Grass 188
Century Plant 162
Ceriman 159
Cerlon Cinnamon 25
Ceylon Elaeocarpus 57
Ceylon Fagraea 107
Ceylon Leucas 144
Ceylon Newlitse 27
Ceylon Pouzolzia 88
Chaffanjon Snakegrape 92
Champaca 22
Champion Magnolia 21
Champion Rhodoleia 81
Champion Slatepenciltree 52
Champion Syzygium 53
Champion Wood Fern 11
Champion's Bauhinia 71
Champion's Millettia 77
Champion's Oak 82
Champion's Orchid 170
Champion's Rhododendron 102
Changeable Rose-mallow 59
Chayote 48
Chee Woodreed 179
Chekiang Machilus 26
Chequer-shape Indocalamus 189
Cherokee Rose 68

Chicken eye Parnassia 37
China Apes Ear-ring 70
China Aster 119
China Eaglewood 44
China Entireliporchis 171
China Fir 18
China Galangal 151
China Heterosmilax 156
China Honeylocust 73
China Hoofrenshen 100
China Ivy 101
China Jasmine 108
China Mallow 59
China Monkeyjoy 57
China Onion 160
China or Bengal Rose 68
China Pistachio 99
China Redbud 72
China Rose Mallow 60
China Saraca 73
China Seaberry 44
China Sida 60
China Snakegrape 92
China Stachyurus 80
China Taxillus 90
China-berry 96
Chinensis Eupatorium 121
Chinese Abelia 117
Chinese Alangium 100
Chinese Albizia 69
Chinese Amaryllis 161
Chinese Anisopappus 118
Chinese Antirhea 112
Chinese Aporusa 61
Chinese Aralia 100
Chinese Arborvitae 18
Chinese Arrow-head 147

C

Chinese Ash   107
Chinese Asystasiella   137
Chinese Aucuba   99
Chinese Beakrush   177
Chinese Bitter-sweet   89
Chinese Box   81
Chinese Brake   6
Chinese Buttercup   28
Chinese Buttonbush   112
Chinese Calogyne   127
Chinese Calophanoides   137
Chinese Chestnut   81
Chinese Chirita   135
Chinese Chives   161
Chinese Chrysanthemum   121
Chinese Cladium   173
Chinese Clematis   27
Chinese Clinopodium   143
Chinese Crinium   161
Chinese Crossostephium   120
Chinese Cryptocarya   25
Chinese Cryptolepis   111
Chinese Cymbidium   170
Chinese Desmos   23
Chinese Dicliptera   137
Chinese Dipteris   14
Chinese Dischidia   111
Chinese Dodder   130
Chinese Egenolfia   13
Chinese Elaeocarpus   57
Chinese Eriosema   76
Chinese Euonymus   89
Chinese Evergreen   157
Chinese Fan-palm   166
Chinese Feverine   116
Chinese Fringelily   156
Chinese Goddess Bamboo   189

Chinese Gomphostemma   144
Chinese Grass   188
Chinese Hackberry   83
Chinese Hatplant   142
Chinese Hicriopteris   3
Chinese Holly   88
Chinese Hollyfern   11
Chinese Honeysuckle   117
Chinese Incense Cedar   18
Chinese Ixonanthes   60
Chinese Ixora   115
Chinese Juniper   19
Chinese Kale   33
Chinese Lantern   129
Chinese Lasianthus   115
Chinese Lemon   94
Chinese Lespedeza   77
Chinese lily   155
Chinese Lipocarpha   176
Chinese Lobelia   127
Chinese Loropetalum   81
Chinese Lovegrass   182
Chinese Mahogany   96
Chinese Meliosma   98
Chinese Mesona   145
Chinese New Year Flower   102
Chinese Nut Rush   177
Chinese Oatchestnut   82
Chinese Osbeckia   55
Chinese Parasol-tree   58
Chinese Pine   18
Chinese Pink   37
Chinese Privet   108
Chinese Radish   34
Chinese Randia   116
Chinese Red Pine   18
Chinese Sacred Lily   161

Chinese Sage   146
Chinese Sassafras   27
Chinese Scaly Seed   176
Chinese Schoepfia   90
Chinese Scolopia   46
Chinese Sedge   173
Chinese Silbergrass   184
Chinese Snapweed   42
Chinese Spicebush   25
Chinese Spinach   41
Chinese Spiraea   69
Chinese Stauntonia   29
Chinese Sumac   99
Chinese Sweet-leaf   106
Chinese Sweetspire   66
Chinese Taro   157
Chinese Thistle   120
Chinese Thorny Bamboo   189
Chinese Three-awn   178
Chinese Tinospora   30
Chinese Trigonostemon   66
Chinese Trumpet-creeper   135
Chinese Tuliptree   21
Chinese Wedelia   125
Chinese White Cabbage   33
Chinese White Olive   95
Chinese Wisteria   80
Chinese Wolfberry   129
Chinese Wood Fern   11
Chinese Yam   162
ChineseTallow-tree   65
Ching Grass Fern   8
Ching Manglietia   21
Ching Wormwood   118
Chingma Abutilon   59
Chittagong Chickrassy   96
Christina Loosestrife   126

Christmas Bush   61
Christmas Cactus   50
Chu's Newlitse   27
Chusan Lip-fern   7
Ciliata Eurya   51
Ciliate Centipede-grass   183
Ciliate Crabgrass   181
Ciliate Falsepimpernel   133
Ciliate Fuirena   176
Ciliate Pseudocyclosorus   9
Ciliate Sasagrass   184
Ciliate Spikemoss   1
Cinderlike Sweet-leaf   106
Cionbag Milkwort   35
Citronella Grass   180
Clammy Hop Seed   97
Clark's Blumea   119
Clethra Loosestrife   126
Climbed Clubmoss   1
Climber Floscopa   148
Climbing Bauhinia   71
Climbing Bird's Nest Fern   15
Climbing Fig   85
Climbing Fig   86
Climbing Groundsel   124
Climbing Lily   154
Climbing Microsorium   15
Climbing Prickly Ash   95
Climbing Seedbox   44
Cloud-leaf Orchid   172
Clubmoss   1
Clustered Casearia   46
Clustered Knotweed   39
Coarseleaf Eargrass   115
Cochinchina Blastus   54
Cochinchina Centranthera   132
Cochinchina Helicia   45

C

Cochinchina Homalium 46
Cochinchina Momordica 47
Cochinchina Randia 112
Cockbillthorn 72
Cockspur Coralbean 76
Coco Magnolia 21
Coconut Palm 165
Coffee Senna 72
Cogniaux Blastus 54
Colorado-grass 178
Commom Calanthe 169
Commom Rush 173
Common Achyranthes 40
Common Adder's Tongue 2
Common Adenostemma 118
Common Allamanda 108
Common Alyxia 109
Common Amentotaxus 20
Common Anthurium 158
Common Apes Ear-ring 70
Common Apes Ear-ring 70
Common Aspidistra 154
Common Atropa 128
Common Bamboo 189
Common Bergia 37
Common Bluebeard 141
Common Bowringia 74
Common Box 81
Common Bulrush 177
Common Bushweed 66
Common Caladium 158
Common Camptotheca 100
Common Cattleya 169
Common Centotheca 180
Common Cerbera Tree 109
Common Championella 137
Common Chickweed 38

Common Christia 74
Common Chrysanthemum 121
Common Claoxylon 62
Common Cockscomb 41
Common Coltsfoot 125
Common Conehead 139
Common Cosmos 120
Common Cowpea 80
Common Cyanotis 148
Common Danceweed 74
Common Dayflower 148
Common Dryoathyrium 8
Common Elaeocarpus 57
Common Elsholtzia 144
Common Eriachine 183
Common Euscaphis 98
Common Fennel 102
Common Freesia 161
Common Fuchsia 43
Common Garden Canna 153
Common Ginger 152
Common Gymnogrammitis 14
Common Holarrhena 109
Common Indian-mulberry 115
Common Isodon 144
Common Jujube 92
Common Knight's Star 161
Common Lantana 142
Common Lepidagathis 138
Common Mappianthus 90
Common Melastoma 54
Common Mesua 56
Common Morning-Glory 132
Common Muntingia 56
Common Nasturtium 42
Common Ottochloa 185
Common Oxwood 56

C

Common Pentaphylax   52
Common Perilla   145
Common Phaius   172
Common Philydrum   168
Common Phymatodes   15
Common Picria   133
Common Pileostegia   66
Common Pokeberry   40
Common Pond Weed   147
Common Pratia   127
Common Pterocypsela   124
Common Ragwort   124
Common Reedgrass   186
Common Rockvine   93
Common Rue   95
Common Salvia   146
Common Sarcandra   33
Common Saurauia   52
Common Screw Pine   167
Common Selfheal   146
Common Sesbania   79
Common Snapdragon   132
Common Snowthistle   124
Common Spinach   40
Common Squill   154
Common Staghorn   15
Common Teak   142
Common Tricalysia   113
Common Triumfetta   56
Common Turmeric   152
Common Verbena   143
Common Violet   35
Common Water Ammannia   42
Common Watershield   28
Common Wildmedia   96
Common Wrightia   110
Common Yellow Stem-fig   85

Common Zinnia   126
Composite Oplismenus   185
Compressed Galingale   174
Concentrate Neoregelia   150
Concolorous Torenia   134
Conform Climbing Fern   3
Consanguine Fimbristylis   175
Conspicuous-nerved Newlitse   27
Copper Leaf   60
Coral Ginger   152
Coral Hibiscus   59
Coralbean Tree   76
Coralgreens   102
Coral-plant   133
Cordate Drymaria   37
Coriander   102
Cork-leaved Snow-bell   105
Corn   188
Corn Chrysanthemum   119
Corner's Eria   170
Corniculata Candleefruit   104
Corniculata Cayratia   93
Corniculate Albizia   70
Corolla Hairorchis   171
Corymb Woodsorrel   42
Corymbose Hedyotis   114
Cowryleafpalm   165
Crab Cactus   50
Crape Ginger   152
Crape Myrtle   43
Crapnell's Camellia   50
Creeping Dichondra   130
Creeping Falsepimpernel   133
Creeping Gentian   126
Creeping Ixeris   122
Creeping Juniper   19
Creeping Launaea   123

Creeping Peanut 73
Creeping Pellionia 87
Creeping Pothos 159
Creeping Psychotria 116
Creeping Sebastiania 65
Creeping Treebine 93
Creeping Woodsorrel 42
Creepy Mallotus 65
Crenate Procris 88
Crenate Reddish Beautyberry 140
Crenateflower Calceolaria 132
Crepe Jasmine 110
Crested Neriumleaf Euphorbia 63
Cretan Brake 7
Crimson Pitcherplant 37
Crisped Common Perilla 145
Crookneck Squash 47
Croton 62
Croton-oil Plant 62
Crown of Thorns 63
Cruciate Sedge 173
Cruian Gall 105
Crystal Anthurium 158
Cuban Bast 59
Cuban Zephyr-lily 161
Cucumber 47
Cudweed 122
Cuming's Oleandra 13
Cuneate Lespedeza 77
Cup Kalanchoe 36
Cup of Gold Vine 129
Cupgrass 183
Curious Dutchmanspipe 31
Curious Kadsura 23
Curved-awn Fimbristylis 176
Cuspidata Olive 108
Cycad-fern 11

Cymose Fimbristylis 176
Cypress-vine (Star Glory) 132

D

Da Yan Bamboo 189
Daisy White Button 119
Dalmatian Pyrethrum 124
Dalziel Strobilanthus 138
Dashen 158
Date Palm 166
Datoucai Leaf-mustard 34
Dawn Redwood 18
Day-Lily 155
Deceive Wood Fern 12
Decumbent Bugle 143
Decurrent Halberd Fern 12
Decurrent Hemigramma 12
Delavay Magnolia 21
Delavay Schefflera 101
Delicate Spikemoss 1
Dense-bract Galangal 152
Dense-flower Elaeagnus 92
Dense-flowered China Laurel 61
Dense-flowered Smithia 79
Dense-flowered Spikesedge 175
Denticulate Paraixeris 123
Derris 75
Desert Rose 108
Desmodium 75
Devil Tree 109
Devil-pepper 110
Dianella 154
Dichotomous Cordia 128
Dichotomous Fimbristylis 175
Dichotomy Forked Fern 3
Diel's Millettia 77
Diels Trema 84

Different Spikemoss 1
Different-leaved Lindsaea 5
Diffuse Coptosapelta 113
Diffuse Dayflower 148
Diffuse Galingale 174
Diffuse Hedyotis 114
Digitate Colysis 14
Digua Fig 86
Dimorphic Crabgrass 181
Dindygule Peperomia 31
Dioecious Olive 108
Discolor Ludisia 171
Disperate Brake 6
Dissected Ladypalm 166
Distant-cleft Brake 6
Distinct-nerved Tickclover 76
Ditch Millet 186
Ditch Millet 186
Divaricate Conehead 139
Divaricate Typhonium 160
Divaricate Velvetplant 122
Diverse Lepidogrammitis 14
Diverseleaf Fig 85
Diverse-leaved Creeper 93
Dixie Silver-back Fern 7
Dockleaved Knotweed 39
Doederlein's Spikemoss 1
Dog's Tail Bean 80
Don Blue Berry 103
Double-fruited Cassia 72
Doublepetalous White Pomegranate 43
Downy Ground Fern 6
Downy Holly 88
Dragon Arum 158
Dragon Juniper 19
Dragon palm 167
Dry Falsepimpernel 133

Dryland Barnyardgrass 181
Drymoglossum-like
  Lepidogrammitis 14
Duck's Tongue Grass 156
Duhat 53
Dumbcane 158
Dusty Miller 119
Dwalf Qiongpalm 165
Dwarf Banana 151
Dwarf Cowlily 28
Dwarf Date-palm 166
Dwarf Galangal 152
Dwarf Holly 88
Dwarf Ixora 115
Dwarf Mitrasacme 107
Dwarf Mountain Pine 52
Dwarf Ophiorrhiza 116
Dwarf Pycreus 177

E

Eagle's Claw 23
Eared Eurya 51
Eared Strangler Fig 85
Earfruit Macaranga 64
East Asian Tree Fern 4
East Indian Holly Fern 11
East-China Many-flowered May-apple
  29
Ebonyshoot Mistletoe 91
Edible Canna 153
Edible Fig 85
Eel Grass 147
Egg Fruit 104
Egyptian Grass 180
Egyptian Water Lily 28
Elatior Begonia 48
Elecampane 122

Elegant Acacia 69
Elegant Ardisia 104
Elegant Cryptocarya 25
Elegant Fig 85
Elegant Phyllodium 79
Elephant Apple 45
Elephant Bush 38
Elephant's Ear 64
Elephants-foot 121
Elliptic Fishvine 75
Elongated Cycad 17
Elongated Lovegrass 182
Emarginate-leaved Ormosia 78
Emerald Philodendron 159
Endospermum 62
English Daisy 119
English Ivy 100
English Primrose 126
Entire Beautyberry 140
Entire Cherry-laurel 67
Entire Snakegourd 48
Entireleaf Dichrocephala 121
Erect Chlorantus 32
Erose Mussaenda Wild Mussaenda 115
Esculent Bracken 6
Esquirol Mussaenda 115
Euonymusleaf Syzygium 53
European Verbena 143
Euryaleaf Tea 50
Evergreen Mucuna 78
Evergreen Viburnum 118
Exaltated Snaketailgrass 185
Excised Spleenwort 10
Eximious Shield Fern 12
Eyelash Begonia 48
Eyre's Chinkapin 82

F

Faber Orchis 170
Faber Snowbell 105
Faber's Phoebe 27
Faber's Chestnut 82
Fairy Bells 154
Fairy Fern 6
Fairy Fig 85
Fairy Webseedvine 90
Fairylake Cycad 17
Falcate Dimeria 181
Falcate Holly Fern 11
Falcate-leaved Spleenwort 10
Falcate-lobed Pseudocyclosorus 9
False Bird of Paradise 151
False Custard 84
False Daisy 121
False Groundnut 75
False Heather 43
False Mallow 59
False Pineapple 167
False Sumac 95
False Tea 51
Fancy-leaved Caladium 158
Fangchi 31
Fan-leaved Maidenhair Fern 7
Farges's Chinkapin 82
Farrer's Azalea 102
Fasciate Aechmea 149
Fascicled-flower Sabia 98
Faurie's Brake 6
Feather Lovegrass 181
Feathery Spike Sawgrass 177
Fenzel's Engelhardia 99
Fernleaf Hedge Bamboo 189
Fern-like Sedge 173

Feruginous-scaled Fimbristylis 175
Fewflower Bladderwort 134
Fewflower Codonacanthus 137
Few-flower Lysionotus 135
Few-spikelet Fimbristylis 176
Fiddleleaf Aster 118
Fiddleleaf Fig 86
Fiddle-leaved Philodendron 159
Field Grass 184
Field Horsetail 2
Field Lacquertree 99
Field Sowthistle 124
Figleaf Goosefoot 40
Fig-leaved Holly 88
Filiform Cassytha 24
Fimbriate-sepal Chirita 135
Fine-nerved Plagiogyria 3
Finet Bogorchis 172
Finet Clematis 27
Finger Citron 94
Fingerleaf Morning Glory 131
Fire Dahlia 120
Fire-cracker Vine 136
Fireplant 63
Fischer Begonia 48
Fish Geranium 41
Fishpole Bamboo 189
Fishtail Palm 165
Fistulose Blumea 119
Fiveleaf Gynostemma 47
Five-leaved Yam 162
Fivepetal Helixanthera 90
Fivestamen Dendrophthoe 90
Flag Grass 173
Flame Bottletree 57
Flame Tree 72
Flameray Gerbera 122

Flaming Glorybower 141
Flaming-sword 150
Flat Hicriopteris 3
Flat-sheath Fimbristylis 175
Fleshy Lady-fern 8
Fleshy Nut Tree 104
Fleury Oak 82
Fleury Podocarpus 19
Flex Lovegrass 182
Flexuose Climbing Fern 4
Floating Bladderwort 134
Floating Heart 126
Florida Waltheria 58
Florists Cyclamen 126
Florists Gentian 126
Flowering Banana 150
Flowering Chinese Cabbage 34
Foetid Eryngo 102
Fokien Angiopteris 2
Fokien Cypress 18
Fokien Microtropis 89
Fool Proof Plant 149
Footleaf Anthurium 158
Ford Checkwood 73
Ford Dutchmanspipe 31
Ford Fishvine 75
Ford's Derris 75
Ford's Manglietia 21
Ford's Meliosma 98
Ford's Phlegmariurus 1
Ford's Sweet-leaf 106
Ford's Yam 162
Forest Gray Gum 53
Forgeirontree 105
Formosa Prive 108
Fortune Firethorn 68
Fortune Holly Fern 11

Fortune Keteleeria 17
Fortune Microsorium 15
Fortune Plumyew 19
Fortune's Cape Jasmine 113
Fortune's China-bell 105
Fortune's Drynaria 15
Fortune's Loosestrife 126
Foully Operculina 132
Four-finger Rattan Palm 164
Fourleaf Devilpepper 110
Four-leaf Peperomia 32
Fourseed Conehead 139
Fourstamen Stephania 30
Four-veined Eulalia 183
Foveolate Michelia 22
Fox-tail Agave 162
Foxtail Palm 167
Fragrant Ainsliaea 118
Fragrant Bullbophyllum 169
Fragrant Dracaena 163
Fragrant Eupatorium 121
Fragrant Gloryberry 141
Fragrant Heteropanax 101
Fragrant Lindsaea 5
Fragrant Litse 25
Fragrant Michelia 22
Fragrant Plantain-lily 155
Fragrant Polygonum 40
Fragrant Rosewood 75
Fragrant Snow-bell 105
Fragrant Syzygium 54
Fragrant Water Lily 28
Fragrant-leaved Geranium 41
Frail Horsetail 2
Franchet Groundcherry 129
Frangipani 109
French Marigold 125

Fringed Iris 162
Fringed Pink 37
Fruitful-grass 38
Fruricose Nailheadfruit 111
Fulvous Fig 85
Funnel-shaped Crossandra 137
Furcate Screwpine 167
Fusiform Melodinus 109

G

Gahnia 176
Gairo Morning Glory 131
Galingale 174
Garden Alternanthera 40
Garden Asparagus 154
Garden Balsam 42
Garden Dahlia 120
Garden Eggplant 130
Garden Lettuce 123
Garden Nicotiana 129
Garden Pansy 35
Garden Petunia 129
Garden Radish 34
Garden Spurge 63
Garden Strawberry 67
Garden Violet 35
Garlic 160
Gaudichaud Heterosmilax 156
Gelsemium 107
General Leucanthemum 123
Giant Alocasia 157
Giant Bean 70
Giant Cabuga 163
Giant Dioon 17
Giant Dumbcane 158
Giant Purple Liparis 171
Giant Reed 179

Giant Spider Lily   161
Giant Spider-flower   33
Giant Star Potato Tree   130
Gibbous Fig   86
Ginger Lily   152
Ginkgo   17
Girald Beautyberry   140
Girasole   122
Glabrous Curculigo   167
Glabrous Elaeagnus   92
Glabrous Greenbrier   157
Glabrous Leea   93
Glabrous Newlitse   27
Glabrous Pittosporum   45
Glabrous Reticulated Leaf-flower   65
Glabrous Sepel Crotalaria   75
Glabrous Tanoak   82
Glabrous Torenia   134
Glabrous Turpinia   98
Glabrous-leaved Uvaria   24
Glabrous-pod Mimosa   70
Glandular Glorybower   141
Glandular Parathelypteris   9
Glandular-leaved Sweet-leaf   106
Glandular-stipe Sweet-leaf   106
Glaucescent Diploclisia   30
Glaucous Diplopterygium   3
Glaucous Echeveria   36
Glaucous Sweet-leaf   106
Glaucousleaf Cyclea   30
Glittering-leaved Millettia   77
Globate Yellowcress   34
Globeamaranth   41
Globose Twinball Grass   184
Globular Spike Pycreus   177
Glossy Privet   108
Glossy Wild Sorghum   187

Glow Vine   136
Gluey Bark Litse   25
Glutene-rice Grass   178
Glycosmis   95
Goat Horns   110
Gold Button   124
Gold Stargrass   168
Gold-dust Dracaena   163
Golden Calla Lily   160
Golden Dewdrops   142
Golden Dock   40
Golden Rain Tree   97
Golden Sweet Osmanthus   108
Golden-eyed-grass   167
Goldenflower Tea   51
Golden-hair Grass   186
Golden-leaved Banyan   86
Golden-leaved Tree   103
Goldenmelon   47
Goldenrod   124
Golden-sheathed Eulalia   183
Golden-shower   72
Goldhair Eargrass   114
Goldthreadweed   39
Goose-no-eat   121
Gourd Tree   136
Gout Stalk   64
Grantham's Camellia   50
Grape Honeysuckle   117
Grassleaf Bladderwort   134
Grassleaf Liriope   155
Grassleaf Sweetflag   157
Gray Nickers   71
Great Burdock   118
Great Galangal   151
Greater Duck-weed   160
Greater Yam   162

Green Ailanthus 95
Green Amaranth 41
Green Bamboo 189
Green brier 156
Green Fox-tail 187
Green Imperial Begonia 49
Green-brown Tainia 172
Greenflower Tainia 168
Greenfruit Fig 86
Grenish Twin-sorus Fern 8
Grested Iris 162
Grey Lemongrass 180
Grey Manglietia 21
Greyblue Ensete 150
Greyblue Spicebush 25
Grey-green Peperomia 31
Greyhair Glorybower 141
Greyhair Kiwifruit 52
Grey-leaved Greenbrier 156
Griffith's Dictyocline 9
Groff's Eurya 51
Gross-dentate Osmanda 2
Groundnut 73
Gryblue Pericampylus 30
Guananbana 23
Guangdong Bauhinia 71
Guangdong Goldleaf 102
Guangdong Half-capitate Hemiboea 135
Guangdong Helicia 45
Guangdong Lily-turf 155
Guangdong Litse 26
Guangdong Osmanda 2
Guangdong Rose 68
Guangdong Waterstarwort 44
Guangxi Newlitse 27
Guangxi Turmeric 152

Guangzhou Yellowcress 34
Guatemalan Bird of Paradise 151
Guava 53
Gugertree 51
Guile's Boeica 135
Guilfoylei Polyscias 101
Guinea Grass 185
Gulf Barn-yard Grass 181

H

Haichow Elsholtzia 144
Hainan Adina 116
Hainan Adinandra 50
Hainan Belltree 136
Hainan Bushbeech 142
Hainan Calophanoides 137
Hainan Chaulmoogratree 46
Hainan Childvine 112
Hainan Cycad 17
Hainan Elaeocarpus 57
Hainan Galangal 152
Hainan Homalium 46
Hainan Hypolytrum 176
Hainan Manglietia 22
Hainan Mucuna 78
Hainan Naupaka 128
Hainan Ormosia 78
Hainan Phoenix Tree 58
Hainan Sarcandra 33
Hair Triumfetta 56
Haired-twig Cinnamon 24
Hairflower Dallisgrass 186
Hairrode Arundinella 178
Hairy Bur-Marigold 119
Hairy Chickrassy 96
Hairy Crabgrass 181
Hairy Crabgrass 181

H

Hairy Fig 85
Hairy Gerbera 122
Hairy Gonostegia 87
Hairy Melon 47
Hairy Merremia 131
Hairy Morning-glory 130
Hairy Nightshade 129
Hairy Persimmon 103
Hairy Polygonum 39
Hairy Randia 112
Hairy Rosary Pea 73
Hairy Spicebush 25
Hairyflower Kiwifruit 52
Hairyfruit Eurya 51
Hairyfruit Musella 151
Hairyfruit Netseedgrass 148
Hairyfruit Pearlsedge 178
Hairy-fruited Abacus Plant 63
Hairy-fruited Croton 62
Hairyleaf South Ailanthus 95
Hairy-scale Rattan Palm 164
Hairy-sepal Lemon Sabia 98
Hairysorus Grammitis 16
Hairy-stemmed Spleenwort 10
Halftooth Camellia 51
Hance Syzygium 54
Hance's Scaly-fern 5
Hance's Ardisia 104
Hance's Pepper 32
Hance's Tanoak 83
Hance's Viburnum 117
Hanging Lobster Claw 151
Hard-fruited Ormosia 78
Hard-stemmed 179
Harland's Balanophora 91
Harland's Box 81
Harland's Chien Fern 11

Harland's Nut Rush 178
Hawaiian Baby Woodrose 130
Hawker Snapweed 42
Hawk's Beard 126
Headed-flowered Adenosma 132
Heartleaf Houttuynia 32
Heartleaf Tubergourd 48
Heart-leaved Arisaema 158
Heart-leaved Philodendron 159
Heart-leaved Sida 60
Heartshape Mikania 123
Hedge Sageretia 92
Helicteres 58
Heliotrope Ehretia 128
Hempleaf Negundo Chastetree 143
Hemsley's Rockvine 93
Henderson Allamanda 108
Henna 43
Henry Clematis 27
Herbaceous Geophila 114
Herbaceous Myrica 69
Herbst Bloodleaf 41
Herklots Cane 189
Herry Roughleaf 115
Heterocarpous Cyclosorus 9
Heterophyllous Tick Clover 76
Heterophyllous Wingseedtree 58
Hidden Homalomena 159
Hilo Grass 185
Hilo Holly 104
Himalayas Creeper 93
Hinds Pseudosasa 190
Hindu Datura 129
Hirsuta Neanotis 116
Hirsute Arundinella 179
Hirsute Carrionflower 111
Hirsute Indigo 77

Hirsute Mussaenda 115
Hirsute Notholaena 7
Hirsute-foot Parathelypteris 9
Hispid Actinidia 52
Hispid Amischotolype 147
Hispid Calabash 47
Hog Plum 98
Hogfennel 102
Hollyhock 59
Hololeaf Fig 86
Holy Thistle 124
Hong Kong Abacus Plant 64
Hong Kong Balanophora 91
Hong Kong Eulophia 171
Hong Kong Hedyotis 115
Hong Kong Lady's-slipper Orchid 172
Hong Kong Machilus 26
Hong Kong Milkwort 35
Hong Kong Pepper 32
Hong Kong Spiranthes 172
Hong Kong Taro-vine 159
Hongkong Azalea 103
Hongkong Dogwood 100
Hongkong Eagle's Claw 23
Hongkong Elaeagnus 92
Hongkong Fissistigma 24
Hongkong Giantarum 158
Hongkong Gordonia 51
Hongkong Hawthorn 68
HongKong Jasmine 107
Hongkong Millettia 78
Hongkong Mucuna 78
Hongkong Orchid Tree 71
Hongkong Pavetta 116
Hongkong Pencilwood 96
Hongkong Rosewood 75
Hongkong Snapweed 42

Hongkong Tainia 168
Hongkong-Canton Cupgrass 149
Honolulu Vine 39
Hooked Clematis 27
Hooked Rattlebox 74
Hooked Spikemoss 2
Hooker Cuphea 43
Hooker Mallotus 64
Hooker Themeda 188
Hooker's Scaly-fern 5
Hoop Pine 19
Horn Nut 44
Hornwort 28
Horny Holly 88
Horny Tanoak 82
Horsetail Tree 83
Horseweed 120
How Eelvine 109
Hu Rustyhair Raspberry 69
Hubei Thistle 120
Humifuse Spurge 63
Hunan Ladybell 127
Hyacinth 155
Hyacinth-bean 77
Hybrid Lantana 142
Hybrid Verbena 142
Hypoglaucous Greenbrier 157

I

Illigera 27
Imbricate Galingale 174
Imitating Cymbidium 170
Imperial Japanese Morningglory 132
Incised Histiopteris 6
India Abutilon 59
India Carallia 55
India Coral Tree 76

India Duck-beak 184
India Locust 73
India Lovegrass 182
India Paspalum 186
India Perotis 186
India Sesbania 79
India Yellowcress 35
Indian Kalimeris 123
Indian Aeginetia 134
Indian Almond 55
Indian Azalea 103
Indian Chickweed 38
Indian Damnacanthus 113
Indian Epimerdei 143
Indian Heliotrope 128
Indian Hibiscus 59
Indian Jujube 92
Indian Lettuce 123
Indian Millettia 78
Indian Pharbitis 132
Indian Pokeberry 40
Indian Polyscias 101
Indian Quassiawood 95
Indian Red Water Lily 28
Indian Rotala 43
Indian Shot 153
Indian Skullcap 146
Indian Staghorn 15
Indian Trumpetflower 136
Indian Wikstroemia 44
India-rubber Tree 85
Indigo Vine 77
Inida Zehneria 48
Insignis Nouelia 123
Insular Bridelia 61
Integrifolious Ainsliaea 118
Integrifolious Chlorantus 32

Intermediate Lilyturf 155
Intermediate Melastoma 54
Interrupted Colysis 14
Interrupted Cyclosorus 9
Involucrate Stahlianthus 152
Involute Spikemoss 1
Iron Olive 104
Iron Plant 163
Iron-cross Begonia 49
Iron-Weed 125
Iron-Weed 125
Island Mahonia 29
Ito Burmannia 168
Ivorywhite Chirita 135
Ivy Tree 101
Ivy-arum 159
Ivy-like Euonymus 89
Ivy-like Merrema 131

J

Jackfruit 84
Jade Plant 36
Jamaica Vervain 142
Janpanese Blyxa 146
Japan Ardisia 104
Japan Camellia 51
Japan Cayratia 93
Japan Cleyera 51
Japan Galangal 152
Japan Korthalsella 90
Japan Pink 37
Japan Sabia 98
Japan Turnjujube 91
Japanese Athyriopsis 8
Japanese Banana 150
Japanese Boneset 121
Japanese Bugle 143

J — K

Japanese Buttercup 28
Japanese Cassia 24
Japanese Cedar 18
Japanese Chain Fern 11
Japanese Clave Fern 7
Japanese Climbing Fern 4
Japanese Dodder 130
Japanese Elaeocarpus 57
Japanese False Hellebore 156
Japanese Fatsia 100
Japanese Felt Fern 15
Japanese Glorybower 141
Japanese Honeysuckle 117
Japanese Ixeris 122
Japanese Lovegrass 182
Japanese Lovegrass 182
Japanese Maesa 105
Japanese Mahonia 29
Japanese Maple 97
Japanese Mazus 133
Japanese Ophiorrhiza 116
Japanese Osmanda 2
Japanese Pittosporum 45
Japanese Polygonum 39
Japanese Premna 142
Japanese Silbergrass 184
Japanese St. Johnswort 56
Japanese Stephania 30
Japanese Supple-jack 91
Japanese Thistle 120
Japanese Velvetplant 122
Japanese Zelkora 84
Japannese Goldenray 123
Jaurez Dahlia 120
Java Apple 54
Java Bishopwood 61
Java Campanumoea 127

Java Fern 15
Java Glorybower 141
Java Staghorn 16
Jensen's Cinnamon 25
Jet-glumed Grass 178
Jiangxi Raspberry 68
Jimson Weed 129
Job's Tears 180
Joyful Knotweed 39
Joyweed 41
Jungle Rice 181
Juteleaf Melochia 58

K

Kaffir Lily 161
Kalanchoe 36
Kamala Tree 65
Kandelia 55
Kapok Ceiba 58
Kariyat 137
Kassod Tree 72
Kawakami Chestnut 82
Kazinoki Papermulberry 84
Kenaf Hibiscus 59
King Palm 164
Kiss Camellia 51
Knotgrass 39
Knotgrass 186
Knotted Lovegrass 182
Knottedflower Phyla 142
Knotweed 40
Koehne Callery Pear 68
Kohlrabi 33
Korean Lawngrass 188
Korean Lovegrass 182
Korsakow Monochoria 156
Kudzu Vine 79

Kurz Alangium 100
Kwai-Fah 108
Kwangtung Beautyberry 140
Kwangtung Glorybower 141
Kwangtung Mussaenda 115
Kwangtung Pine 18
Kwangtung Rehdertree 105
Kwangtung Scolopia 46

L

Laciniate Kalanchoe 36
Lacy Tree Philodendron 159
Ladder Brake 7
Ladder Orchid 168
Ladies Tresses 172
Lady Palm 166
Laevigate Bowfruitvine 111
Lamont Castanopsis 82
Lance Asiabell 127
Lance Coreopsis 120
Lancebract Kopsia 109
Lanceleaf Litse 26
Lanceleaf Wingseedtree 58
Lanceolate Cymibidum 170
Lanceolate Raphiolepis 68
Lanceolobed Rustyhair Raspberry 69
Langyu Elm 84
Lao Mytilaria 81
Large Gymnosphaera 5
Large Hare's-foot Fern 13
Largebract Murdannia 148
Largeflower Cape Jasmine 113
Largeflower Carrionflower 111
Largeflower Woodsorrel 42
Large-flowered Honeysuckle 117
Large-flowered Thunbergis 139
Large-flowered Uvaria 24

Large-fruit Hodgsonia 47
Large-fruit Manglietia 22
Largefruit Murdannia 148
Largefruit Pachira 58
Largeleaf Caesalpinia 72
Largeleaf Hydrangea 66
Large-leaf Kalanchoe 36
Large-leaf Manglietia 22
Largeleaf Markingnut 99
Largeleaf Philodendron 159
Largeleaf Uvaria 24
Largeleaf Wildmedia 96
Large-leaved Abacus Plant 63
Large-leaved Beauty-berry 140
Large-leaved Curculigo 167
Large-leaved Flemingia 77
Large-leaved Iron-weed 125
Large-pinna Spleenwort 10
Latter Pine 18
Laucel Magnolia 21
Laurel Clockvine 139
Laurel Sweet-leaf 106
Leaf-mustard 33
Leafy Cactus 50
Leatherleaf Millettia 78
Leathery Leaf Litse 26
Lebbek Tree 70
Lemon 94
Lemon-scented Gum 53
Lemon-yellow Cattleya 169
Lens Coldwaterflower 88
Lenticel-bearing Embelia 105
Leopard Camphor 26
Leopard Plant 122
Leschenault Rush 173
Lesser Duck-weed 160
Lesser Galangal 152

L

Lesser Sedge 173
Lesser Thorny Bamboo 188
Leutescent Pycnospora 79
Levine's Newlitse 27
Li's Privet 108
Liangguang Adinandra 50
Licuala Palm 166
Lily Magnolia 21
Lily Turf 155
Limbated Spikemoss 2
Lindley Butterflybush 107
Lindley's Boneset 121
Linear Forked Fern 3
Linearleaf Thistle 120
Linear-spike Sedge 173
Lineate Brake 6
Lineate Supple-juck 91
Lingnan Artocarpus 84
Linguanramie 87
Lionhead Thelocactus 50
Lipstick Palm 165
Lipstick Vine 134
Litseleaf Tanoak 83
Little Groundcherry 129
Littleflower Melaleuca 53
Little-fruited Grape 93
Littleleaf Indian-mulberry 115
Little-leaf Lemmaphyllum 14
Littleleaf Rourea 99
Littleleaf Sida 59
Littoral Spinegrass 187
Lizard's Tail 32
Lobed-leaf Borecole 34
Lobster Claw 151
Lobster Claws 150
Loddiges's Dendrobium 170
Lollypops 138

Lollypops Super Goldy 137
Long Arrowleaf Knotweed 39
Long Stephania 30
Long Stipe Dunbaria 76
Longan 97
Longbeak Eucalyptus 53
Long-calyx Rattlebox 74
Longcapsuled Falsepimpernel 133
Long-ear Cane 189
Longear Water Hymenachne 183
Long-flower Crabgrass 181
Longflower Ehretia 128
Long-flowered Honeysuckle 117
Longhair Sweet-leaf 106
Long-hairy Clematis 27
Longleaf Bolbitis 13
Longleaf Croton 62
Long-leaf Magnolia 21
Long-leaf Pepper 32
Long-leaved Bird-nest Fern 10
Long-leaved Greenbrier 157
Long-leaved Xylosma 46
Longpeduncule Kadsura 23
Longpetiole Elaeocarpus 57
Longsepal Violet 35
Longspike Lovegrass 182
Longstalk Crepidomanes 4
Longstalk Fimbristylis 175
Longstalk Glinus 38
Longstalk Sida 60
Longstalk Silvertree 58
Longstalked Phyllanthus 65
Longstem Liparis 171
Longtube Ground Ivy 144
Looking-glass Tree 58
Loose-flowered Euonymus 89
Loose-flowered Pottsia 110

Loosehairy Leucas 144
Loosespike Lovegrass 182
Lopez Root 95
Loquat 67
Loquatleaf Beautyberry 140
Lotung Parakmeria 23
Loureiro Elaeagnus 92
Lovable Pseudosasa 190
Lovely Azalea 103
Lovely Stringbush 44
Low Galingale 174
Lucidleaf Maple 97
Lucidleaf Sageretia 92
Lueddemann Aechmea 149
Luofu Maple 97
Luofushan Joint-fir 20
Lychee 97
Lyrata Hemistetea 122

M

Macadamianut 45
Macao Nightshade 130
MacArthur Palm 166
Macartney's Eurya 51
Macropodous Tigernanmu 66
Madacaru 49
Madagascar Grass 185
Madagascar Terminalia 55
Magnific Tigridiopalma 55
Magnificent Medinilla 54
Maguan Manglietia 22
Maguey 163
Maiden's Jealousy 60
Maidenhair Fern 7
Majesty Palm 166
Maki Podocarpus 19
Makino Pleurosoriopsis 10

Malabar Ottochloa 185
Malabar-Nightshade 41
Malacca Galingale 174
Malachite Stonecrop 36
Maluanshan Maple 97
Mamillate Ardisia 104
Mangium Acacia 69
Mango 99
Mangrove Brake 7
Mangrove Date Palm 166
Manila Grass 188
Manila Palm 167
Manila Tamarind 71
Manybranched Armgrass 179
Manyflower Childvine 112
Manyflower Fissistigma 24
Manyflower Landpick (Fragrant Landpick) 155
Manyflower Melastoma 54
Manyflower Paris 156
Many-flower Tickclover 76
Many-flowered Garcinia 56
Many-flowered Gardneria 107
Many-flowered May-apple 29
Many-flowered Polygonum 40
Manyhead Stairweed 87
Many-petaled Mangrove 55
Many-pinnate Phymatodes 15
Manyspike Chlorantus 32
Manyspike Galingale 174
Manyspine Fagerlindia 113
Manystem Cusweed 122
Maomian Azalea 103
Marchand Doellingeria 121
Marginate Rockbell 127
Marginate Scaly-fern 5
Marginate Spanish Bayonet 163

M

Mariana Maiden Fern 9
Marine Cowpea 80
Marsh Fleabane 124
Mascarene Grass 188
Mate Spikemoss 2
Matthew Twin-sorus Fern 8
Maud's Michelia 22
Maxican Sunflower 125
Mayflower Gloryberry 141
McClure's Michelia 22
Meadow-rue 28
Medicinal Citron 94
Medicinal Fatheadtree 116
Melia-leaved Evodia 94
Membranaceus Beautyleaf 56
Memorial Rose 68
Menten Twinballgrass 183
Metal Palm 165
Mexican Fire Plant 63
Mexican Stonecrop 36
Meyen's Clematis 27
Midribsorus Wood Fern 11
Mile-a-minute Weed 123
Milk-bush 63
Milky Mangrove 63
Mimosa-leaved Cassia 72
Ming Fern 154
Mini Hairorchis 171
Miniature Umbrella Plant 101
Minireed 179
Minute Gonocormus 4
Mioga Ginger 152
Miracle Fruit 104
Mission Grass 186
Mistletoe 90
Mitre-flower Aloe 153
Mock Elaeocarpus 57

Mock Ginseng 38
Mock Jasmine 107
Mock Lime 96
Moellendorff Spikemoss 2
Molucca Albizia 70
Money Tree 160
Moneywort 102
Mongolian Snakegourd 48
Monostyle St. Johnswort 56
Montain Burmareed 185
Montane Kudzu 79
Montane Turpinia 98
Moon Valley 88
Morris's Persimmon 103
Moslem Garlie 28
Mosquito Fern 16
Moss Crassula 36
Moth Orchid 172
Mother-of-mosquitoes Tree 80
Moto Manglietia 22
Mountain Begonia 48
Mountain Fig 85
Mountain Hemp 84
Mountain Kumquat 94
Mountain Lichi 80
Mountain Orange 109
Mountain Yam 162
Mousetail Cupscale 187
Moutain Pomegranate 113
Moutain Tallow 65
Moutain Wampi 112
Mume Plant 67
Mung Bean 80
Muricate Eurya 51
Musk Mallow 59
Musk Melon 47
Muskmallow 58

Muyou Oiltung 66
Myrobalan 65
Myrobalan Plum 67
Myrsinaleaf Oak 82
Mysore Thorn 72

N

Nagai Podocarpus 19
Nai Bamboo 188
Nakedanther Ternstroemia 52
Nakedflower Beautyberry 140
Nakedflower Murdannia 148
Nakedleaf Wildjute 84
Napier Grass 186
Narrow Habeuaria 171
Narrowed-leaf Chinese Aucuba 100
Narrow-flowered Poisonnut 107
Narrowleaf Box 81
Narrowleaf Chirita 135
Narrowleaf Eurya 51
Narrowleaf Falsepimpernel 133
Narrow-leaf Fevervine 116
Narrowleaf Goosefoot 40
Narrowleaf Java Paraphlomis 145
Narrowleaf Laurel 61
Narrow-leaf Osmanda 2
Narrowleaf Reddish Beautyberry 140
Narrowleaf Stairweed 87
Narrowleaf Wildjute 84
Narrow-leaved Borreria 113
Narrow-leaved Cat-tail 160
Narrow-leaved Euonymus 89
Narrow-leaved Gardenia 113
Narrow-leaved Lip-fern 7
Narrow-leaved Merrema 131
Narrow-leaved Sword Fern 16
Narrow-leaved Taiwan Fig 85

Native Cobblers Pegs 122
Necklace Plant 65
Needle Spikesedge 175
Neilingding Diploclisia 30
Neilingding Twin-sorus Fern 8
Nepal Buckthron 92
Nepal Cycad 17
Nepal Twinballgrass 183
Neple Pennywort 102
Nerve Plant 138
Nettleleaf Falsepimpernel 133
Nevin Lovegrass 182
New Nymph Flower 162
New-Zealand Spinach 38
Night Jessamine 129
Night-closing Leaf 65
Night-fragrant Flower 111
Nipple Fruit 130
Noble Bottle-tree 58
Noble Dendrobium 170
Noble Sanchezia 138
Nodalflower Synedrella 124
Nodding Clubmoss 1
Nodding Wikstroemia 44
Norfolk Island Pine 19
Normal Spleenwort 10
Norrow-leaved Grape 93
North China Grape 93
Notchleaf Statice 126
Nude Fern 1
NutGrass Galingale 174

O

Oakleaf Goosefoot 40
Oakleaf Tanoak 83
Obcordata Christia 74
Obliqueswollen Adhatoda 138

Oblong Gymananthera 112	Orchid Cactus 49
Oblongleaf Blumea 119	Orchid Canna 153
Oblongleaf Euonymus 89	Oriental Blechnum 10
Oblongleaf Garcinia 56	Oriental Buckthorn 92
Oblong-leaf Sweetspire 66	Oriental Cattail 160
Oblong-leaved Eustigma 80	Oriental Chain Fern 11
Obscure Morning-glory 131	Oriental Mistletoe 91
Obscure Spleenwort 10	Oriental Ostrich Fern 10
Obtuse Didymostigma 135	Oriental Pickling Melon 47
Obtuse-leaf Erycibe 131	Oriental Stephania 30
Obtuse-leaf Peperomia 31	Ornamental Pepper 128
Obtuseleaf Premna 142	Ornamental Pumpkin 47
Obtuse-leaved Crateva 33	Oval Kumquat 94
Obvious Crepidomanes 4	Ovate Eargrass 114
Officinal Breynia 61	Ovate Tylophora 112
Oil Sindora 73	Ovate-leaved Stauntonia 29
Oil Tea 51	Ovate-leaved Stauntonia 29
Oiltung 66	Oxtail Greenbrier 157
Old Man Palm 165	Oyster Plant 149
Old World Adder's-tongue 2	
Oldham Tigernanmu 66	**P**
Oldham's Fissistigma 24	
Oleander 109	Paddle Plant 36
Oleanderleaf Rapanea 105	Paddy Galingale 174
Oleander-leaf Seatung 45	Painted Drop-tongue 157
Omoto Nipponlily 156	Painted Graphistemma 111
One-flowered Abelia 117	Pale Purple Eulophia 171
One-spike Chlorantus 32	Paleleaf Echinodorus 147
Onion Grass 149	Palette Flower 158
Opaque Greenbrier 157	Pallid Rattlebox 74
Operculate Waterfig 53	Palmate Begonia 49
Oppenheim's Ctemanthe 153	Palm-grass 187
Oppositeleaf Fig 85	Panic Grass 185
Orange Lantana 142	Panic Grass 185
Orange Sweet Osmanthus 108	Panic Grass 187
Orange-jessamine 95	Panicle Tanoak 83
Orbicular Lindsaea 5	Paniculate Jacquemonia 131
	Paniculate Microcos 56

Papaya 49
Paper Flower 45
Paper Galingale 174
Paper Mulberry 84
Paper-bark Tree 53
Papery Daphne 44
Parado Spride 72
Parasitic Cyclosorus 9
Parlor Palm 165
Pasquier Madhuca 104
Passion flower 46
Patenthairy Melastoma 54
Pauho Machilus 26
Peach 67
Peachform Mango 99
Peacock Plant 153
Peacock-plume Grass 180
Pea-like Fruit Popowia 24
Pearl Haworthia 155
Pearleaf Microglossa 123
Pearleaf Raspberry 69
Pearl-orchid 32
Pectinate Rungia 138
Pedateleaf Snakegourd 48
Peduncle Acronychia 94
Pedunculated Adder's Tongue 2
Peepul Tree 86
Peking Cabbage 34
Peltate Coldwaterflower 88
Pendent Japanese Pagodatree 79
Pendulous Orchis 170
Pennsylvania Bittercress 34
Pensylvanian Cudweed 122
Pentas 116
Peppermint 145
Perennial Fox-tail 187
Perennial Lovegrass 182

Perfoliate Knotweed 40
Periwrinkle 109
Perpetual Begonia 49
Persimmon 103
Petalless Yellowcress 34
Petersen Athyriopsis 8
Petticoat Palm 167
Phantom Orchid 171
Philippine Almond 55
Philippine Flemingia 77
Philippine Garcinia 56
Philippine Glorybower 141
Philippine Hackberry 83
Philippine Maidenhair Fern 7
Philippine Violet 137
Philippines Abacus Plant 64
Photinia 67
Physic Nut 64
Pigeon Pea 74
Pikeleaf Ungeargrass 178
Pileate Morning-glory 131
Pilose Lovegrass 182
Piloselike Drymoglossum 14
Pilosus Galingale 174
Pilous Purslane 38
Pindo Palm 164
Pineapple 149
Pine-leaf Bottle-brush 52
Pineleaf Eargrass 114
Pink Bone-wort 114
Pink Plumepoppy 33
Pink Quill 150
Pink Reineckia 156
Pink Silk Cotton Tree 58
Pink Trumpet-tree 136
Pinkleaf Seedbox 44
Pinnatevein Christmas Bush 60

Pitaya 49	Prostrate Spurge 63
Pitcher Plant 31	Pubescent Violet 35
Pittier Philodendron 159	Pubescent Water Lily 28
Plains Coreopsis 120	Puer Tea 50
Plantain 127	Pummelo 94
Playfair's Spicebush 25	Punting Pole Bamboo 189
Plum 68	Purple Bauhinia 71
Plume Grass 186	Purple Beauty-berry 140
Plumed Celosia 41	Purple Doubled Snow Azalea 103
Poinsettia 63	Purple Granadilla 46
Pointed-leaf Camellia 50	Purple Heart 148
Poisonousroot Ironweed 125	Purple Hypoestes 138
Polycephalous Adina 112	Purple Justicia 138
Pomegranate 43	Purple Star Liparis 171
Pond Cypress 18	Purple Tephrosia 80
Pongam Tree 79	Purple Toona 96
Pop-gum Seed 62	Purpleflower Angelica 101
Portia Tree 60	Purpleflower Elsholtzia 144
Potato 130	Purpleflower Violet 35
Potmarigold Calendula 119	Purple-flowered Crotalaria 74
Potted Dumbcane 158	Purple-fruited Spikesedge 174
Poverty Grass 187	Purslane 38
Powdery Alligator-flag 153	Pygmy Water Lily 29
Praxelis 121	Pygmy Water-lily 28
Prayer-beads 73	**Q**
Prettynerved Slugwood 24	
Prickly Ash 95	Qiongpalm 165
Prickly Snowthistle 124	Queen Crape Myrtle 43
Prickly-pear 49	Queen Palm 167
Primrose Willow 43	Queen Victoria Agave 163
Prince's Feather 40	Queensland Bottletree 57
Privetleaf Memecylon 55	Queensland Kauri 19
Procumbent Falsepimpernel 133	Queensland Woodrose 132
Procumbent Indian Mallow 60	**R**
Prolific Chain Fern 11	
Prolongated Spleenwort 10	Racemose Cyclea 30
Prostrate Cyathula 41	Racemose Sweet-leaf 10ƒ

Radiate Sawgrass 177
Rainbow Heliconia 151
Raintree 71
Rambutan 97
Ramie 87
Ramose Scouring Rush 2
Rangoon Creeper 55
Raspberry 69
Rat's Eye Bean 79
Rattan Palm 165
Rattlesnake Orchid 172
Reactiongrass 42
Red Azalea 103
Red Beautyberry 140
Red Bract Sedge 173
Red Frangipani 109
Red Globeamaranth 41
Red Hair Grammitis 16
Red Latan Palm 165
Red Machilus 26
Red Oatchestnut 82
Red Orchid Cactus 49
Red Pouzolzia 88
Red Poverty Grass 187
Red Powderpuff 70
Red Pronephrium 9
Red Psychotria 116
Red Sandalwood 69
Red Scale Ctenitis 12
Red Water-lily 28
Redback Osmanthus 63
Redbird Cactus 65
Redbracted Lysidice 73
Redcalyx Glorybower 141
Red-flower Loropetalum 81
Red-flower Manglietia 22
Redflower Ragleaf 120

Redglandular Raspberry 69
Redhot Cat-tail 60
Redleaf Ainsliaea 118
Redleaf Hibiscus 59
Redleaf Rourea 99
Red-leaved Mussaenda 115
Redlip Sage 146
Redlip Stonebean-orchis 168
Red-man Orchid 171
Red-scaled Pycreus 177
Reed-like Sugarcane 186
Reflexpetal Calanthe 169
Rehder Syzygium 54
Repent Humate 14
Reticulate Helicia 45
Reticulate Oreocharis 135
Reticulated Leaf-flower 65
Retuse Ash 107
Rheed Cansjera 90
Ribbed Bush Fig 86
Ribbed-sorus Ctenitis 12
Ribbon Fan Palm 166
Rice 185
Rice Galingale 174
Ricepaperplant 101
River Oak 83
Rivier Giantarum 158
Rock Seed Nut Rush 178
Rock Vine 159
Rocket Consolida 28
Rock-ginger Fern 15
Romana Salat 123
Root Crabgrass 181
Rooted Pellionia 87
Rosary Pea 73
Rose 68
Rose Mallow 60

Rose Myrtle 53
Rose Natal Grass 186
Rose of Sharon 59
Rose-apple 54
Rose-leaved Strawberry 69
Rose-moss 38
Rose-painted Calathea 153
Rostrate Cleisostoma 169
Rosula Ardisia 104
Rote Stern Ludwigia 44
Rottboell's Grass 186
Rough Knotweed 40
Rough Pellionia 87
Rough Sword Fern 13
Rough Tibouchina 55
Roughleaf Ironweed 125
Roughleaf Raspberry 68
Rough-leaved Borreria 112
Rough-leaved Holly 88
Roulett Rose 68
Round Kumquat 94
Round Leaf Codariocalyx 75
Roundleaf Bittersweet 89
Roundleaf Bladderwort 134
Roundleaf Litse 26
Round-leaved Dunbaria 76
Round-leaved Rotala 43
Roundpetal Coldwaterflower 88
Roundpod Jute 56
Roxburgh Goldlineorchis 168
Royal Palm 167
Rue Lemongrass 180
Rugged Pronephrium 9
Rugose-fruited Razorsedge 178
Running Montaingrass 185
Rush-like Bulrush 177
Rushlike Dopatrium 133

Rustyhairy Litse 26
Rutabaga 34

S

Sacred Bamboo 29
Sacred Lily 28
Sago-palm 17
Salamander Tree 61
Salamander-tree 61
Salomonia 35
Sampson Coilthrum 90
Sampson Macaranga 64
Sampson's Aster 119
Sand Ammannia 42
Sand Pear 68
Sander's Dracaena 163
Sandpaper Vine 45
Sandwort 37
Sanfendan Childvine 112
Santa Cruz Water Lily 29
Sapodilla 104
Sappan Caesalpinia 72
Sapphire-berry Sweet-leaf 106
Sargent gloryvine 30
Sarmentose Fig 86
Sasagrass 184
Sassanqua 51
Satsuma Orange 94
Saucer Magnolia 21
Saucerwood 66
Sausage Tree 136
Sawgrass 177
Sawtooth Oak 83
Scabrous Aphanathe 83
Scabrous Doellingeria 121
Scabrous Mosla 145
Scaly Tree-fern 5

Scandent Rosewood 75
Scapose Sedge 173
Scarab-like Cajanus 74
Scargrass 183
Scarlet Bush 114
Scarlet Kadusra 23
Scarlet Rosemallow 59
Scarlet Sage 146
Scarlet Sterculia 58
Scott Wood Fern 12
Scurfy Cycad 17
Scutate-dentate Bittercress 34
Sea Fig 86
Sea Sword Bean 74
Sea-lavender 126
Seashore Dropseed 188
Seashore Sumac 99
Seaside Purslane 38
Securidaca 35
Semiliquidambar 81
Semi-pinnated Brake 7
Senegal Khaya 96
Sensitive Plant 71
Sentry Palm 165
Serissa 117
Serpentgourd 48
Serrate Clubmoss 1
Serrulate Pendent-bell 102
Sessile Alternanthera 41
Sessile Blumea 119
Seven Sisters Rose 68
Severalflower Dewflower 148
Seville Orange 94
Shade Meadow-rue 28
Shag Mucuna 78
Shame Plant 71
Sharpleaf Acmena 52

Sharpleaf Fighazel 81
Sharpleaf Galangal 152
Sharp-veind Litsea 25
Sheathed Murdannia 148
Sheathflower 90
Shell Ginge 152
Shepherd's Needle 119
Shepherd's Purse 34
Shilu Michelia 22
Shining Eurya 51
Shining Hypserpa 30
Shining Sweet-leaf 106
Shiningfruit Nightshade 129
Shiningleaf Elaeocarpus 57
Shiningleaf Millettia 77
Shiny Peperomia 31
Shiny-leaved Prickly Ash 95
Short Glume Crabgrass 181
Short Spur Brachycorythis 168
Shortcalyx Lysidice 73
Short-inflorescence Machilus 26
Shortleaf Dyckia 150
Short-leaved Killinga 176
Short-leaved Malacca Galingale 174
Shortpetiole Cleidion 62
Shortscape Curculigo 167
Shortspike Pogostemon 146
Short-spiked Pepper 32
Shortstalk Beautyberry 139
Short-stalk Grammitis 16
Shortstalk Lobelia 127
Short-toothed White-hairy Paraphlomis 145
Shot-awn Foxtail 178
Shoulang Yam 162
Showy Millettia 78
Shrimp Claw Plant 124

Shrimp Plant 138
Shrotstyle Camellia 50
Shrubby Argyranthemum 118
Shrubby Parasiticvine 91
Shrubby Wood Nettle 87
Siam Lily 154
Siamense Selenodesmium 4
Siberian Cocklebur 125
Sickle Senna 72
Sida Hemp 60
Siebold Pipewort 149
Si-lao Bamboo 190
Silk Grass 173
Silk Oak 45
Silk-tree 70
Silky Kangaroo Grass 188
Silky-haired Fimbristylis 176
Silvemargin Spurge 63
Silver Jade Plant 36
Silver-back Artocarpus 84
Silverbell 105
Silverflower Globeamaranth 41
Silverline Calanthe 169
Silverspike Lakemelongrass 177
Similar Diploclisia 30
Similar-logania Eargrass 114
Simond Closedspurorchis 169
Simple Coelachne 180
Simple Pronephrium 9
Simpleleaf Shrub Chastetree 143
Simplex Glorybower 141
Simplex Murdannia 148
Singlespike Fishtailpalm 165
Sisal Agave 163
Sixangular Pipewort 149
Six-seeded Cycad 17
Skinner Michelia 23

Skullcaplike Coleus 144
Skybluewing Passionflower 46
Slash Pine 17
Slender Clinopodium 144
Slender Hairy Polygonum 39
Slender Ixeridium 122
Slender Ixeris 122
Slender Palm 166
Slender Thryallis 60
Slenderbranch Eurya 51
Slender-leaved Pholidota 172
Slender-pedicel Sedge 173
Slenderstyle Acanthopanax 100
Small Allamanda 109
Small Centipeda 119
Small Coleus 144
Small Cowpea 80
Small Fishtail Palm 164
Small Knife Bean 74
Small Lovegrass 182
Small Microchloa 133
Small Persimmon 103
Small Pipewort 149
Small Razorsedge 178
Small Stylidium 128
Small Yellow Bladerwort 134
Small-breact Sanchezia 139
Smallflower Aspidistra 154
Small-flower Bracket Plant 154
Small-flower Camphor Tree 25
Small-flower Grewia 56
Smallflower Ixora 115
Smallflower Rush 173
Smallflower Seaberry 44
Small-flowered Capillipedium 179
Small-flowered Holly 88
Small-flowered Urceola 110

Smallfruit Banyan 86
Smallfruit Chinese Holly 89
Smallfruit Greenbrier 157
Smallfruit Slatepentree 52
Small-fruited Honeylocust 73
Small-hooked Hedyotis 114
Smallleaf Caesalpinia 72
Small-leaf Carmona 128
Smallleaf Childvine 112
Small-leaf Climbing Fern 4
Small-leaf Pepper 32
Smallleaf Silvertree 58
Small-leaved China Laurel 61
Small-leaved Desmodium 76
Small-leaved Embelia 105
Small-leaved Holly 89
Small-leaved Joint-fir 20
Small-leaved Lasianthus 115
Small-sorus Mecodium 4
Small-spiked Climbing Fern 4
Smartweed 39
Smooth Knotweed 39
Smooth Lawn Grass 183
Smooth Sticherum 3
Smooth-branched Hooktea 91
Smooth-fruited Ventilago 92
Smooth-leaved Sweet-leaf 106
Smooth-lipped Cymbidium 170
Snail Plant 138
Snail Seed 30
Snake Acacia 69
Snake Aroid 158
Snake Caesalpinia 72
Snake Plant 163
Snake Strawberry 67
Snakegourd 48
Snakeking Vine 46

Snake's-eye Fern 14
Snow Bush 61
Snow Flower 159
Snow of June 117
Snowbell-leaved Tick Clover 76
Soap Berry 97
Sodaapple Nightshade 129
Soft-fruited Ormosia 78
Softhair Asiaglory 130
Softstem Bulrush 177
Solena 48
Song of India 163
Sour Creeper 110
South American Royal Palm 166
South China Elaphoglossum 13
South China Hackberry 83
South China Hare's-foot Fern 13
South China Honeylocust 73
South China Maple 97
South China Pepper 32
South China Redcarweed 87
South China Rosewood 75
South China Screwpine 167
South China Telosma 111
South China Wood Fern 12
South Crape Myrtle 43
South Sea-Blite 40
Southern China Barthea 54
Southern China Davallia 13
Southern China Homalium 46
Southern China Milkwort 35
Southern China Spleenwort 10
Southern Pteridrys 12
Southern Viburnum 117
Soy Bean 77
Spanish Dagger 164
Spanish-cherry 104

Sparraw Herb 133
Spathulate Sundew 37
Spathulate Sundew 37
Spatulate leaf Sauropus 65
Spear Sansevieria 163
Spear-Leaved Lady-fern 8
Speckled Toadlily 156
Spicate Clerodendranthus 143
Spice snowball 34
Spicy Jatropha 64
Spider Aralia 101
Spider Tree 33
Spike Gayfeather 123
Spike Indigo 77
Spiked Loosestrife 43
Spikeflower Axispalm 166
Spine Aralia 100
Spine Axispalm 166
Spineless Yucca 164
Spiny Alsophila 5
Spiny Amaranth 41
Spiny Bears Breech 137
Spiny Date-palm 166
Spiny Sesbania 79
Spiny-tooth Holly Fern 11
Splash-of-white 116
Splendid Forked Fern 3
Splitlip Peristyle 172
Spotted Ardisia 104
Spotted Begonia 49
Spotted-leaf Begonia 49
Spotted-leaf Creeping Peperomia 32
Spreading Cyrtococcum 180
Spreading Violet 35
Spring Caesalpinia 72
Spring Orchis 170
Spurge 63

Squarelea China Laurel 61
Squarestem Eargrass 114
Srilanka Hunteria 109
Srilanka Hydrolea 128
Stalk-leaf Wood Fern 12
Star Begonia 49
Star Jasmine 110
Starshape Sweet-leaf 107
Staunton's Ragwort 124
Stellatehair Vatica 52
Stemless Cryptanthus 150
Stick Rattanpalm 164
Sticky Adenosma 132
Sticky Germander 146
Stiff Bottle-brush 52
Stiff-leaved Meliosma 98
Stiff-spike Fimbristylis 175
Stink Grass 33
Stipefruit Mistletoe 91
Stipulate Breynia 61
Stock 34
Stoneblood 110
Stonecrop 36
Stonegarlic 161
Straight-lip Cirrhopetalum 169
Strawberry Tree 81
Strength-vine 87
Striate Kummerowia 77
Stringy Stonecrop 36
Striolate Ginger 152
Stripe Bamboo 189
Striped Abutilon 59
Styraxleaf Artocarpus 84
Subcordate Bolbitis 13
Subcordate Sida 60
Suberet Spatholobus 79
Suberous Greenstar 24

Sublobate Hewittia 131
Subshrub Indigo 77
Subshrub Nightshade 130
Sugar Cane 187
Sugar Palm 164
Sugar-apple 23
Sulphur Cosmos 120
Summer Fimberstylis 175
Sumuntia Sweet-leaf 107
Sunflower 122
Sunshine Tree 72
Sunshing Fimbristylis 175
Superb Fig 86
Surinam Calliandra 70
Surinam Cherry 53
Swallow Wort 111
Swallowtail Pinangapalm 166
Swamp Bladderwort 134
Swamp Cypress 18
Swamp Mahogany 53
Sweet Acacia 69
Sweet Basil 145
Sweet Broomwort 133
Sweet Jute 56
Sweet Orange 94
Sweet Potato 131
Sweet Viburnum 117
Sweet Wormwood 118
Sweet-Flag 157
Sweetgumshoot Mistletoe 91
Sweetpalm 164
Sweetscented Basil 145
Swinhoe Raspberry 69
Swinhoe's Sabia 98
Swiss Chard 40
Swodr-leaved Dendrobium 170
Sword Brake 6

Swordleaf Cymbidium 170
Sword-leaved Lindsaea 5
Sword-leaved Spikemoss 2
Szechwan Taxillus 90

T

Tabletform Aeonium 35
Tail Themeda 188
Tailleaf Eargrass 114
Tail-leaved Camellia 50
Taiwan Acacia 69
Taiwan Chloris 180
Taiwan Cycad 17
Taiwan Dallisgrass 186
Taiwan Fig 85
Taiwan Goldraintree 97
Taiwan Leea 93
Taiwan Osmanthus 108
Taiwan Plumegrass 183
Taiwan Rhynchotechum 135
Talipot Palm 165
Tall Bottle-brush 53
Tall Brake 6
Tall Nut Rush 178
Tall Reed 186
Tall-culm Galingale 174
Tamarind 73
Tamarisklike Spikemoss 2
Tapered Tarenna 117
Tapering Cyclosorus 8
Taro 158
Tassel-Flower 121
Tatch Screwpine 167
Tea 51
Tea Rose 68
Teafruit Holly 89
Tender Bothriospermum 128

T

Tender-clavate Fimbristylis 175		Three-leaved Pronephrium 9	
Tender-flowered Hedyotis 114		Three-lobed Morning-glory 131	
Tenstamen Biggingseng 101		Threestamen Waterwort 37	
Tenstamen Maple 97		Threevein Aster 118	
Tenuous-leaved Plagiogyria 3		Thunberg Leek 161	
Terete Cleisostoma 169		Thunberg's Lepisorus 14	
Ternstroemialike Anisetree 23		Thyme-leaved Spurge 63	
Terrateleaf Macadamia 45		Thyrselike negundo Chastetree 143	
Tetragonal Fimbristylis 176		Tibet Oatchestnut 82	
Thalia Lovegrass 182		Tien-tai Spicebush 25	
Thalia Lovegrass 182		Tiger Lily 155	
Thickbark Sweet-leaf 106		Tiger-grass 188	
Thickcupsule Tanoak 82		Tigerstick 39	
Thickleaf Oak 82		Titimo 43	
Thick-leaved Abacus Plant 63		Tiwan Beautyberry 140	
Thick-leaved Clematis 27		Tobacco 129	
Thick-leaved Croton 62		Tobacco Tree 129	
Thick-nerved Cinnamon 25		Tomato 129	
Thick-pericarped Millettia 78		Tomentose Elephantfoot 121	
Thickrostrate Begonia 48		Tomentose Feverine 116	
Thin Evodia 94		Tomentose Grape 93	
Thin Hedgehogcactus 49		Tomentose Machilus 26	
Thinfruit Monkeyjoy 57		Tomentose Pileostegia 66	
Thin-leaved Machilus 26		Tonguaeleaf Flower 38	
Thin-nerved Microtropis 89		Tongue Habenaria 171	
Thitmin 19		Tonkin Artocarpus 84	
Thomson Vionet 35		Tonkin Neohusnotia 184	
Thomson's Fimbristylis 176		Tonkin Sindora 73	
Thomson's Kudzu 79		Tonkin Snowbell 105	
Thorny Elaeagnus 92		Toothed Dentate Black Tree-fern 5	
Thorny Wingnut 91		Toothleaf Snowbell 105	
Thread Sprangletop 184		Topeng Pygeum 68	
Three-flowered Desmodium 76		Tougue-fern 15	
Three-leaf Arrow-head 147		Tow-leaves Morning-glory 131	
Threeleaf Chastetree 143		Trailing Lantana 142	
Three-leaved Acanthopanax 100		Translucent Haworthia 155	
Threeleaved Halberd Fern 13		Traveller's Palm 151	

Tree Cotton 58
Tree Ginseng 100
Trian Cattleya 169
Triandra Palm 164
Triangle Palm 165
Triangular Euphorbia 63
Tri-awned Minireed 179
Tricolored Stromanthe 153
Tricuspid Cudrania 85
Tridax 125
Trifoliolate Indigo 77
Trilobate Wedelia 125
Trilobed Typhonium 160
Triplicata Calanthe 169
Triquetrous Tadehagi 80
Trumpet-flower 136
Truncate Cyclosorus 9
Truncate-glume Sedge 173
Trunk Rosewood 75
Tsang Raspberry 69
Tsang's Beilschmiedia 24
Tsiang Persimmon 103
Tso Michelia 22
Tsoong's Tree 23
Tube-bract Bristle Fern 4
Tuber Oatgrass 178
Tuber Stemona 162
Tuberous Begonia 49
Tuberous Sword Fern 13
Tubular Kalanchoe 36
Tulip 156
Tuliptree 21
Turbinate Dillenia 45
Turk's Cap 59
Turn-in-the-wind 64
Tutcher Psychotria 116
Tutcher's Maple 97

Tutcher's Persimmon 103
Tweaksheath Lemongrass 180
Twelvestamen Melastoma 54
Twiggy Buckthorn 91
Twiggy Phyllanthus 65
Twig-hanging Embelia 105
Twinleaf Zornia 80
Twin-sorus Fern 8
Twisted Rosewood 75
Twoanther Mosla 145
Twocolor Arrowroot 153
Two-color Cattleya 169
Twocolored Velvetplant 122
Two-dotted Crepidomanes 4
Twoflower Lycianthes 129
Twoflower Pearlsedge 177
Twoflower Wedelia 125
Two-flowered Hedyotis 114
Twolobed Officinal Magnolia 21
Two-ranked-leaf Eurya 51
Two-spiked Signal-grass 179
Two-spikelet Fimbristylis 176
Twotooth Achyranthes 40

U

Umbel-flowered Poisonnut 107
Umbellate Merremia 132
Umbellate Rockjasmine 126
Umbrella Grass 176
Umbrella Plant 174
Unarmed Glorybower 141
Undulateleaf Oplismenus 185
Unigemmate Chain Fern 11
Uni-spike Killinga 176
Unulate Speedwell 134
Uvariaformleaf Tanoak 83
Uvarialeaf Wendlandia 117

V

Vachell's Interrupted Fern 3
Vagabondage Microstegium 184
Variant Wood Fern 12
Variedleaf Fig 86
Variegated Benjamin Fig 85
Variegated Spider Plant 154
Vegetable Sponge 47
Veined Fig 86
Veitch Screwpine 167
Velvet Leaved Desmodium 76
Venus Fly-trap 37
Verdant Bamboo 189
Veriegated Snake Plant 163
Verschaffelt Agave 163
Verticillate Kalanchoe 36
Vest Greenbrier 157
Vewrticillate Hydrilla 147
Vicary Golden Privet 108
Vietnam Fakesnaketailgrass 183
Vietnam Leaf-flower 65
Vietnam Sweet-leaf 106
Villous Armgrass 179
Villous Casearia 46
Villous Fatoua 85
Vimineous Microstegium 184
Vine Panic Grass 185
Vinebamboo Panicgrass 185
Violet Crabgrass 181
Virid-leaved Boehmeria 87
Voodoo Lily 157

W

Walking Maidenhair Fern 7
Waller Snapweed 42
Wallich Gugertree 52
Wallich's Wedelia 125
Wampi 94
Wandering Jew 149
Wandering Jew Zebrina 149
Water Ball-fruit 187
Water Cape Jasmine 113
Water Celery 102
Water Chestnut 175
Water Cress 34
Water Fern 7
Water Hyacinth 156
Water Hyssop 132
Water Lettuce 159
Water Melon 47
Water Nightshade 130
Water Pine 18
Water Plantain 147
Water Shamrock 16
Water Smartweed 39
Water Spangles 16
Water Spinach 131
Water Starwort 44
Waterbamboo Syzygium 53
Water-dragon 43
Water-meal 160
Watermelon Peperomia 31
Water-plantain Ottelia 147
Wax Gourd 47
Wax Plant 111
Waxy Leaf 61
Weak Philodendron 159
Weakleaf Yucca 164
Weaver's Bamboo 189
Wedge-shaped Spleenwort 10
Weeping Cypress 18
Weeping Fig 85
Weeping Willow 81

Welsh Onion 160
West Indies Mahogany 96
Westerhout's Sugar Palm 164
Westland's Birthwort 31
White Azalea 103
White Bauhinia 71
White Champak 22
White Fig Tree 87
White Flag 160
White Flower Leadwort 127
White Gironniera 83
White Heart-leaf 178
White Ixora 115
White Justicia 138
White Loosestrife 126
White Mulberry 87
White Ox Creeper 114
White Popinac 70
White Powdery Bamboo 188
White Smartweed 39
White Thunbergia 139
White Wormwood 118
White Yam 162
White Zephyr Flower 161
White-back Mallotus 64
Whiteback Sumac 99
Whitedrumnail 38
Whiteflower Conyza 120
Whiteflower Raspberry 68
Whiteflower Tarenna 117
White-flowered Derris 75
White-flowered Embelia 105
White-fruited Randia 112
Whitehairy Beautyberry 139
White-leaved Fissistigma 23
Whiteligulate Aster 118
Whitescale Galingale 174

White-veined Anthurium 158
Whitish Rattlebox 74
Whorl-leaf Litse 26
Whorlleaf Syzygium 54
Wideleaf Croton 62
Wideleaf Cymbidium 170
Wideleaf Razorsedge 178
Wideleaf Statice 126
Wide-leaved Borreria 113
Wide-leaved carpetgrass 179
Widened Twin-sorus Fern 8
Wight's Toxocarpus 111
Wild Asparagus 153
Wild Banana 150
Wild Bittertea 64
Wild Cane 187
Wild Ginger 152
Wild Himalayan Cherry 67
Wild Honeysuckle 117
Wild Kudzu Vine 79
Wild Lettuce 123
Wild Loquat 67
Wild Nightshade 129
Wild Oat Grass 180
Wild Perilla 145
Wild Sensitive-plant 72
Wild Siris 70
Wild Spikenard 144
Wild Syzygium 54
Wild Violet 35
Wild Vitex 143
Wilford's Pennywort 102
Williams Elder 117
Willowleaf Hygrophila 138
Willowleaf Machilus 26
Willowleaf Maesa 105
Willowleaf Raphiolepis 68

Willow-leaved Camellia 51
Willow-leaved Iron-weed 125
Winded Cassia 72
Windmill Palm 167
Wine Grape 94
Wine Palm 165
Wingceltis 84
Winged Laggera 123
Winged-bean 79
Wingedstem Treebine 93
Winit Pleocnemia 12
Witches's Broom 90
Wolly Eria 171
Wood Ceropegia 110
Wood Gossip Caesalpinia 71
Wooded Hypolytrum 176
Wooden Tortoise 48
Woodland Elaeocarpus 57
Woolly Fimbristylis 175
Woolly-axle Bracken 6
Woolly-flowered Persimmon 103
Wooly Grass 183
Wormwoodlike Mother-wout 144
Wright's Abacus Plant 64
Wrinkled Duck Beak 184
Wrinkledleaf Bristlegrass 187
Wrinkled-leaf Peperomia 31

Y

Yam Bean 78
Yanmin 99
Yaoping Photinia 67
Yaoshan Cupgrass 149
Yard Grass 181
Yellow Basket-willow 99
Yellow Bramble 143
Yellow Bristlegrass 187

Yellow Camphor Tree 25
Yellow Canna 153
Yellow Eye Grass 149
Yellow Flax 41
Yellow Grass 183
Yellow Hair Aralia 100
Yellow Hairy Dunbaria 76
Yellow Hibiscus 59
Yellow Ixora 115
Yellow Jasmine 107
Yellow Machilus 26
Yellow Oleander 110
Yellow Poinciana 73
Yellow Pui 136
Yellow Spider-flower 33
Yellow Velvetleaf 146
Yellow Water Lily 28
Yellow-bow Dendrobium 170
Yellowflower Lanatan 142
Yellowflower Milkwort 35
Yellow-flower Pea 74
Yellowflower Torenia 134
Yellow-flowered Calanthe 169
Yellowleaf Eargrass 115
Yellow-vein Coralbean Tree 76
Yuanjiang Cycad 17
Yulan Magnolia 21
Yunnan Michelia 23
Yunnan Parakmeria 23
Yunnan Roughleaftree 83
Yuyuan Manglietia 22

Z

Zebra Haworthia 155
Zedoary Turmeric 152
Zhang Arundinella 179
Zhejiang Camellia 50

Zigzag Asparagus 154
Zippel Microsorium 15

Zonate Cryptanthus 150
Zululand Cycad 17

参考文献

于法钦，廖文波，周海旋等．2005．深圳市笔架山公园维管植物编目．中山大学学报（自然科学版）44（增刊）：92－114．

中国科学院植物研究所．1996．新编拉汉英植物名称．北京：航空工业出版社．

中国植物志编辑委员会．1978~1998．中国植物志．北京：科学出版社．

王发祥，梁惠波，罗蒙．1998．深圳园林植物．北京：中国林业出版社．

王伯荪，余世效，胡玉佳．1986．深圳宝安的银叶树林．生态科学（2）：89－91．

王勇进，张寿洲，李勇等．2003．深圳市国家重点保护野生植物的区系特点与分布状况．华南农业大学学报（自然科学版）24（1）：63－66．

冯志坚，李镇魁，吴永彬等．2002．广东惠东莲花山盘珠保护区植物资源调查．华南农业大学学报23（4）：49－52．

叶华谷，邢福武．2005．广东植物名录．广州：世界图书出版公司．

仲铭锦，王晓明，廖文波，等．2002．深圳莲花山公园的植物多样性编目．中山大学学报（自然科学版）41（增刊2）：69－81．

刘胜祥．1994．植物资源学．武汉：武汉出版社，44－51．

孙卫邦．2003．乡土植物与现代城市园林景观建设．中国园林（7）：63－65．

孙延军，梁嘉声，黄康有等．2003．深圳围岭公园植物多样性编目．中山大学学报（自然科学版）增刊（2）：62－81．

邢福武，余明恩．2000．深圳野生植物．北京：中国林业出版社．

邢福武，余明恩，张永夏．2002．深圳植物物种多样性及其保育．北京：中国林业出版社．

邢福武，周远松，龚友夫，张永夏．2004．深圳市七娘山郊野公园植物资源与保护．北京：中国林业出版社．

邢福武．2004．澳门植物名录．澳门：澳门民政总署园林绿化部．

严岳鸿，邢福武，黄向旭等．2004．深圳的外来植物．广西植物24（3）：232－238．

吴世捷，高力行．2002．不受欢迎的生物多样性：香港的外来植物物种．

生物多样性 10（1）：109－118.

吴章文，陈就和，吴楚材．2003．广东象头山国家自然保护区科学考察集．北京：中国林业出版社．

吴德邻．1994．海南及广东沿海岛屿植物名录．北京：科学出版社．

吴德邻主编．2002．香港植物名录（第七版）．香港：香港特别行政区政府渔农自然护理署出版．

李沛琼，张寿洲，王勇进．2003．耐荫半耐荫植物．北京：中国林业出版社．

李佩琼．1998．深圳仙湖植物园植物名录．北京：中国林业出版社．

李振宇，解炎．2002．中国外来入侵种．北京：中国林业出版社．

李烨，陈锡沐，李镇魁等．2001．深圳市重要药用植物资源调查．中国野生植物资源 20（4）：26－30．

李镇魁，陈涛，冯志坚等．2001．广东深圳野生观赏植物资源调查．亚热带植物科学 30（4）：40－44．

杨永川，达良俊．2005．上海乡土树种及其在城市绿化建设中的应用．浙江林学院学报 22（3）：286－290．

杨际明，何仲坚，冯志坚等．2005．深圳塘朗山郊野公园的植物资源．亚热带植物科学 34（1）：56－59．

陈飞鹏，汪殿蓓，暨淑仪，等．2001．深圳南山区天然植物群落的聚类分析．武汉植物学研究 19（5）：385－390．

陈灵芝，马克平．2001．生物多样性原理与实践．上海：上海科学技术出版社．P1－308．

陈封怀，吴德邻，等．1987－2005．广东植物志．广东：广东科技出版社．

陈涛．2006．深圳森林景观生态构建．北京：中国林业出版社．

范慧芬，杨远攸，汤仲之．1984．深圳梧桐山主要森林类型的垂直分布及其特征．广西植物 3（3）：215－224．

郑万均，傅立国．1978．中国植物志（第七卷）．北京：科学出版社．

秦仁昌．1978．中国蕨类植物科属系统排列和历史来源．植物分类学报，16（3）：1－19；16（4）：16－37．

深圳市人民政府城市管理办公室．1998．深圳园林植物．北京：中国林业出版社．

深圳市人民政府城市管理办公室等．1997．深圳特区古树名木．北京：中国林业出版社．

深圳市人民政府城市管理办公室等．2003．梧桐山植物．北京：中国林业出版社.

深圳市人民政府城市管理办公室等. 2004. 深圳市园林植物续集（一）. 北京：中国林业出版社.

黄智明. 1991. 广东南昆山野生花卉资源（四）：种类丰富的兰科植物. 广东园林（3）：14 – 16.

黄辉宁，李思路，朱志辉. 2005. 珠海市外来入侵植物调查. 广东园林（6）：24 – 27.

蓝崇钰，王勇军，等. 2001. 广东内伶仃岛自然资源与生态研究. 北京：中国林业出版社.

黎盛臣主编. 1991. 中国植物园参观指南. 北京：金盾出版社.

HUTCHINSON J. 1959. The families of flowering plants. Vol. I – II. Oxford：Clarendon.